# ESSENTIAL REPRODUCTION

TO BOB EDWARDS AND
JOE HERBERT

who first stimulated
our interest in
the science of reproduction

# Essential Reproduction

## Martin H. Johnson
MA, PhD
Fellow Christ's College, Cambridge
Reader in Experimental Embryology
University of Cambridge

## Barry J. Everitt
MA, BSc, PhD
Fellow and Director of Studies in Medicine
Downing College, Cambridge
Lecturer, Department of Anatomy
University of Cambridge

THIRD EDITION

OXFORD

**BLACKWELL**

**SCIENTIFIC PUBLICATIONS**

LONDON EDINBURGH BOSTON

MELBOURNE PARIS BERLIN VIENNA

© 1980, 1984, 1988 by
Blackwell Scientific Publications
Editorial offices:
Osney Mead, Oxford OX2 0EL
25 John Street, London WC1N 2BL
23 Ainslie Place, Edinburgh EH3 6AJ
3 Cambridge Center, Cambridge,
    Massachusetts 02142, USA
54 University Street, Carlton
    Victoria 3053, Australia

Other Editorial Offices:
Arnette SA
2, rue Casimir-Delavigne
75006 Paris
France

Blackwell Wissenschaft
Meinekestrasse 4
D-1000 Berlin 15
Germany

Blackwell MZV
Feldgasse 13
A-1238 Wien
Austria

First published 1980
Reprinted 1983
Second edition 1984
Third edition 1988
Reprinted 1988, 1990, 1991

Set by Times Graphics
Singapore
Printed and bound in Great Britain
by Hartnolls Ltd, Bodmin, Cornwall

DISTRIBUTORS

Marston Book Services Ltd
PO Box 87
Oxford OX2 0DT
(*Orders:* Tel. 0865 791155
        Fax: 0865 791927
        Telex: 837515)

USA
Mosby-Year Book, Inc.
11830 Westline Industrial Drive
St Louis, Missouri 63146
(*Orders:* Tel. (800) 633-6699)

Canada
Mosby-Year Book, Inc.
5240 Finch Avenue East
Scarborough, Ontario
(*Orders:* Tel. (416) 298–1588)

Australia
Blackwell Scientific Publications
(Australia) Pty Ltd
54 University Street
Carlton, Victoria 3053
(*Orders:* Tel. (03) 347–0300)

British Library
Cataloguing in Publication Data

Johnson, Martin H.
    Essential reproduction.—3rd ed.
    1. Mammals—Reproduction
    I. Title   II. Everitt, Barry J.
    599.01'6                          QP251

    ISBN 0-632-02183-7

# Contents

# Preface to Third Edition

The science of reproduction is progressing rapidly. In the few years since our second edition was written several grey areas have become more boldly black and white and several totally novel discoveries have been made. We have substantially restructured some chapters to take these new developments into account, to clarify the more opaque parts of our earlier text and to simplify some diagrams and arguments. We have received many helpful comments and letters of advice from our students and from teachers all over the world and have attempted in this third edition to provide a sound and comprehensive text for students of reproduction.

# How to Use this Book

This book represents an integrated approach to the study of reproduction. There can be few subjects which so obviously demand such an approach. During our teaching of reproduction at Cambridge University, the need for a book of this kind was clear to both ourselves and our colleagues. We hope this volume goes some way towards filling this need.

We have written the book for medical, veterinary and science students of mammalian reproduction. Throughout we have attempted to draw out the general, fundamental points common to reproductive events in all or most species. However, a great range of variation in the *details* of reproduction is observed amongst different species, and in some respects very *fundamental* differences are observed. Where the details differ, we have attempted to indicate this in the numerous tables and figures rather than clutter the general emphasis and narrative of the text. Where the fundamentals differ, an explicit discussion is given in the text. These fundamental differences should not be ignored. For example, preclinical medical students may consider the control of luteal life in the sheep or ovarian cyclicity in the rat to be irrelevant to their future interests. However, as a result of inappropriate extrapolation between species, the human female has been treated both as a ewe and as a rat in the past decade (much to her discomfort and detriment). If on finishing this book the student appreciates the dangers of wholesale extrapolations between species, we will have achieved a major aim.

Confinements of space have unfortunately necessitated omission from the text of the subject of embryonic development. We felt that to give only passing reference to this subject would be an injustice, to treat it fully would require a text of similar length to the present one. We recommend that the interested student seeks this information elsewhere.

We suggest that first you read through each chapter with only passing reference to tables and figures. In this way, we hope that you will grasp the essential fundamentals of the subject under discussion. Then reread the chapter, referring extensively to the tables and to the figures and their legends, in which a great deal of detailed or comparative information is located. Finally, because we have adopted an integrated approach to the subject, the book needs to be taken as a whole, as it is more than the sum of its constituent chapters.

M.H.J.
B.J.E.

# Acknowledgments

For help at many stages of the preparation of the book we owe particular thanks to many people: to our students of former years for their interest, stimulation and responsiveness; to our colleagues at Blackwell Scientific Publications for all their help and advice; to Shirley French, Jane Hugh, Tracy Kelly, Susan Currie and, especially, Caroline Hunt for their patience and help in typing the many manuscript drafts; to Tim Crane, Roger Liles and Chris Burton for their advice and help with photographic illustrations; to Alex Everitt for help with illustrations; to Dr Peter Braude, Dr Graham Burton, Dr Ruth Kleinfeld, Professor Tomas Hökfelt, J. Moeselaar, Dr Bernard Maro, Dr Tony Plant and Dr J. M. Tanner for allowing us to use their original photographs and data; and to our many colleagues who read and criticized our drafts and encouraged us in the preparation of the book, especially: Peter Braude, Graham Burton, Jerry Eberhart, Norma Lynn Fox, Stefan Hansen, Alan Findlay, Michael Hastings, Brian Heap, Tomas Hökfelt, Barry Keverne, Ruth Kleinfeld, Charlie Loke, Ken Lowe, Alan McNeilly, Nick Martensz, Rachel Meller, Bob Moor, Hester Pratt, Idwal Rowlands, Brian Setchell, Erica Stanfield, Don Steven, Pat Tate and Barbara Weir. Finally, the stimulating environment generated by the research workers in our laboratories has been extremely important in providing us with the help and motivation to write this book, and to them we are very grateful.

# Figure Acknowledgments

*Chapter 1*

Figs 1.2, 1.3, 1.4, 1.6 all redrawn from Langman J. *Medical Embryology*. Williams and Wilkins, 1969.

Fig. 1.7a reproduced with permission from Money J, Ehrhardt A. *Man and Woman, Boy and Girl*. Johns Hopkins University Press, 1972.

Fig. 1.7b redrawn with permission from Overzier C. *Intersexuality*. Academic Press, 1963.

Figs 1.9, 1.10 redrawn from Goy RW. *Phil Trans Roy Soc B* 1970; **259**: 149–162.

Figs 1.13, 1.17 from Tanner JM. *Growth at Adolescence*. Blackwell Scientific Publications, 1962.

Figs 1.14, 1.15, 1.16 from van Wieringen JC, Wafelbakker F, Verbrugge HP, de Haas JH. *Growth Diagrams 1965 Netherlands: Second National Survey on 0–24 year olds*. Netherlands Institute for Preventative Medicine TNO Leiden and Wolters Noordhoff, 1971.

*Chapter 5*

Figs 5.4a, b redrawn from Heimer L. *The Human Brain and Spinal Cord*. Springer-Verlag, 1983.

Fig. 5.5 redrawn from Nikolics K *et al*. 1983. *Nature* 1985; **316**: 511–7.

Fig. 5.7 redrawn from Clarke IJ, Cummins JT. *Endocrinology* 1982; **111**: 1737–1739.

Fig. 5.8 redrawn from Belehetz PE *et al*. *Science* 1978; **202**: 631–633.

Figs 5.9, 5.10, 5.13, 5.14, 5.21, 5.25, 5.26, 5.28 from Yen SSC, Jaffe R (Eds) *Reproductive Endocrinology*. WB Saunders and Co., 1978.

Fig. 5.11 redrawn from Martin GB *et al*. *J Endocrinol* 1986; **111**: 287–296.

Fig. 5.12 redrawn from Rivier C *et al*. *Science* 1986; **234**: 205–208.

Fig. 5.15 from Clayton RN in *Neuroendocrinology* (Eds Lightman SL, Everitt BJ). Blackwell Scientific Publications, 1986.

Fig. 5.16 redrawn from Clarke IJ, Cummins JT. *Endocrinology* 1985; **116**: 2376–2383.

Fig. 5.17 redrawn from Ferin M *et al*. *Rec Progr Horm Res* 1984; **40**: 441–481.

Fig. 5.19 redrawn from Dorner G in *Sex, Hormones and Behaviour*. Ciba Foundation Symposium 62. Elsevier, 1979.

Figs 5.18, 5.29 from Lincoln GA. *Br Med Bull* 1979; **35**(2): 167–172.

Fig. 5.23 redrawn from Moore RY in *Reproductive Endocrinology* (Eds Yen SSC, Jaffe R). WB Saunders, and Co., 1978.

Fig. 5.30 redrawn from Bittman EL *et al*. *Biol Reprod* 1984; **30**: 585–593.

Figs 5.31, 5.24 from Keverne EB in *Sex, Hormones and Behaviour*. Ciba Foundation Symposium 62. Elsevier, 1979.

Figs 5.33, 5.34, 5.35 from Everitt BJ, Keverne EB in *Neuroendocrinology* (Eds Lightman SL, Everitt BJ). Blackwell Scientific Publications, 1986.

*Chapter 6*

Figs 6.1, 6.7 from Tanner JM. *Growth at Adolescence*. Blackwell Scientific Publications, 1962.

Fig. 6.2 redrawn from Grumbach M in *Reproductive Endocrinology* (Eds Yen SSC, Jaffe R). WB Saunders and Co., 1978.

Fig. 6.3 reproduced with permission from Weitzmann ED. *Recent Progress in Hormone Research*. Academic Press, 1975.

Fig. 6.4 drawn from Boyer RM *et al. J clin Invest* 1974; **54**: 609.

Figs 6.6, 6.12, 6.13 from Grumbach M in *Control of the Onset of Puberty* (Eds Grumbach M, Grave GD, Mayer FE). John Wiley and Sons, 1974.

Fig. 6.8 from Frisch RE. *Pediatrics* 1972; **50**: 445–450.

*Chapter 7*

Figs 7.7, 7.8 redrawn with permission from Bermant G, Davidson JM. *Biological Bases of Sexual Behaviour*. Harper and Row, 1974.

Fig. 7.9 redrawn from Bancroft J. *Human Sexuality and its Disorders*. Churchill Livingstone, 1983.

Figs 7.10b, 7.13 redrawn with kind permission of Herbert J.

Fig. 7.12 from Udry JR, Morris NM, Waller I. *Arch Sex Behav* 1973; 2(3): 205–214.

Fig. 7.14 data adapted from Bermant G, Davidson JM. *Biological Bases of Sexual Behaviour*. Harper and Row, 1974.

*Chapter 8*

Fig. 8.7 from Maro B *et al.* Changes in action distribution during fertilization of the mouse egg. *J Embryol Exp Morphol* 1984; **81**: 211–237.

*Chapter 11*

Table 11.2, Figs 11.1, 11.3, 11.4, 11.5 after Biggers JD in *Med Physiol* (Ed. Mountcastle V). Mosby, 1979.

*Chapter 12*

Fig. 12.2 adapted from Liggins GC. *Br Med Bull* 1979; **35**(2): 145–150.

Fig. 12.5 reproduced with kind permission of Niswander KR.

*Chapter 13*

Fig. 13.1 redrawn from Netter FH. *The Ciba Collection of Medical Illustrations*. Ciba, 1954.

Figs 13.2, 13.4 redrawn from Cowie AT in *Reproduction in Mammals* (Eds Austin CR, Short RV). Cambridge University Press, 1972.

Fig. 13.5 redrawn and adapted with kind permission of McNeilly A.

Fig. 13.9 redrawn from Hinde RA. *Biological Bases of Human Social Behaviour*. John Wiley and Sons, 1974. With the kind permission of Hinde RA.

# Chapter 1
# Sex

The reproduction of mammals involves sex. The essential feature of *sexual reproduction* is that the new individual receives its genetic endowment in two equal portions—half carried in a *male gamete*, the *spermatozoon*, and half carried in a *female gamete*, the *ovum*. These gametes come together at fertilization to form the new *zygote*. In order to reproduce itself subsequently, the individual must transmit only half its own chromosomes to the new *zygotes* of the next generation. In sexually reproducing species, therefore, a special population of *germ cells* is set aside. These cells undergo a *reduction division* known as *meiosis*, in which the chromosomal content of the germ cells is halved and the genetic composition of each chromosome modified as a result of chromosomal exchange (Fig. 1.1). The generation of genetic diversity within a population that results from sexual reproduction may mean that the population as a whole provides a richer and more varied source of material on which natural selection may operate. The population would, therefore, be expected to show greater resilience in the face of environmental challenge.

However, sex is not by any means an essential component of reproductive processes. Asexual (or vegetative) reproduction occurs continuously within the tissues of our own bodies as individual cells grow, divide *mitotically* (Fig. 1.1) and generate two offspring which are genetically identical to

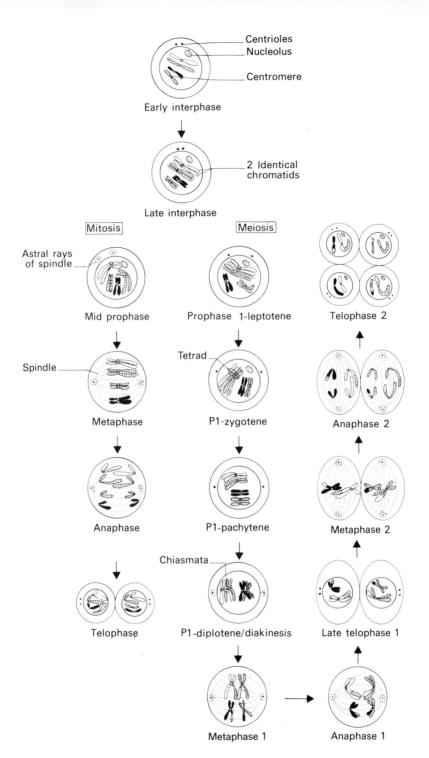

**Fig. 1.1.** Mitosis and meiosis, illustrated by a cell containing two homologous pairs of chromosomes (each human cell contains 23 such pairs or 46 chromosomes in total; it is thus said to be *diploid*—having two complete sets of chromosomes). Prior to division in interphase the chromosomes each reduplicate their DNA in the

each other and to their single parent. Many unicellular organisms reproduce themselves mitotically just like the individual cells of the body. Amongst multicellular organisms, several reproduce themselves by setting aside a population of 'germ cells' which divide not meiotically but mitotically, and differentiate to generate a complete new organism which is genetically identical to its parent. This process of reproduction, as a result of simple division, is not used at all by mammals, or indeed most phyla, for perpetuating the species. Each time a mammal produces new offspring, they will be genetically unique organisms differing from their parents and their siblings. The generation of such genetic diversity is achieved as a result of sexual reproduction.

Whilst the reasons for sex may be sought in terms of population genetics, the consequences of this mode of reproduction permeate all aspects of mammalian life. At the core of the process lies the creation and fusion of gametes.

---

synthetic (S) phase of the cell cycle. Each chromosome then consists of two identical chromatids joined at the centromere. In *mitotic* prophase the two chromatids first become distinctly visible as each shortens and thickens by a spiralling contraction; the nucleoli and nuclear membrane break down. In metaphase, microtubules form a spindle radiating from the centrioles and the chromosomes lie on the equator. In anaphase, the centromere of each chromosome splits and each chromatid migrates to opposite poles of the spindle. Telophase sees the reformation of nuclear membranes and nucleoli, cytokinesis (or division of the cytoplasm), breakdown of spindle and decondensation of chromosomes. Two genetically identical daughter cells now exist where one existed before. Mitosis is a nonsexual or vegetative form of reproduction. In meiosis, prophase is extended and is divided into several sequential steps: leptotene chromosomes are long and thin; in zygotene homologous pairs of chromosomes come to lie side by side along parts of their length; in pachytene chromosomes start to thicken and shorten and are associated in pairs along their entire length at which time synapsis, crossing over and chromatid exchange take place, nucleoli disappear; in diplotene and diakinesis chromosomes shorten further and show evidence of being closely linked to their homologue at the chiasmata where crossing over and chromatid exchange has occurred, giving a looped or cross-shaped appearance. Metaphase: nuclear membrane breaks down, pairs of chromosomes lie on equator of spindle. Anaphase: homologous chromosomes move in opposite directions drawn by centromeres. Telophase: cytokinesis and reformation of nuclear membrane yields two daughter cells each with half the number of chromosomes (only one member of each homologous pair), but each chromosome consisting of two unique chromatids. In a second division, these chromatids then separate as in mitosis, to yield two *haploid* cells, i.e. each containing only one complete set of chromosomes. Due to chromatid exchange and the random segregation of homologous chromosomes, each haploid cell is genetically unique. At fertilization, two haploid cells come together to yield a new diploid zygote.

3    *Sex*

The gametes take distinctive male or female forms (to prevent self-fertilization) and are made in distinctive male and female *gonads*—the *testis* and *ovary*, respectively. In addition, the gonads elaborate a group of hormones—the *sex steroids*—which modify the tissues of the body to generate distinctive male and female somatic phenotypes suited to maturing and transporting their respective gametes. The steroids also affect the behaviour and physiology of the individuals of each sex to ensure that, in most mammals, mating will only occur between opposite sexes at times of maximum fertility. Finally, in mammals, not only do the steroids provide conditions to facilitate the creation of new individuals, they also prepare the female to carry the growing embryo for prolonged periods (*viviparity*), and to nurture it after birth through an extended period of *maternal lactation* and *parental care*. Thus, sex has come to occupy a central position in mammalian biology, shaping not just anatomy and physiology, but also aspects of behaviour and social structure. This ramification of sex throughout a whole range of biological and social aspects of mammalian life is mediated largely through the actions of the gonadal hormones.

# 1 The genesis of two sexes
## a Genetic determinants

Sex has a genetic basis in mammals. Examination of human chromosomes reveals a consistent difference between the sexes in *karyotype* (or pattern of chromosomal morphologies). Man has 46 chromosomes, 22 pairs of *autosomes* and one pair of *sex chromosomes* (Fig. 1.2). Human females, and indeed all female mammals, are known as the *homogametic sex* because the sex chromosomes are both *X chromosomes* and all the gametes (ova) are similar to one another in that they each possess one X chromosome. Conversely, the male is termed the *heterogametic sex*, since his pair of sex chromosomes consist of *one X* and *one Y*, so producing two distinct populations of spermatozoa, one bearing an X and the other a Y chromosome (Fig. 1.2). Examination of a range of human patients with chromosomal abnormalities has shown that if all, or part, of a Y chromosome is present then the individual develops the *male* gonads (testes). If the Y chromosome is absent the *female* gonads develop (ovaries). The number of X chromosomes or autosomes present appears to be without effect on the *primary determination of gonadal sex* (Table 1.1). Similar studies on a whole range of other mammals provide substantial support for the hypothesis that Y-chromosome activity alone determines gonadal sex.

The Y chromosome itself is small and has very little DNA which actively synthesizes RNA. Therefore, it seems unlikely that all the structural genes required to make an organ as complex as the testis are located on the Y chromosome itself. It is more probable that these genes lie on other autosomal

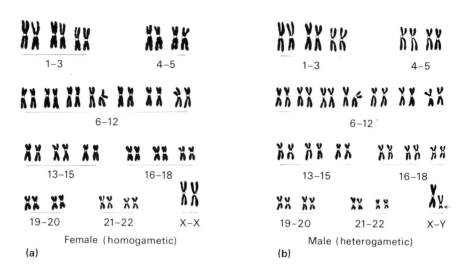

**Fig. 1.2.** Karyotypes of two human cells, obtained by arresting mitosis with colchicine in metaphase when the chromosomes are compacted and clearly visible. The chromosomes are arranged and classified according to the so-called 'Denver' system. It can be seen that while the 44 autosomes are grossly similar in size in either sex, the two sex chromosomes are distinguishable by size, being XX in a female (a) and XY in a male (b). Arrow indicates site of testis gene(s).

chromosomes, or even on the X chromosome, and that a 'switching' controller gene (or genes) on the Y chromosome regulates their expression.

Evidence from deletions and translocations involving the Y chromosome suggests that the crucial gene(s) is on the short arm near to the centromere (see Fig. 1.2b). An intense research effort to locate, map and clone the gene(s) is underway. Success in this venture should reveal the precise

**Table 1.1** Effect of human chromosome constitution on development of the gonad.

| Chromosomes | | | |
|---|---|---|---|
| Autosomes | Sex chromosomes | Gonad | Syndrome |
| 44 | XO | Ovary | Turner's |
| 44 | XX | Ovary | Normal female |
| 44 | XXX | Ovary | Super female |
| 44 | XY | Testis | Normal male |
| 44 | XXY | Testis | Klinefelter's |
| 44 | XYY | Testis | Super male |
| 66 | XXX | Ovary | Triploid |
| 66 | XXY | Testis | (non-viable) |
| 44 | XX$^{sxr}$ | Testis | Sex reversed* |

*An X$^{sxr}$ chromosome carries a small piece of Y chromosome translocated onto the X.

mechanism by which the Y chromosome coordinates the construction of a testis.

It is already clear that sex steroid hormones (see Chapter 2) are *not* likely to be involved, since the injection of 'male' hormones into the female fetus does not produce a testis; similarly neutralization of 'male' hormones within, or injection of 'female' hormones into, a male fetus does not suppress testis formation or produce an ovary. In recent years some clues have accumulated to suggest that a protein called the *H–Y antigen* might be involved in testis construction. This antigen can be recognized by injecting male cells into a female, who then mounts an immune response to the foreign protein and generates specific antibodies. Use of these antibodies has shown that there is a correlation between the presence of the antigen and the possession of a testis (however rudimentary or malfunctioning). However, recently some mice have been found which have testes but lack the antigen. It seems that the gene 'make a testis' and that encoding *H–Y* are closely linked but separable. Thus, expression of *H–Y* is not required for a testis to be formed.

*b Gonadal dimorphism*

The early development of the gonad proceeds identically in both males and females. In both sexes the gonads are derived from two distinct tissues: *somatic mesenchymal* tissues which form the matrix of the gonad; and the *primordial germ cells* (PGCs) which migrate into and colonize this matrix to form the gametes.

The *genital ridge primordia* are two knots of mesenchyme overlain by a columnar *coelomic (or germinal) epithelium*. The primordia develop at about 3.5–4.5 weeks in human embryos, on either side of the central dorsal aorta on the posterior wall of the embryo in the lower thoracic and upper lumbar region (Fig. 1.3c). The genital ridges are superficial and medial to the developing *mesonephric tissue* (Fig. 1.3). Columns of cells derived from proliferation and inward migration of both the mesonephros and the coelomic epithelium form the *primitive sex cords* (Fig. 1.3d).

The primordial germ cells arise outside the genital ridge region, and are first identifiable in the human embryo at about 3 weeks in the epithelium of the yolk sac near the developing allantois (Fig. 1.3a). There they may be recognized by morphological criteria and by a distinctive alkaline phosphatase activity. By the 13- to 20-somite stage the PGC population, expanded by mitosis, can be observed migrating to the connective tissue of the hind gut and from there into the gut mesentery (Fig. 1.3b). From about the 25-somite stage onwards, 30 days or so after fertilization, the majority of cells have passed into the region of the developing kidneys, and thence into the adjacent gonadal primordia (Fig. 1.3). This migration of PGCs occurs primarily by amoeboid movement,

6    *Chapter 1*

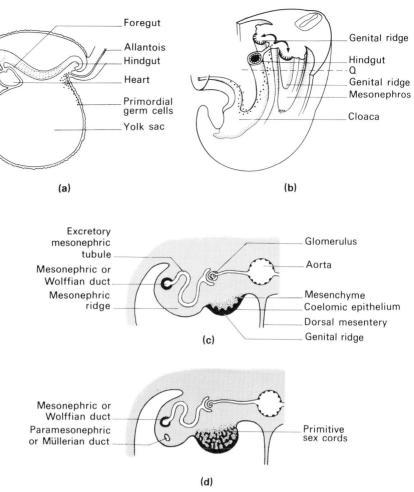

**Fig. 1.3.** Drawings of a 3-week human embryo showing: (a) origin of primordial germ cells; (b) the route of their migration. Section Q is plane of transverse section shown in (c) and (d) which are highly schematic representations of a transverse section lumbar region of 4- and 5-week human embryos at the 'indifferent gonad' stage.

and the cells have appropriately well-defined pseudopodia. There is a suggestion that the genital ridges may produce a chemotactic substance which attracts the PGCs, and so gives direction to their amoeboid movement. Thus, PGCs cultured *in vitro* display well-directed, active amoeboid movement towards the genital ridge tissue placed in the culture dish. In addition, gonadal tissue grafted into abnormal sites within the embryo stimulates germ cells to migrate towards it and to colonize it quite normally.

The dividing PGCs are joined in the gonad primordium by an ingress of cells from both the coelomic germinal epithelium (the sex cords) and the mesonephros (the medullary cords). Throughout this phase of PGC migration and early

7    *Sex*

colonization it is not possible to discriminate between male and female gonads, which are therefore said to be *indifferent*. The Y-chromosomal determination of gonadal sex manifests itself only when colonization is completed during the sixth week in the human embryo. At this time, and in male embryos only, the cells of the cortical mesenchyme condense to form a fibrous layer—*the tunica albuginea*. Meanwhile, a vigorous proliferation of the sex cord cells deep into the medullary region of the gonad establishes contact with ingrowing cords of mesonephric tissue (*the rete blastema* or *rete testis cords*) and leads to the formation of the definitive testis cords (Fig. 1.4a,b). These cords incorporate most of the PGCs (which cease mitosis) within their columns, and the cords then secrete an outer basement membrane. They are then known as the *seminiferous cords*, and will give rise to the *seminiferous tubules* of the adult (Fig. 1.4b). Of the two cell populations within the cords, the PGCs, now known as *prospermatogonia*, will give rise to spermatozoa, and the mesodermal cord cells, now known as *sustentacular cells*, will give rise to *Sertoli cells*. Between the cords, the loose mesenchyme vascularizes and develops as *stromal tissue*, within which cells condense in clusters to form specific endocrine units—the *interstitial glands of Leydig*.

Whilst the male gonad undergoes these marked changes in organization the female gonad continues to appear indifferent. The primitive sex cords of the human female are ill-defined, although in sheep and pigs well formed cords do appear transiently. Unlike the situation in the male, the ingrowth of coelomic epithelial cells is followed by their condensation in the more *cortical* regions of the gonad. Small cell clusters then surround the germ cells to initiate formation of the *primordial follicles* characteristic of the ovary (Fig. 1.4d). In these follicles the mesenchymal cells secrete an outer basement membrane—*membrana propria*. The mesenchymal cells will give rise to the granulosa cells of the follicle, whilst the germ cells, which have now ceased their mitotic proliferation, give rise to oocytes (see Chapter 4). Clusters of *interstitial gland* cells form around the primordial follicles.

The morphological divergence of the ovary and testis is dependent upon whether or not the Y chromosome is expressed. Almost certainly it is the Y-chromosome activity within the mesodermal cells of the genital ridge which is critical, since germ cells do not appear to play an active or essential role in the *initiation* of gonadal differentiation. Thus, functional germ cells may be depleted either experimentally or pathologically, but gonadal differentiation is still initiated.

However, subsequent development of the gonad, particularly of the ovary and its follicles, *is* dependent upon a population of normal germ cells. For example, women

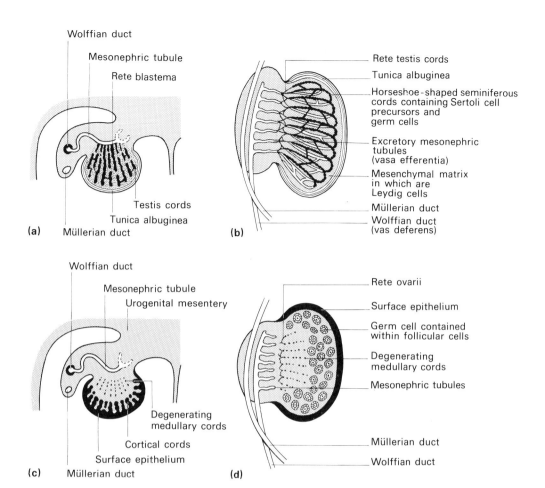

**Fig. 1.4.** Testicular development during (a) the eighth, and (b) the sixteenth to twentieth week of human fetal life. (a) The primitive sex cords proliferate in the medulla and establish contact with the rete testis. The tunica albuginea (fibrous connective tissue) separates the testis cords from the coelomic epithelium, and eventually forms the capsule of the testis. (b) Note the horseshoe shape of the seminiferous cords and their continuity with the rete testis cords. The vasa efferentia, derived from the excretory mesonephric tubules, connect the seminiferous cords with the Wolffian duct (see text). Comparable diagrams of ovarian development around (c) the seventh, and (d) the twentieth to twenty-fourth weeks of development. (c) Any primitive medullary sex cords degenerate to be replaced by the well-vascularized ovarian stroma. The cortex proliferates, and mesenchymal condensations later develop around the arriving PGCs. (d) In the absence of medullary cords and a true persistent rete ovarii, no communication is established with the mesonephric tubules. Hence, in the adult, ova are shed from the surface of the ovary, and are not transported by tubules to the oviduct (cf. male, see Chapters 3 and 4).

suffering from *Turner's syndrome* (see Table 1.1), who have a normal autosomal complement but only one X chromosome, develop an ovary. Subsequently, however, normal oocyte

9     *Sex*

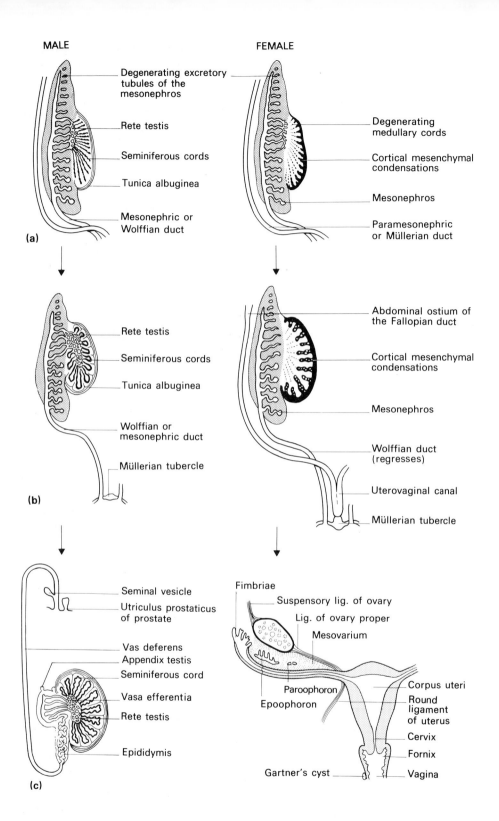

MALE

Degenerating excretory tubules of the mesonephros

Rete testis

Seminiferous cords

Tunica albuginea

Mesonephric or Wolffian duct

(a)

FEMALE

Degenerating medullary cords

Cortical mesenchymal condensations

Mesonephros

Paramesonephric or Müllerian duct

Rete testis

Seminiferous cords

Tunica albuginea

Wolffian or mesonephric duct

Müllerian tubercle

(b)

Abdominal ostium of the Fallopian duct

Cortical mesenchymal condensations

Mesonephros

Wolffian duct (regresses)

Uterovaginal canal

Müllerian tubercle

Seminal vesicle

Utriculus prostaticus of prostate

Vas deferens
Appendix testis
Seminiferous cord

Vasa efferentia

Rete testis

Epididymis

(c)

Fimbriae

Suspensory lig. of ovary

Lig. of ovary proper

Mesovarium

Paroophoron

Epoophoron

Corpus uteri

Round ligament of uterus

Cervix

Fornix

Gartner's cyst

Vagina

growth requires the activity of both X chromosomes, and the activity of only one X in individuals with Turner's syndrome leads to death of the oocyte. Secondary loss of the follicle cells follows, *ovarian dysgenesis* (abnormal development) occurs, and a highly regressed or *streak* ovary is present in such women. Conversely, men with *Klinefelter's syndrome* (see Table 1.1) have a normal autosomal complement of chromosomes but *three* sex chromosomes, two X and one Y. Testes form normally in these individuals. However, all the germ cells die off later in life when they enter meiosis and their death is the result of the activity of *two* rather than *one* X chromosomes. These syndromes provide us with two important pieces of clinical evidence. First, *initiation* of gonad formation can occur when *mesenchymal cells* have *one* Y (testis) or *one* X (ovary) chromosome *only*. Second, *completion* of normal gonad development requires that the *germ* cells have two X chromosomes in an *ovary* but *do not have* more than one X chromosome in a *testis*.

We have established that Y-chromosomal activity converts an indifferent gonad into a testis, whereas its absence results in an ovary. *Genetic maleness* leads to *gonadal maleness*. This primary step in sexual differentiation is remarkably efficient, and only rarely are individuals found that have *both* testicular *and* ovarian tissue. Such individuals are called *true hermaphrodites* and arise, in most if not all cases, because of the presence of a mixture of XY and XX (or XO) cells. The way in which such *chimaerism* (mixture of cells with different genotypes) can arise will be discussed in more detail in Chapter 9.

The main role of the sex chromosomes in sexual determination is completed with the establishment of the fetal gonad. From this point onwards genetic sex is unimportant, rather the gonads assume a pivotal position about which the rest of sexual differentiation occurs both pre- and post-natally. In the embryo the testis takes over the 'baton of masculinity' in this sexual relay, and through the endocrine activity of its stromally derived tissue stimulates somatic or *phenotypic maleness*.

**Fig. 1.5.** Differentiation of the internal genitalia in human male and female from: (a) week 6 through (b) the fourth month to (c) the time of descent of testis and ovary. Note the Müllerian and Wolffian ducts are present in both sexes early on, the former eventually regressing in the male and persisting in the female, while the converse is true of the Wolffian ducts. The appendix testis and utriculus prostaticus in the male, and epoophoron, paroophoron and Gartner's cyst in the female are remnants of the degenerated Müllerian and Wolffian ducts, respectively.

**2 The differentiation of
two sexes**

*a The genitalia*

Endocrine activity in the ovaries is *not* essential to sexual differentiation during fetal life. In contrast, the testes actively secrete two *essential* hormones. The mesenchymal interstitial cells of Leydig secrete steroid hormones, *the androgens*, and

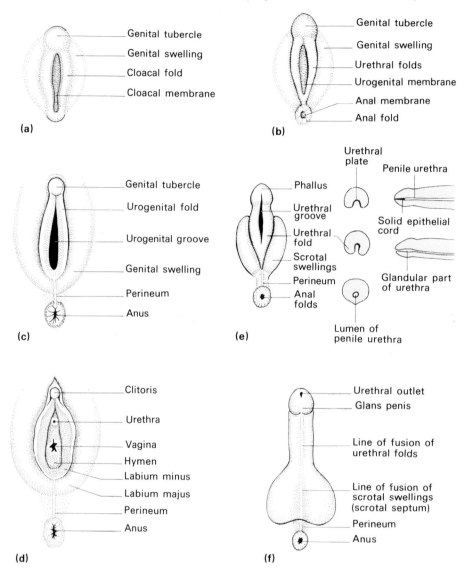

**Fig. 1.6.** Differentiation of the external genitalia in the female (left) and male (right) from common primordia at the indifferent stage: (a) at 4 weeks; (b) at 6 weeks. The changes are more pronounced in the male, with enlargement of the genital tubercle to form the glans penis and fusion of the urethral folds to enclose the urethral tube and the shaft of the penis (the genital swellings probably contribute cells to the shaft). The definitive external genitalia of the male at birth are shown in (f). The changes in the female are much less pronounced, formation of the labia minora from the urethral folds and elongation of the genital tubercle to form the small clitoris being the major changes between the fifth month (c) and birth (d).

the sustentacular, or Sertoli, cells of the seminiferous cords secrete a dimeric glycoprotein hormone (molecular weight 123 000) called *Müllerian inhibiting hormone* (MIH; also called anti-müllerian hormone or AMH). It is these hormones that are the messengers of masculine sexual differentiation sent out by the testis. In their absence feminine sexual differentiation occurs. Sexual differentiation must be actively diverted along the male line, whereas differentiation along the female line reflects an inherent trend requiring no active intervention.

Examination of the primordia of the male and female *internal genitalia* (see Fig. 1.5). shows that instead of one indifferent primordium, as was the case for the gonad, there are two separate sets of primordia, each of which is *unipotential*. These are termed the *Wolffian* or mesonephric (male) and *Müllerian* or paramesonephric (female) ducts. In the female, the Wolffian ducts regress naturally and the Müllerian ducts persist and develop to give rise to the *oviducts, uterus, cervix* and possibly the *upper vagina* (Fig. 1.5). If a male or female fetus is *castrated* (gonads removed), its internal genitalia develop in a normal *female* pattern. This observation demonstrates that ovarian activity is not required for development of the female tract.

In the normal male the two testicular hormones prevent this inherent trend towards the development of female genitalia. Thus androgens, secreted in considerable amounts by the testis, actively induce the Wolffian ducts to develop and give rise to the *epididymis, vas deferens* and *seminal vesicles*. If androgen secretion by the testes should fail, or be blocked experimentally, then the Wolffian duct system will regress and these organs will fail to develop. Conversely, exposure of female fetuses to androgens will cause the development of male internal genitalia.

Testicular androgens have no influence on the Müllerian duct system, however, and the normal regression of this duct in males is under the control of the second testicular hormone, MIH. Thus, incubation of MIH *in vitro* with the primitive internal genitalia of female embryos provokes abnormal regression of the Müllerian ducts.

The primordia of the *external genitalia*, unlike those of the internal genitalia, are *bipotential* (Fig. 1.6). In the female the *urethral folds* and *genital swellings* remain separate, thus forming the *labia minora* and *majora*, whilst the *genital tubercle* forms the *clitoris* (Fig. 1.6). If the ovary is removed these changes still occur, indicating their independence of ovarian endocrine activity. In contrast, androgens secreted from the testes in the male cause the urethral folds to fuse (so enclosing the urethral tube and contributing together with cells from the genital swelling to the shaft of the penis), the genital swellings to fuse in the midline (so forming the

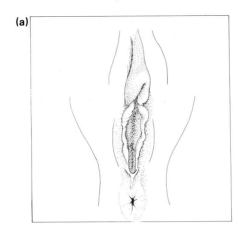

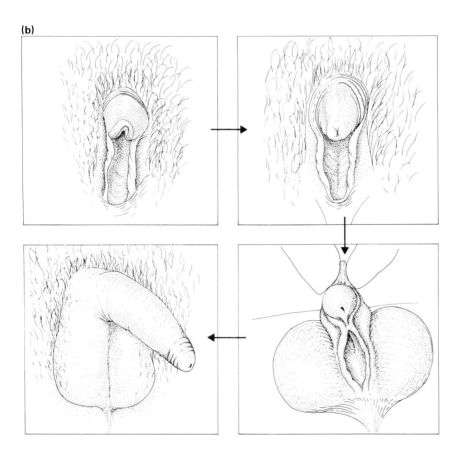

**Fig. 1.7.** (a) External genitalia of an XY adult with testicular feminization syndrome and a testis. The genitalia are indistinguishable from those of a 'normal' female, and at birth the child would be classified as a girl (see Chapter 2 for more details). Copyright Johns Hopkins University Press. (b) The external genitalia from XX girls with the adrenogenital syndrome to varying degrees of severity, from merely an enlarged clitoris to development of a small penis and (empty) scrotum. Ovaries are present. The adrenal cortex has

14　　*Chapter 1*

*scrotum*) and the genital tubercle to expand (so forming the *glans penis*) (Fig. 1.6). Exposure of female fetuses to androgens will 'masculinize' their external genitalia, whilst castration, or suppression of endogenous androgens, in the male results in 'feminized' external genitalia.

Failure of proper endocrine communication between the gonads and the internal and external genitalia can often lead to a dramatic dissociation between gonadal and phenotypic sex. Such individuals are called *pseudohermaphrodites*. For example, in the syndrome of *testicular feminization* (Tfm) the genotype is XY (male), and the testes develop in the normal way and secrete androgens and MIH. However, the fetal genitalia are insensitive to the action of androgens (see detailed discussion in Chapter 2), which results in complete regression of the androgen-dependent Wolffian ducts and the development of female external genitalia. Meanwhile the MIH secreted from the testes exerts its action fully on the Müllerian ducts which regress. Thus, this genetically male individual bearing testes appears female with labia, clitoris and a vagina, but totally lacks other components of the internal genitalia (Fig. 1.7a).

A naturally occurring counterpart to testicular feminization is the *adrenogenital syndrome* in female fetuses. In this syndrome the XX female develops ovaries in the usual way. However, the fetal adrenal glands are hyperactive and secrete large quantities of steroids, some with strong androgenic activity. These androgens stimulate development of the Wolffian ducts, and also cause the external genitalia to develop along the male pattern. The Müllerian system remains as no MIH has been secreted. Thus the individual appears male with penis and scrotum, but is, nonetheless, genetically and gonadally *female* and possesses the internal genitalia of *both* sexes (Fig. 1.7b).

Individuals with Müllerian duct syndrome present as males in whom either MIH production, or responsiveness to it, is inadequate. They will therefore be genetic males, with testes that stimulate external genitalia and Wolffian ducts via androgens but *retain* Müllerian duct structures. They are thus genetically and gonadally *male* and possess the internal genitalia of *both* sexes.

Apart from the problems of immediate clinical management raised by diagnosis of syndromes such as these, abnormalities of development of the external genitalia may have

---

inappropriately secreted androgens at the expense of glucocorticoids during fetal life and directed development of the genitalia along the male line. (If this occurred early enough, the Wolffian ducts would also persist.) Clearly, the more severe cases could lead to assignment as a boy, or to indecision, the consequences of which are discussed later. Copyright Academic Press.

important long-term consequences. The single, most important event in the identification of sex of the new-born human is examination of the external genitalia. These may be unambiguously male or female in type, regardless of whether the genetic and gonadal constitution corresponds. They may also be ambiguous as a result of partial masculinization during fetal life. Since sex assignment at birth is one important step which contributes to the development of an individual's *gender identity*, errors at this early stage can have major consequences for an individual's sexual orientation and attitudes in later life, as we shall now see.

*b The brain and behavioural dimorphism*

Males and females differ from each other during childhood, adolescence and adulthood not only physically, but also behaviourally. How do *sexually dimorphic behaviours develop*? Two major influences have been proposed, but the balance of their relative importance appears to vary depending on the species. It is clear that the action of steroid hormones on the developing brain may result in sexually dimorphic behaviour patterns just as it results in sexually dimorphic genitalia. However, there is also evidence that in primates the influence during childhood of other individuals and society may affect the development of sexual behaviour. We will consider the evidence for both primates and non-primates that bears on the relative significance of the *endocrine* and *social determinants* of behavioural dimorphism.

i Non-primates

The differentiation of neural mechanisms that underlie many types of sexually dimorphic behaviour patterns occurs late in fetal life (e.g. guinea-pig, sheep) or neonatally (e.g. rat, mouse, hamster). Such patterns of behaviour displayed in adulthood include, for example, the distinctive urination patterns shown by the dog (cocked leg) and bitch (squatting). However, the most obvious of sexually dimorphic behaviours are those concerned with reproduction—the patterns of male and female sexual responses which bring the two sexes together and result in coitus. Thus, species-typical patterns of *courtship*, *mounting*, *intromission* and *ejaculation* may be recognized in adult males (Fig. 1.8), while *courtship* (*soliciting*) and *acceptance* patterns, such as *lordosis postures* (Fig. 1.8), may be recognized in adult females. These behaviours are *predominantly*, but not *exclusively*, typical for each sex. Thus, normal males will occasionally accept mounts by other males, i.e. show patterns of behaviour usually associated with females, while females in *heat* will mount one another. *The differences in behaviour are not absolute but quantitative.* Furthermore, it is clear that the differences in behaviour patterns are due to differences in the brains of the male and female animals that result from the irreversible actions of androgens in late fetal or early neonatal life.

16    *Chapter 1*

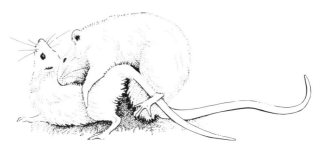

**Fig. 1.8.** Sexually dimorphic patterns of sexual behaviour in male and female rats. Note the immobile *lordosis posture* shown by the *receptive* female which enables the male to *mount* and achieve *intromissions* which will result in *ejaculation*. Lordosis is predominantly shown by females, while mounts, intromission and ejaculation patterns of behaviour are predominantly shown by males. Thus, they are *sexually dimorphic*.

Experimental evidence in support of this view is abundant. Treatment of normal female rats with androgens during the first 5 days of life (the so-called *critical period* of neural sensitivity to hormones) enhances their capacity to display masculine patterns of sexual behaviour in adulthood and also reduces their· capacity to display feminine patterns. Conversely, removal of androgens by castration of a normal male during this critical period results in an enhanced capacity to display feminine patterns of sexual behaviour in adulthood while reducing the capacity for masculine patterns of behaviour. Thus, '*masculinization*' in the rat (and other non-primate mammals) is accompanied by '*defeminization*'.

These changes in adult sexual behaviour induced by neonatal hormone manipulations are clearly due to effects on the brain. For example, neonatal implantation of androgens directly into a region of the neonatal female's brain called the *anterior hypothalamus* (see Chapter 5) results in adults that are both defeminized and masculinized behaviourally. Moreover, the androgens elicit permanent structural changes in the brain, particularly in a region of the anterior hypothalamus called the *medial preoptic area* (see Chapter 5 for details of hypothalamic organisation), which is considerably larger in the male than the female, and where there are also sex differences in the types of synapses made with the dendrites of preoptic neurons. Treatment of neonatal females with androgens produces similar structural changes, and if the neonatal anterior hypothalamus is cultured *in vitro*, addition of androgens stimulates growth and differentiation.

Thus, as we saw for the genitalia, the presence of androgens over a critical period of brain development ensures a masculine pattern of development. In the absence of any steroid influence over this period, female behaviour patterns will develop.

To what extent do androgens exert the same effect on the development of sexually dimorphic behaviour in primates? Experiments on rhesus monkeys suggest some similarities, since young females exposed to high levels of androgens during fetal life display levels of sexually dimorphic behaviours in childhood intermediate between normal males and females, for example, patterns of play (Fig. 1.9) and mounting behaviour (Fig. 1.10). Infant male *and* female monkeys will mount other infants (Fig. 1.10a). However, normal maturing males, but not normal females, progressively display mounts of a mature pattern. (Fig. 1.10b). Androgenized females *also* develop this mounting pattern. Moreover, as adults they attempt to mount other females at a higher frequency than non-androgenized monkeys. Thus, neonatal androgenization does produce persistent 'masculinization'. However, the androgenized females, unlike androgenized female rats, may as adults show normal menstrual cycles and become pregnant. Presumably then they display patterns of adult feminine sexual behaviour at least *adequate* for them to interact successfully with males, suggesting that they are not totally or permanently 'defeminized'.

These results suggest a less complete or persistent effect of androgens on the development of primate bahaviour than on non-primates. However, this apparent difference may be

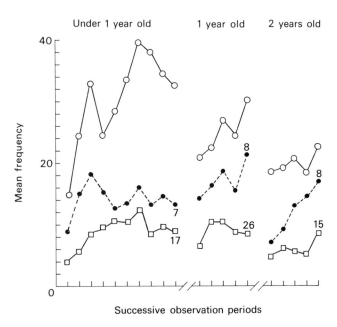

**Fig. 1.9.** Frequency of 'rough-and-tumble play' in normal male (O), female (□) and prenatally testosterone-treated, pseudohermaphroditic 'female' (●) rhesus monkeys during their first, second and third years of life. Note that males display this behaviour at a higher frequency than females and that pseudohermaphroditic females are intermediate.

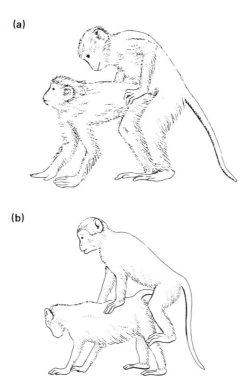

(a)

(b)

**Fig. 1.10.** Sexually dimorphic patterns of mounting behaviour in young rhesus monkeys. Early in life both males and females show 'immature' mounts (a) by standing on the cage floor. During development males show progressively more 'mature' mounts in which (b) they clasp the female's calves so that she entirely supports his weight. Pseudohermaphroditic females display more of the latter type of 'mature' mounts than do normal females.

simply a consequence of the degree to which rats and monkeys are exposed to androgens during development. Thus, the 'sensitive period' for the rat's brain is neonatal, after genital phenotype is established, and is easily accessible to manipulation. If a critical period exists in primates it occurs during fetal life, and may be prolonged. Attempts to androgenize primate fetuses *in utero* often lead to abortion if doses of administered androgens are too high and also tend to masculinize genitalia. Thus, a clear and specific effect of a high dose of androgen on the brain may not yet have been achieved.

iii Humans

Not surprisingly, the difficulty in studying primates is exacerbated further when the human is considered. Sexually dimorphic behaviour patterns analogous to those studied in monkeys can be identified and quantified in humans. For example, (i) *energy expenditure during play* is higher for boys than girls; (ii) *parental rehearsal* patterns in children, such as doll playing and fantasies about adulthood, differ; (iii)

19    *Sex*

*attention to personal appearance* such as clothes, hair and body decoration differ in ways that may be analogous to grooming differences in monkeys: (iv) explicitly *sexual behaviour patterns* also differ. In addition, humans may also be studied uniquely by verbal or questionnaire techniques to determine (v) *gender identity and role* (Table 1.2). It is important to stress again that by all these criteria, the sexes differ quantitatively and not exclusively. Moreover, it is possible for any given individual to show behavioural discordance; for example, a genetically (*XY*) and gonadally male individual who may also be typically male by criteria i–iv, having a male gender identity in the sense of feeling himself to be male, nevertheless may have as the object of all his sexually orientated activity another man. Thus, extrapolation from the demasculinized rat to the homosexual man is naive and quite unjustifiable.

Using these various criteria, the influence of prenatal hormones on subsequent behaviour has been investigated, and two clinical conditions have proved particularly useful. Genetic females with the adrenogenital syndrome (AGS) are nature's counterpart to experimental animals treated exogen-

**Table 1.2** Definition of some elements of human sexuality and behaviour.

| | |
|---|---|
| Gender identity | The self awareness of one's identity or self-image as male, female, or ambivalent. It is a *personal experience*, but which can be expressed socially as gender role. |
| Gender role | Is everything an individual says and does to convey (consciously or subconsciously) to other people his/her gender identity as male, female or ambivalent. It includes sexual arousal, preference and response but is not restricted to them. |
| Sexual preference | Is the preference shown by an individual for members of the same or opposite sex in a sexual context. A heterosexual woman shows sexual preference for a man. A homosexual man also shows partner preference for a man. |
| Sexual arousal | Is the state of sexual excitement or desire at any point in time, but particularly during sexual interaction. |
| Sexual arousability | Is the ability to respond to sexual stimuli with an increase in sexual arousal. |
| Libido | Means, in Freudian terms, 'psychic drive' or 'energy', but has become especially associated with *sexual drive*. The term is often used interchangeably with sexual desire, arousal and receptivity. |

ously with androgens during the critical period. Two sub-groups of these patients have been studied: those exposed to continuing high levels of androgens after birth, so-called 'late-treated' (this group dates back to the 1950s before the simple and effective therapy for this condition was discovered); and 'early-treated' subjects in whom the hyper-secretion of androgens was controlled by cortisol therapy soon after birth. Another group of subjects exposed inadvertently to androgens during gestation also dates to the 1950s when mothers were treated with synthetic progestagens to prevent a threatened abortion. It was subsequently discovered that these progestagens have metabolites with androgenic activity which were responsible, therefore, for 'masculinizing' effects in the fetus, yielding *progestagen-induced hermaphrodites*. The converse of these syndromes is the condition of testicular feminization (Tfm) which occurs in genetic males with a target-organ insensitivity to androgens, the counterpart of experimental castration of neonatal males.

Behavioural findings from the studies of girls with AGS or progestagen-induced hermaphroditism suggest *increased* levels of energy expenditure and athletic interests, associated with a *decreased* incidence of 'rehearsals' of parental behaviour and doll-play activities, as well as diminished interest in dresses, jewellery and hairstyles. This spectrum of behaviour has been termed '*tomboyism*' and is well-recognized and accepted in Western culture. It provides few, if any, problems for children so affected. Furthermore, this 'tomboyism' seems to represent the consequence of the effects of androgens on the fetal brain, the behaviour appearing to be more characteristic of boys than girls, rather like the changes in rough-and-tumble play in infant monkeys exposed pre-natally to androgens. However, if these human subjects are again examined as adults, there is little evidence of behav-ioural consequences of these abnormal endocrine events in their fetal life. Thus, their sexual orientation is heterosexual, they have boyfriends (although they begin dating a little later than controls), they get married, they have children and they prove to be unexceptional mothers. There is no evidence of a higher incidence of dissatisfaction with their female gender identity than in controls and no evidence of obligatory homosexuality, although occasional lesbian contacts were reported by some of the 'late-treated' AGS subjects. Thus, the *early androgen exposure*, although recognizable in childhood play behaviour, *did not obviously determine gender identity* in the adult.

These results apply only if one important condition is fulfilled. It is critical that these children be *assigned as girls at birth and unambiguously reared as such*. However, if the 'masculinization' of their external genitalia is severe enough for them to be assigned as boys at birth, and they are subsequently *reared as boys*, then their adult *gender identity is masculine*. They are attracted to and date girls and have no desire for sex reassignment. Of course, a 'masculinizing' puberty must be achieved by giving exogenous androgens, and their ovaries must be removed, although an enduring problem for these subjects in sexual encounters is the presence of a very small penis and the absence of testes.

Subjects with the syndrome of complete or partial testicular feminization provide evidence for similar conclusions. In the complete form of Tfm, subjects reared as girls, because their external genitalia were unambiguously female at birth, display a clearly feminine gender identity despite their XY genotype. They are attracted to, date and marry men. Conversely, Tfm subjects in which the syndrome is incomplete, and who have partially masculinized external genitalia, are reared as boys and develop a masculine gender identity. This presents a more serious problem at puberty since the androgen insensitivity which characterizes the syndrome ensures that a masculinizing puberty is impossible—indeed, these subjects develop breasts and female body form under the action of their own oestrogens.

These clinical cases have been taken as evidence that *sex of rearing* is the most important single event determining subsequent differentiation of gender identity, since it establishes the framework for sexually specified social interaction during childhood and adolescence.

Clinical psychologists and psychiatrists generally believe that this process of differentiation of male or female gender identity is probably completed by about 2–3 years of age, after which it is an irrevocable part of a boy's or girl's personality. Ambiguity on the part of parents about the sex of their child may then represent one risk to the subsequent, safe development of gender identity. Thus, it has been suggested, but not universally agreed, that cases of transexualism are particularly closely associated with a sexually ambiguous childhood; for example, parents treating their son more as a girl, clothing him in dresses and not reinforcing 'boyish' activities in play and sport. If this is indeed the case, much depends on the genitalia at birth!

However, it will be evident from what has gone before that this postulated environmental determination of sexual behaviour can only be a tentative conclusion. Clearly, only the less severely androgenized females with AGS will be brought up as girls, and only the more severely as boys. The converse applies with Tfm syndrome. Thus, a clear separation of endocrine and social factors has not been achieved. Powerful evidence *against* a major role for social factors has come from the recognition of a syndrome first identified among a subgroup of the population of the Dominican Republic, but now identified in Europe also. These individuals are genetic males with normal functional testes; however, they are born with external genitalia sufficiently diminutive as to be classifiable as female. They are reared as girls but at puberty the 'clitoris' rapidly enlarges to form a penis, the 'labia' enlarge to receive the testes and the body hair pattern and phenotype assume the characteristics of a man. What is especially important is that at puberty these individuals, the so-called '*Guevodoces*' (or 'penis-at-twelves'), are reported to switch their psychosexual identity to that of a male. The syndrome is caused by the presence of a recessive gene on an autosome that results in the lack of a specific enzyme, 5*a*-reductase (see Chapter 2 for details). This enzyme normally converts the weak androgens secreted by the testes to a very potent form that is more active in the tissues. Thus, the individuals have weak androgens during development, but only at puberty when the output of testicular androgens rises dramatically are levels of these weak androgens sufficiently high to overcome the lack of enzyme and thereby to complete the full process of masculinization. Thus, the immature individuals are masculine but minimally and ambiguously so, the adults are fully masculine and despite being reared as girls, they nonetheless can function behaviourally as men. Careful studies of these patients may permit the balance of social and endocrine influences on human sexual behaviour to be assessed more accurately.

iv Summary

It is perhaps not surprising that the relatively simple rules governing the development of sex differences in the behaviour of rodents cannot be applied easily to primates and man. Thus, the finding that exposure of the rodent's brain to androgens during a critical period subsequently *masculinizes* while reciprocally *defeminizing* patterns of sexual behaviour in adulthood, does not find a simple counterpart in monkeys or humans. In both the latter species, behavioural evidence of exposure of the fetal brain to androgens is abundant in the form of childhood behaviour, the so-called 'tomboy syndrome'. However, it is not clear whether such masculinization of behaviour is accompanied by defeminization, and affected individuals can display apparently normal patterns of feminine sexual behaviour as adults. This has led to a dogma that sex assignment at birth and subsequent gender-specific patterns of social interaction are the most important determinants of behavioural dimorphism. This view is undoubtedly too simple and the sex hormone environment of the developing human's brain may well be an important determinant of adult sexual orientation. The interaction between prenatal hormonal, postnatal hormonal and social factors in determining behavioural sexual dimorphism thus remains an area of intense interest and investigation, with the relative importance of each factor yet to be determined.

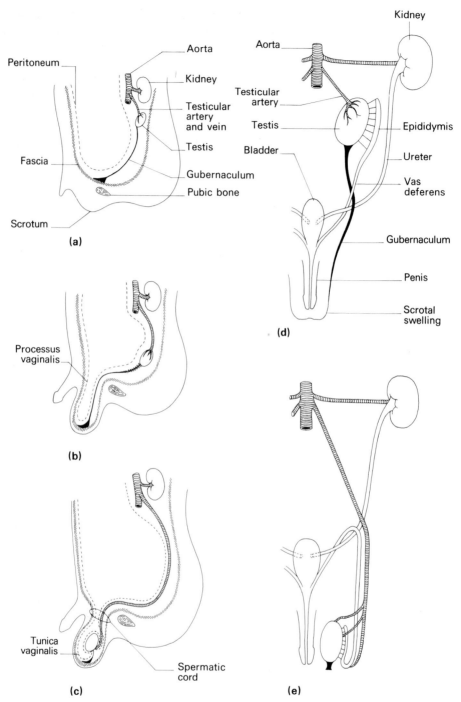

**Fig. 1.11.** (a)–(c) Parasagittal sections through developing abdomen. Initial retroperitoneal, abdominal position of testis shifts to pelvis, extending blood supply (and also Wolffian duct derivatives not shown) as gubernaculum becomes shorter. Musculo-fascial layer evaginates into scrotal swelling accompanied by peritoneal membrane which forms the processus vaginalis. At about 8 months of pregnancy in the human, the testis migrates over pubic bone behind

## 3 Pre- and post-natal growth of the gonads

We have seen how many of the essential phenotypic and behavioural features of the male and female are laid down in the embryo. The male arises by active genetic and then endocrine activities, the female develops passively in the absence of these activities. These same rules operate during the remainder of prenatal life and during postnatal life up to puberty. Over this period, further divergence of sexual phenotypes occurs only at a very slow pace. In the male, this process is dependent upon low levels of gonadal androgens. The female grows and retains a feminine phenotype in the *absence* of such androgens or indeed of any gonadal steroids.

### a The testis

The genital ridge develops in the upper lumbar region of the embryo, yet in the adults of most species of mammals the testes have descended through the abdominal cavity and over the pelvic brim to arrive in the scrotum (Fig. 1.11). Evidence of this extraordinary migration is found in the nerve and blood supplies to the testis, which retain their lumbar origins and pass on an extended course through the abdomen to reach their target organ. Testicular migration may, in certain pathological conditions, be arrested at some point on the migratory route resulting in *cryptorchidism* (hidden gonad). Examples of 'evolutionary cryptorchidism' are provided by a few species in which testes normally do not descend at all from the lumbar site (monotremes, elephants and hyraxes), migrate only part of the route to the rear of the abdomen (armadillos, whales and dolphins), lodge in the inguinal canal (hedgehogs, moles and some seals) or retain mobility in the adult, migrating in and out of the scrotum to and from inguinal or abdominal retreats (most rodents, wild ungulates).

It is not clear *how* or *why* the testis undergoes this migration in most mammals, although there is evidence that both androgens and MIH are required for successful completion of the process (the latter probably acting indirectly by removing the Müllerian duct structures which would otherwise obstruct descent; cryptorchid testes are frequently associated clinically with residual Müllerian duct structures). Migration may be rather too active an expression since much

---

processus vaginalis, which wraps around it forming a double-layered sac. The fascia and peritoneum become closely apposed above the testis, obliterating the peritoneal cavity leaving only a tunica vaginalis around the testis below. The fascial layers, obliterated stem of processus vaginalis, vas deferens and testicular vessels and nerves form the spermatic cord. In some species, and pathologically in man, the closure may not be complete and the testis may be able to move back up into a pelvic location. (d) and (e) Front view of migration showing extended course ultimately taken by the testicular vessels and vas deferens.

of the descent within the abdomen may occur simply by a process of differential growth. Thus, the testis is firmly attached caudally to the posterior abdominal wall by the fibrous remains of the mesonephric tissues, which form a *posterior gonadal ligament*, and by a fibrous structure, the *gubernaculum* (Fig. 1.11). As the body increases in size, the gubernaculum does not, and thus the relative position of the testis becomes increasingly caudal.

The consequences of cryptorchidism in those mammals with scrotal testes demonstrate that a scrotal position is essential for normal testicular function. Although endocrine activity is not affected in any major way, spermatogenesis is arrested, testicular metabolism is abnormal and the risk of testicular tumours rises. These effects can be simulated by prolonged warming of the scrotal testis, either experimentally or with thick tight underwear! The normal scrotal testis functions best at temperatures 4–7°C lower than abdominal 'core' temperatures. Cooling of the testis is improved by copious sweat glands in the scrotal skin and by the circulatory arrangements in the scrotum. The *internal spermatic arterial* supply is coiled, or even forms a rete in marsupials, and passes through the spermatic cord in close association with the draining venous *pampiniform plexus*, which carries peripherally cooled venous blood (Fig. 1.12). Therefore, a heat exchange is possible, cooling the arterial and warming the venous blood. However, although the scrotal testis clearly requires a lower ambient temperature, this requirement may be a secondary *consequence* of its scrotal position rather than the original evolutionary *cause* of its migration. Those species in which testes remain in the abdomen survive, flourish and reproduce despite the high testicular temperature. It remains unclear as to why testes are scrotal in so many species.

By the sixteenth to twentieth week of human fetal life, the testis consists of an outer fibrous tunica albuginea enclosing vascularized stromal tissue, which contains condensed Leydig cells and solid seminiferous cords comprised of a basement membrane, sustentacular cells and germ cells. These cords connect to the cords of the rete testis, the vasa efferentia and thereby to the epididymis (see Figs 1.4 and 1.5). The Leydig cells in the human testis actively secrete testosterone from at least 8–10 weeks of fetal life onwards, blood levels peaking at 2 ng/ml at around 13–15 weeks. Thereafter, blood levels decline and plateau out by 5–6 months at a level of 0.8 ng/ml (cf. adult male level of *c*. 9 ng/ml: Chapter 2). This transient *prenatal peak* of blood testosterone is a feature of many species although in some, for example rat and sheep, the peak may approach, or span, the period of parturition, and only begin its decline postnatally. This testosterone secretion is, as we saw, essential for establishing the male phenotype. In addition, the males of some primate species, including man, show a second

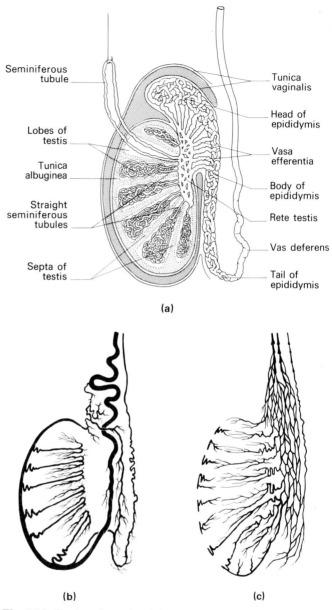

Seminiferous tubule

Lobes of testis

Tunica albuginea

Straight seminiferous tubules

Septa of testis

Tunica vaginalis

Head of epididymis

Vasa efferentia

Body of epididymis

Rete testis

Vas deferens

Tail of epididymis

(a)

(b)

(c)

**Fig. 1.12.** Section through adult human testis to show: (a) general structure; (b) arterial supply; (c) venous drainage.

*postnatal* peak in plasma testosterone concentrations reaching 2–3 ng/ml by 2 months postpartum, but declining to around 0.5 ng/ml by 3–4 months.

Throughout pregnancy and early postnatal growth, testis size tends to increase slowly but steadily, but the germ cells do not contribute to this by proliferating, and indeed, in some species, many of them die. At puberty, however, there is a sudden increase in testicular size, endocrine secretion by the

Leydig cells increases sharply, the solid seminiferous cords canalize to give rise to tubules, the sustentacular cells increase in size and activity and become *Sertoli cells*, and the germ cells resume mitotic activity. The causes of this sudden growth at puberty will be discussed in detail in Chapter 6. The consequences are the subject of much of the remainder of this chapter.

*b The ovary*

The ovary, unlike the testis, retains its position within the abdominal cavity, shifting slightly in some species, such as man, to assume a pelvic location. It is attached to the posterior abdominal wall by the ovarian mesentery or mesovarium. The ovary, like the testis, grows slowly but steadily in size during early life. Over this period its output of steroids is minimal and, indeed, removal of the ovary does not affect prepubertal development. However, at puberty marked changes in both the structure and endocrine activity of the ovary occur, and for the first time the ovary becomes an essential and positive feminizing influence on the developing individual.

**4 Puberty**

Puberty is best regarded as a collective term which encompasses all the physiological, morphological and behavioural changes which occur in the growing individual as the gonads change from an infantile to an adult condition. Puberty has been studied in most detail in the human. A fairly definitive sign that this has occurred in girls is *menarche*, the first *menstrual bleeding*, which indicates that the ovary has secreted sufficient of its steroids, *oestradiol* and *progesterone*, to induce development of the uterus. An indication of a similar stage of maturity in boys is the first *ejaculation*, which often occurs nocturnally and is thus much more difficult to date precisely. The first menstruation and ejaculation do not signify fertility. Indeed, in early puberty the ovary does not ovulate and the ejaculate consists of small quantities of seminal plasma lacking spermatozoa. Rather, these two dramatic events are signs that the gonads have been awakened and are beginning to assume adult levels of activity.

Some 2–4 years prior to these obvious signs of sexual maturation, a series of other changes in most of the organs and in the structure of the body is initiated. These changes are dependent on, and orchestrated by, the increasing level of sex steroids from the gonads and also from the adrenal glands. This sequence of maturational changes does not begin at the same chronological age or take the same length of time to reach completion in all children, even though the sequence in which these changes occur varies but little.

*a Physical changes*

The *adolescent growth spurt* is an acceleration, followed by a deceleration, of growth in most skeletal dimensions, and can be divided into three stages: the time of minimum growth velocity (or 'age at take-off'); the time of peak height velocity (PHV); and the time of decreased growth velocity and cessation of growth at epiphyseal fusion. In Fig. 1.13 it can be seen that boys commence their growth spurt about 2 years later than girls, they are therefore taller at the age of take-off and reach their PHV 2 years later. The height gain of boys and girls between take-off and cessation of growth is similar, about 28 and 25 cm respectively, indicating that the 10 cm difference in mean height between adult males and females is due more to the height difference at take-off than gain during the spurt. Age at take-off and PHV are poor indicators of adult height, and in addition they show a poor correlation with the rate of passage through the various stages of puberty described below. Virtually every muscular and skeletal dimension is involved in the adolescent growth spurt; however, sex differences in the growth rate of different regions occur which enhance sexual dimorphism in the adult, for example the shoulders (greater in boys) and hips (greater in girls). This dynamic phase of growth is dependent not only on

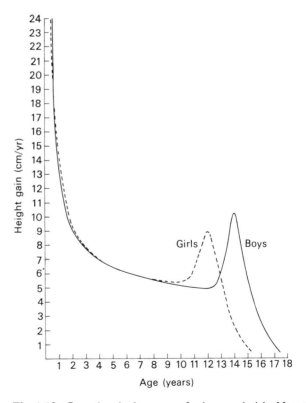

**Fig. 1.13.** Growth velocity curves for boys and girls. Note the later 'take off' in boys which generally ensures a greater height at the start of the adolescent growth spurt. Also note that average peak height velocity (PHV) is 9 cm/year for girls and 10 cm/year for boys.

sex steroids but also on *growth hormone* from the anterior pituitary. Thus, hypopituitary patients must be given both growth hormone and steroids if a pubertal growth spurt is to occur.

As well as growth, considerable changes in body composition occur during puberty. Lean body mass and body fat, components of total body mass, are virtually identical in prepubertal girls and boys. In adulthood, men have about 1.5 times the lean body mass of women, while women have twice as much body fat as men. In addition, the skeletal mass of men is 1.5 times that of women. These alterations in body mass commence at about 6 and 9 years of age in girls and boys re-

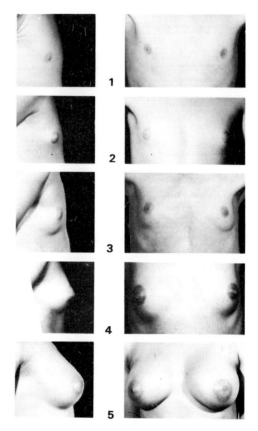

**Fig. 1.14.** Stages of breast development. 1 pre-adolescent stage during which the papilla (nipple) alone is elevated; 2 breast bud stage in which papilla and breast are elevated as a small mound; 3 continued enlargement of the breast and areola, but without separation in their contours; 4 further breast enlargement but with the papilla and areola projecting above the breast contour; 5 mature stage in which the areola has become recessed, and forms a smooth contour with the rest of the breast. Only the papilla is elevated. Classification of this stage is independent of breast size, which is determined principally by genetic and nutritional factors.

spectively, and represent the earliest changes in body composition at puberty. The greater strength of men compared to women reflects a greater number of larger muscle cells.

In addition to growth and changes in body composition, development of the *secondary sexual characteristics* occurs at puberty, e.g. breasts (Fig. 1.14), genitalia and pubic hair (Figs 1.15 and 1.16), beard growth and voice change. Ovarian oestrogens regulate growth of the breast and female genitalia, but *androgens* from both ovary and adrenal (see Chapter 6) control the growth of female pubic and axillary hair. Testicular androgens not only control development of the genitalia and body hair in boys, but by enlarging the *larynx* and *laryngeal muscles* lead to *deepening* of the voice.

These various characteristics develop at very different chronological ages in different individuals (see for example

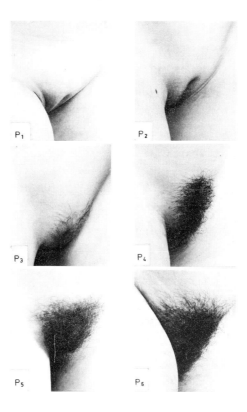

**Fig. 1.15.** Stages of pubic hair development in girls. 1 No pubic hair is visible; 2 sparse growth of long, downy hair which is only slightly curled and situated primarily along the labia; 3 appearance of coarser, curlier and often darker hair; 4 hair spreads to cover labia; 5 hair spreads more over the junction of the pubes and is now adult in type but not quantity. No spread to the medial surface of the thighs. 6 Adult stage in which the classical 'inverse triangle' of pubic hair distribution is seen, with additional spread to the medial surface of the thighs.

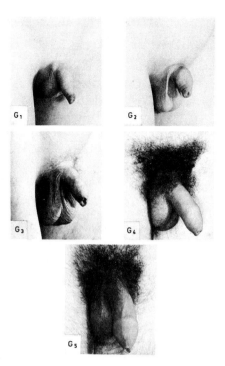

**Fig. 1.16.** Stages of genitalia development in boys. 1 Pre-adolescent stage during which penis, testes and scrotum are of similar size and proportion as in early childhood: 2 scrotum and testes have enlarged; texture of the scrotal skin has also changed and become slightly reddened; 3 testes and scrotum have grown further but now the penis has increased in size—first in length and then in breadth; facial hair appears for the first time on upper lip and cheeks; 4 further enlargement of the testes, scrotum (which has darkened in colour) and penis; the glans penis has now begun to develop; 5 adult stage; facial hair now extended to lower lip and chin.

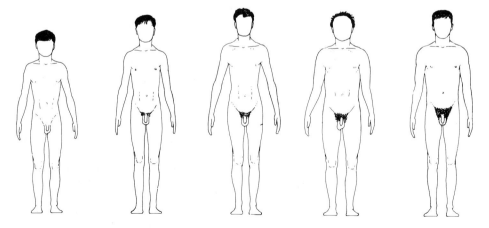

**Fig. 1.17.** Five boys aged 14 to illustrate the marked differences in maturation at the same chronological age.

Fig. 1.17). However, the *sequence* in which the changes occur is quite *characteristic* for each sex. This is important for the clinician, who has *staging criteria* (summarized in Figs 1.14, 15 and 16) by which he can detect abnormalities and make comparisons among individuals, populations and cultures. For example, a boy with advanced penile and pubic hair growth but small testes must have a nongonadal source of excess androgen, such as congenital adrenal hyperplasia or an adrenal tumour.

*c Puberty in other mammals*

All mammals undergo pubertal changes equivalent to those described in detail for the human. However, the details differ as do the timings. For example, whereas in the human female the first evidence of fertility occurs between 12 and 14 years postnatally, in most other species it occurs much earlier, e.g. ewe (6–7 months), cow (12 months), pig (7 months), mouse (30–35 days). Prepubertal development is prolonged in higher primates.

## 5 Summary

In this chapter we have seen that sexual differentiation is an enduring process of divergence, which begins with the expression of a genetic message that establishes the structure and nature of the fetal gonad, and which then extends from the gonad, via its hormonal secretions and over a period of several years, to many tissues of the body. Thus, sex may be defined at several levels and by several parameters. Concordance at all levels is often incomplete, and the medical, social and legal consequences of this 'blurring' of a clear, discrete sexual boundary may pose problems. However, in this chapter we have been able to define, by a broad set of criteria, how the two sexes are established and attain sexual maturity. In subsequent chapters we will look at the physiological processes regulating fertility and sexual behaviour which result in conception, pregnancy, parturition, lactation and maternal care.

**Further reading**

Austin CR & Edwards RG. *Mechanisms of Sex Differentiation in Animals and Man*. Academic Press, 1981.

Bancroft J. *Human Sexuality and its Problems*. Churchill Livingstone, 1983.

Byskov AG. Differentiation of mammalian embryonic gonad. *Physiol Rev* 1986; **66**: 71–117.

Goodfellow PN, Craig IW, Smith JC & Wolfe J. The mammalian Y chromosome: molecular search for the sex-determining gene. *Development* 1987; **101** (Suppl).

Green R. *The 'Sissy Boy Syndrome' and the Development of Homosexuality*. Yale University Press, 1987.

Hughes IA. Precocious puberty and its management. *Br Med J* 1983; **66**: 664–665.

Josso N & Picard JY. Anti-Müllerian hormone. *Physiol Rev* 1986; **66:** 1038–1091.

McLaren A & Wylie C. *Current Problems in Germ Cell Differentiation.* Cambridge University Press, 1983.

Money J. *Man and Woman, Boy and Girl.* Johns Hopkins University Press, 1972.

Tanner JM. *Growth at Adolescence.* Blackwell Scientific Publications, 1966.

# Chapter 2
# The Sex Steroids

Two major conclusions were drawn in Chapter I: first, the gonads were seen to be pivotal organs in the reproductive process, translating an individual's genetic sex into phenotypic and behavioural sex; second, the gonads were seen to achieve this function largely through the mediation of their endocrine secretions—the steroid hormones. In this chapter, we will examine more closely some general properties of the steroid hormones themselves. The complexity of the steroid hormone system can appear daunting at first, but in grasping a few *general* points about steroids, the complexities evaporate.

## 1 Steroidogenesis
### a Cholesterol

The steroid hormones comprise a large group of molecules all derived from a common sterol precursor–cholesterol (Fig. 2.1). Throughout this book we will illustrate the pathway for biosynthesis of steroids from cholesterol in the format shown in Fig. 2.2 (see folder, facing p. 36); continued reference to this figure will give a visual framework of the molecular relationships.

Cholesterol is a lipid, and it is synthesized from acetate in many tissues of the body. It is a highly fat-soluble molecule with only limited solubility in water, and accordingly it localizes in the lipid regions of cell membranes. Indeed, it has become clear in recent years that cholesterol is a molecule of major importance in maintaining the structural integrity of the plasma membrane of the cell. If the cholesterol content of mammalian cells is manipulated to abnormal levels experimentally, or varies pathologically, then the cell membranes cease to function properly, are destabilized and rupture more easily. *Cholesterol has a widespread structural role in*

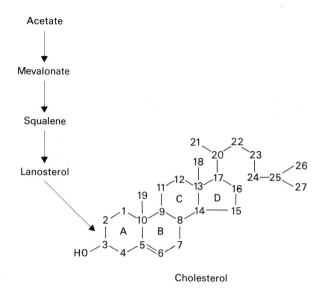

Acetate

↓

Mevalonate

↓

Squalene

↓

Lanosterol

Cholesterol

**Fig. 2.1.** The basic structure of the cholesterol molecule. Each of the 27 carbon atoms is assigned a number and each ring a letter. The individual carbon atoms are simply carrying hydrogen atoms unless otherwise indicated, e.g. C1 and C19 are $-CH_2-$ and $-CH_3-$ residues whereas C3 is $-CHOH-$. Cholesterol is converted to pregnenolone by cleavage of the terminal six carbons leaving a steroid nucleus of 21 carbons. This conversion occurs within the mitochondria, and requires NADPH and oxygen.

the body, and is not merely a substrate for steroidogenesis.

In steroidogenic tissues, it is probable that the bulk of the steroids produced are derived primarily from acetate with cholesterol as an intermediate product (Fig. 2.1). However, direct synthesis of steroids from cholesterol also occurs, and the cholesterol substrate utilized may be derived from one of two sources. Most cells have reserves or pools of esterified cholesterol to draw on but, in addition, cholesterol circulates in the blood and tissue fluids, and may be taken up and used by cells. The relatively poor aqueous solubility of cholesterol means that much of the blood cholesterol is complexed with

**Table 2.1.** Steroid binding proteins in human plasma.

| Protein | Rough percentage of non-conjugated steroids bound* | | | |
|---|---|---|---|---|
| | Progestagens | Androgens | Oestrogens† | Cortisol |
| Albumin | 48 | 40 | 48 | 20 |
| Cortisol binding globulin | 50 | 1 | — | 70 |
| Sex steroid binding globulin | — | 57 | 51 | — |
| Free steroid | 2 | 2 | 1 | 10 |

* Steroids conjugated as sulphates or glucosiduronates bind weakly to albumin only.
† Oestrone and oestriol bind mainly to albumin.

carrier protein molecules, like albumin and lipoproteins. These proteins carry a large reservoir of cholesterol which is in equilibrium with much lower levels of free cholesterol in aqueous solution—rather in the way that haemoglobin carries reservoirs of bound oxygen.

*b General properties of steroids*

The conversion of cholesterol to pregnenolone marks the first and common step in the formation of all the major steroid hormones. The conversion is rate-limiting and is therefore a most important point in the biosynthetic path at which the *rate* of steroid synthesis may be *regulated*. The conversion occurs on the inner mitochondrial membrane and requires NADPH, oxygen and cytochrome P-450. Pregnenolone is then converted to the active sex steroids. It can be seen from Fig. 2.2 that the overall basic structure of all steroids is very similar to that of cholesterol. Therefore, it follows that the steroids too are highly fat-soluble but only poorly water-soluble. This relatively low aqueous solubility has two consequences. First, like cholesterol, much of the steroid present in the blood is also carried on binding proteins. There are proteins such as albumin, which, although they have a relatively low affinity for steroids, are present in abundance and therefore their total capacity for carrying steroids is high. In addition, there are also special binding proteins which have high affinities for steroids (Table 2.1).

Second, like cholesterol the steroid hormones in the fluids of the body have ready access to the interiors of cells. The lipid cellular boundaries present no barrier to the steroids, which will partition selectively to these hydrophobic areas of the cell. This ubiquity of steroids has important physiological consequences for, just as insertion of large numbers of cholesterol molecules into cell membranes affects the membrane properties, so does insertion of high levels of steroids. For example, high levels of progesterone affect the stability and ionic permeability of cell membranes in the central nervous system, and this underlies progesterone's anaesthetic properties. Similarly, high levels of corticosteroids stabilize lysosomal membranes and thereby reduce inflammation. It is important to remember when considering steroids that if they are present in large enough amounts, they have the capacity to act as 'structural units' of the cell. In order to exert more restricted reproductive effects, rather than these widespread general effects, the blood levels of the sex steroids are maintained at relatively low levels and well below those of cholesterol (see Table 2.2). The specific actions of the sex steroid are achieved by special 'concentrative' mechanisms. Thus, although the steroids permeate all the tissues and enter the cells of the body, there are some tissues which have the capacity to select, concentrate and retain certain steroids (Fig. 2.3). These tissues contain binding proteins known as

**Table 2.2.** Mean blood levels of cholesterol and of some major steroids (human).

| | Blood level (ng/ml) | | | |
|---|---|---|---|---|
| | Men | Women | | |
| | | Prior to ovulation | After ovulation | Pregnant |
| Cholesterol | $15 \times 10^5$ | $15 \times 10^5$ | $15 \times 10^5$ | $15 \times 10^5$ |
| Progesterone | 0.3 | 1.0 | 11.0 | 160.0 |
| Testosterone | 6.6 | 0.4 | 0.4 | 1.4 |
| Androstenedione | 1.2 | 1.6 | 1.6 | 0.8 |
| 5 $\alpha$-Dihydro-testosterone | 0.5 | 0.2 | 0.2 | 0.3 |
| Oestradiol | 0.02 | 0.1 | 0.2 | 15.0 |
| Oestrone | 0.03 | 0.05 | 0.2 | 52.0 |
| Oestriol | — | — | — | 113.0 |
| Cortisol | 70.00 | 50.0 | 50.0 | 50.0 |

*receptors* with which the steroids can form noncovalent bonds. A variety of receptors exists within the tissues of the body, each having a distinct affinity for binding a specific steroid. A tissue may, therefore, selectively concentrate highlevels of a certain steroid within its cells. The ability of a tissue to bind a steroid correlates well with the ability of the steroid

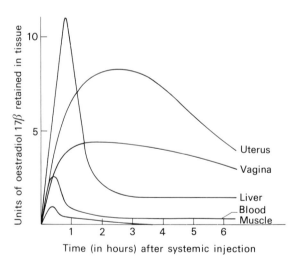

Time (in hours) after systemic injection

**Fig. 2.3.** Fate of oestrogens within various tissues of the body. Radioactive oestradiol 17$\beta$ was injected into female rats, and tissues were sampled for radioactivity at time intervals thereafter. Note how only the target tissues, the uterus and vagina, retain radioactivity due to the presence of binding proteins (molecular weight 100 000–200 000; binding affinities $10^{-9}$ to $10^{-10}$ M). The transient high levels of oestradiol within the liver reflect its role as the principle organ for metabolic degradation of steroids (see later).

to act on that tissue. It seems that tissues become *targets* for specific steroid action only if they possess receptors. We must view the actions of steroids as reflecting not simply their structure and properties, but also the structure and properties of the receptors available to them.

*c Classes of steroid*

Four major classes of steroid are derived via pregnenolone from cholesterol—the *progestagens*, the *androgens*, the *oestrogens* (also spelt *estrogens*) and the *corticosteroids* (Fig. 2.2). Only the first three classes are regarded as the *sex steroids*—the fourth class of steroids is made and released by the adrenal cortex in abundance. An important general point to grasp is that, although there are three distinctive classes of sex steroids, they are related to each other. Indeed, they can be seen as different generations of a family, the progestagens being 'parental' to androgens on the biosynthetic pathway and the oestrogens being 'filial' to the androgens (Fig. 2.2). Interconversion from one class of steroids to another is undertaken by a series of enzymes. It seems probable that all the enzymes required within one cell for synthesis of a specific steroid are arranged together as a 'biosynthetic unit', taking in substrate and passing the molecule along an enzyme production line. For example, the enzymes, 17 α-hydroxylase, 17,20-desmolase, 17-ketosteroid reductase and 3β-hydroxysteroid dehydrogenase would form an enzyme package for the synthesis of testosterone from pregnenolone. They are organized such that there would be little 'leakage' of any intermediates from the production line. Whilst the conversion from cholesterol to pregnenolone occurs in the mitochondria, the subsequent steps in steroidogenesis take place in the adjacent smooth endoplasmic reticulum.

The close relationship of the different classes of steroids means that an enzymic defect at one point in the synthetic pathway may have far-reaching effects. For example, quite common genetic deficiencies in the fetal adrenal are reduced activity of 21-hydroxylase (which converts 17 α-hydroxyprogesterone to 11-deoxycortisol; Fig. 2.2) or of 11β-hydroxylase (which converts 11-deoxycortisol to cortisol). In each case the absence of adequate levels of corticosteroids leads to greater stimulation of the biosynthetic path (*congenital adrenal hyperplasia*) and the accumulation of high levels of 17 α-hydroxyprogesterone. These are converted by 17,20-desmolase to androgens which can masculinize female fetuses (see Chapter 1; adrenogenital syndrome). Conversely, *genetic deficiency* of 17,20-desmolase activity results in *depressed* androgen output, and the failure of male fetuses to masculinize.

Within each class of sex steroids there are several natural members. The criteria for membership of a class are twofold. There are general *functional* criteria that characterize the

39    *The Sex Steroids*

**Table 2.3.** Principal properties of natural progestagens.

| Progestagens (relative potency)* | Properties (either some or all) |
|---|---|
| Progesterone ($=$P) (100%)<br>17 $\alpha$-Hydroxyprogesterone ($=$17$\alpha$OHP) (40–70%)<br>20 $\alpha$-Hydroxyprogesterone ($=$20 $\alpha$-Dihydroprogesterone or 20 $\alpha$-OHP) (5%) | 1 Prepare uterus to receive conceptus<br>2 Maintain uterus during pregnancy<br>3 Stimulate growth of mammary glands, but suppress secretion of milk<br>4 Mild effect on $Na^+$ loss via distal convoluted tubule of kidney<br>5 General mild catabolic effect<br>6 Regulate secretion of gonadotrophins |

* In Tables 2.3–2.5 the relative potencies are only approximate since they vary with species and with the assay used. This variation is due partly to differences in the relative affinity of receptors in different tissues, partly to differences in local enzymic conversions of steroids within tissues and partly due to differences in systemic metabolism—see Section 2.ii for discussion of these factors.


**Table 2.4.** Principal properties of natural androgens.

| Androgens (relative potency)* | Properties (either some or all) |
|---|---|
| 5 $\alpha$-Dihydrotestosterone ($=$DHT) (100%)<br>Testosterone ($=$T) (50%)<br>Androstenedione ($=\Delta$ 4) (8%)<br>Dehydroepiandrosterone ($=$DHA) (4%) | 1 Induce and maintain differentiation of male somatic tissues<br>2 Induce secondary sex characters of males (deep voice, body hair, penile growth) and body hair of females<br>3 Induce and maintain some secondary sex characters of males (accessory sex organs)<br>4 Support spermatogenesis<br>5 Influence sexual and aggressive behaviour in males and females<br>6 Promote protein anabolism, somatic growth and ossification.<br>7 Regulate secretion of gonadotrophins (testosterone) |

* As for Table 2.3.


**Table 2.5.** Principal properties of natural oestrogens.

| Oestrogens (relative potency)* | Properties (either some or all) |
|---|---|
| Oestradiol 17$\beta$ ($=$Estradiol or $E_2$) (100%)<br>Oestriol ($=$Estriol or $E_3$) (10%)<br>Oestrone ($=$Estrone or $E_1$) (1%) | 1 Stimulate secondary sex characters of female<br>2 Prepare uterus for spermatozoal transport<br>3 Increase vascular permeability and tissue oedema<br>4 Stimulate growth and activity of mammary gland and endometrium<br>5 Prepare endometrium for progestagen action<br>6 Mildly anabolic; stimulates calcification<br>7 Active during pregnancy<br>8 Regulate secretion of gonadotrophins<br>9 Associated with sexual behaviour in some species |

* As for Table 2.3.

members of a class, that is, they show certain similarities of action. There are also *chemical* criteria that characterize all natural members of the class, that is, they look alike structurally. Each of these criteria will be examined separately.

i Functional features

Tables 2.3–2.5 summarize the principal natural progestagens, androgens and oestrogens, together with some of the functional criteria by which each class is characterized. In essence, the progestagens are characterized by activities associated with the preparations for pregnancy and its maintenance; the androgens by activities associated with development and maintenance of masculine traits; and the oestrogens by activities associated with the development and maintenance of feminine traits, including preparation for egg production and fertilization. Each of these properties will be examined more closely in later chapters. For the present, however, it is important to note that these are *general* but *not exclusive* characteristics. Common abbreviations or alternative names encountered in the literature are also recorded in Tables 2.3–2.5.

ii Structural features

The functional similarities shown by members of the same steroid class reflect their ability to combine with tissue receptors of similar specificity. Since steroid–receptor interaction depends upon a good stereochemical fit between the two molecules, as a key must fit a lock, it is not surprising that the general shapes and therefore primary structure of members of the same class are similar. Natural progestagens are characterized by 21 carbons (C21 steroids), a double bond between C4 and C5, a $\beta$-acetyl at C17 and a $\beta$-methyl at C13. Natural androgens are characterized by 19 carbons (C19 steroids); strong androgenic activity is associated with a $\beta$-hydroxylated C17 and a ketone structure ($\supset$C=O) at C3. Natural oestrogens have 18 carbons (C18 steroids) with an aromatized A ring, hydroxylated at C3 and with a $\beta$-hydroxyl group at C17. Not surprisingly, some steroid structures 'fit' a particular receptor better than others; thus, we can say that oestradiol 17$\beta$ is a *stronger* oestrogen than oestriol and oestrone, which are very *weak* oestrogens. This concept of strong- and weak-acting members of a steroid class is important when assessing the likely effect of hormones on target tissues. In Tables 2.3–2.5 the natural members of each steroid class are ranked in order of decreasing potency.

**2 How do steroids act?**

When a steroid encounters a compatible receptor within a target tissue and combines with it, the receptor undergoes a conformational change. After being *activated* in this way, the *steroid–receptor complex* is able to bind to specific DNA sequences in the chromatin—the so called *acceptor sites*. Neither free steroid nor free receptor alone is effective at binding to the acceptor sites. The interaction between the complex and the chromatin acceptor sites is associated with a rapid rise in activity of RNA polymerase II, which increases the production of mRNA. Following this early rise in mRNA production, and in the continuing presence and binding of the complex to the chromatin, a more general stimulation of nucleolar and transfer RNA synthesis occurs, and the synthetic machinery of the cells increases. Protein synthesis rises within 30 minutes of steroid stimulation, and there is evidence to suggest that synthesis of specific proteins may be activated selectively. In this way the steroid hormone can exert a specific effect on its target tissue.

We are building up a picture of hormone action in which the steroids are secreted into the blood, permeate the tissues of the body and enter cells, where they encounter receptors with which they combine to form a complex. The complex then interacts with chromatin to induce specific tissue synthetic responses. Steroid activity may thus be regulated at two levels—by regulation of the concentration of *either* the receptor *or* the steroid *within* the target tissue.

*a Regulation of tissue receptors*

Non-target tissues lack, or have very low levels of, receptors for steroids. However, even within target tissues, the sensitivity of the tissue to the steroid can be modified by regulating the quantity of receptors available. The consequences of this *receptor regulation* can be seen in both physiological and pathological conditions. In Chapter 1, we discussed the genetic deficiency condition of *testicular feminization* (Tfm) in which individuals present as phenotypically normal females who are often married with normal external genitalia, well-developed breasts and a female gender identity. On examination these women are found to have an XY-chromosome constitution and abdominal testes secreting almost normal male levels of androgens into the blood. Examination of the androgen target tissues of these women shows an absence or deficiency of androgen receptors. Thus, although the androgens are produced and present, they are not 'seen' by the tissues, blinded by their absence of receptors, and masculinizing activity is lost. This striking example demonstrates the importance of receptor regulation which will be reinforced by many examples later in the book. *It is of critical importance to realize that the detection of a steroid in the blood does not necessarily mean that the steroid will be active on a target organ.*

*b Regulation of tissue steroids*

The levels of a steroid within a tissue will depend upon: (i) the circulating blood level of the steroid; (ii) local factors within the tissue that affect the equilibration of blood and tissue

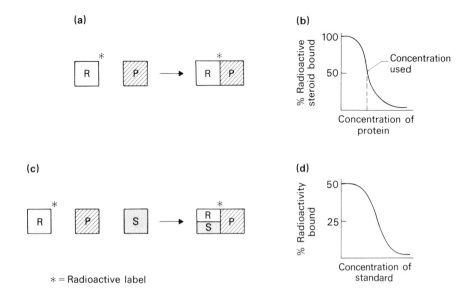

(a)

(b)

(c)

(d)

∗ = Radioactive label

**Fig. 2.4.** Radioassay of steroids. (a) A sample of pure, radioactive steroid (R) is mixed with different amounts of a binding protein (P), which may be a specific antibody (*radioimmunoassay*) or an isolated plasma-binding protein or cell receptor (*radioreceptor assay*). The bound and free steroid are then separated. This may be accomplished in several ways: the protein can be used fixed to the side of the sample tube (solid phase assay), in which case the free steroid is easily washed away; alternatively, proteins in solutions may be precipitated by adding salts or a second antibody directed against the protein. The proportion of the steroid binding to the protein can then be plotted (b). Usually protein levels binding about half the radioactive steroid are selected for use as shown in (c), in which the selected level of protein is incubated with radioactive steroid in the presence of various dilutions of a sample of plasma containing a known level of non-labelled standard steroid (S). The unlabelled standard steroid competes for the binding protein, and so the percentage of bound radioactive steroid declines. In this way a calibration curve can be constructed (d). Dilutions of samples of plasma containing *unknown* levels of steroid can now be used instead of the standard steroid and the concentration of steroid thereby determined from the standard curve.

steroids, such as local blood flow and local enzymes which metabolize or modify the steroids.

i What factors affect blood levels of steroids?

To the clinician, the steroids most directly *visible* are those measured in a sample of systemic blood by various protein-binding radioassays (Fig. 2.4). What interpretation can be placed on such measurements?

Observations on one sample of blood represent a single static measurement taken from a highly dynamic system. Fig. 2.5 shows a series of values of blood testosterone taken from a man over a 24-hour period. Two important points emerge. First, testosterone and many other hormones are secreted not continuously but in a *pulsatile manner*. The magnitude of these pulses, which occur usually at intervals of between 1 and 3 hours depending on the species, is often very large. Thus, as Fig.2.5 shows, levels at a trough are often only one-third or one-quarter of those at a peak. It is therefore very important to remember that *single blood samples taken at one point in time do not take account of the existence of considerable temporal fluctuations and may be misleading*. Second, blood levels of testosterone vary over a 24-hour period, that is to say they show *circadian rhythmicity*. In this case (Fig. 2.5), night-time levels are significantly greater than those in the daytime and very different conclusions about an individual's blood testosterone levels would be drawn if values from samples taken during day or night were compared.

Furthermore, values obtained in a systemic blood sample represent the balance between the rate of secretion of the steroid and its rate of removal and degradation by various routes. There is little evidence for extensive storage of steroids at their site of production and thus the rate at which a steroid is released from a tissue approximates to its rate of synthesis. The mean levels of a steroid in the systemic blood will include steroids from all synthetic sources. For the principal active steroids, for example oestradiol 17$\beta$ or testosterone, most, if not all, of their production will be from one source, namely the gonad. Other steroids, like oestrone, arise from several sources. Thus, in women the ovaries produce only 10–30% of total oestrone by direct secretion. The remainder of the oestrone is derived from metabolic conversion of ovarian oestradiol 17$\beta$ and of adrenal andro-stenedione by peripheral tissues, notably the liver. This distinction between primary secretion and metabolic conversion can be important in interpreting changes in the mean

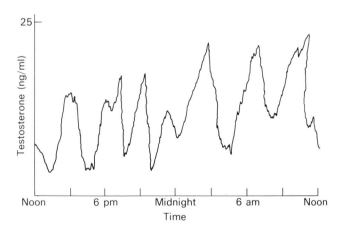

**Fig. 2.5.** Testosterone levels in blood samples taken from a male patient at 45-min intervals over a 24-hour period.

44    *Chapter 2*

**Table 2.6.** Clearance of steroids from the blood in humans (blood volume 4–5 litres).

| Steroid | Clearance* (litres/min) |
|---|---|
| Oestradiol 17β | 1.0 |
| Oestrone | 1.5 |
| Progesterone | 1.5 |
| 17 α-Hydroxyprogesterone | 1.4 |
| 20 α-Dihydroprogesterone | 1.6 |
| Testosterone | 0.5 (women) 0.7 (men) |
| Androstenedione | 1.3 (women) 1.6 (men) |
| Dehydroepiandrosterone | 1.1 |

* Clearance is defined as that volume of blood which would be totally cleared of a substance per unit time by the liver, kidney and any other steroid-metabolizing tissue.

blood levels of a steroid, particularly in pathological conditions.

The rate at which a steroid is removed from the blood is reflected in its *clearance*. Steroids have high clearance rates, that is to say they do not survive for long after secretion, but have a rapid turn-over rate (Table 2.6). Removal of steroids from the blood is affected only marginally by utilization in receptor complexes. The bulk of the steroid produced is rendered ineffective, either by metabolic conversion to a range of biologically inactive steroid derivatives, or by conjugation with glucuronic acid or sulphate at the third and/or seventeenth carbons (Fig. 2.1) to form inactive mono- or diglucosiduronates and sulphates, and sulphoglucosiduronates—all water-soluble conjugates that lack steroidal activity. Most of the conjugated steroids are excreted in the urine and bile but some, e.g. DHA sulphate, may be converted by steroid sulphatases to regenerate active steroids. In some circumstances, e.g. pregnancy, this conversion is of great importance (Chapter 10). Conjugation occurs primarily in the liver, but metabolic degradation also occurs at a variety of extrahepatic sites. The relative rates of conjugation and metabolic degradation as a route for steroid clearance vary with the steroid (Table 2.7). In assays on blood and urine samples it is important to know whether total conjugated and non-conjugated steroid is being measured, since the former are biologically inactive.

In addition to relative rates of primary secretion and secondary conversion and clearance, the mean level of a steroid in the blood is also influenced by the level of steroid-binding proteins present in plasma. In cases of malnutrition and protein deficiency, in which levels of binding proteins are reduced, the *total* level of sex steroids within the blood falls without any reduction in the level of *free* steroid. Conversely, during pregnancy, the levels of steroid-binding proteins rise

**Table 2.7.** Principal metabolic derivates of some steroids (human).

| Steroid | Excreted as |
|---|---|
| Oestradiol 17$\beta$ | Converted to oestrone (see below). Metabolic hydroxylation at various carbons including C16 (to give oestriol) and reduction of C17 ketone—both followed by conjugation. About 10% excreted as urinary oestrone glucosiduronates |
| Oestriol | Little metabolic conversion; most excreted as sulphates, glucosiduronates or sulphoglucosiduronates |
| Progesterone | Mostly as the reduction product pregnanediol (Fig. 2.2), much of it as urinary or biliary conjugates |
| Testosterone | C3 and C5 reduction of either testosterone or its C17 oxidized derivative androstenedione. Conjugation of metabolites. Some conversion to oestrogens |

(see Chapter 10 for details), leading to a corresponding increase in total, measurable steroid levels. It is actually rather difficult to determine accurately the ratio of free to bound steroid in blood samples. However, samples of body secretions, such as saliva, contain levels of steroids which reflect only the levels of *free* blood steroid available for equilibration. Analysis of saliva for steroid levels is also useful because frequent sampling on an outpatient basis is possible.

The blood levels of the steroid, together with the rate of local blood flow, will determine the quantity of steroids entering the tissues. Before encountering receptors, however, the local steroid level is frequently altered by local steroid interconversions.

**ii Steroid interconversions in target tissues**

The gonads are the primary steroidogenic organs and they possess the enzymic apparatus to convert acetate and cholesterol to biologically active sex steroids. However, it has become clear recently that many target organs, although not able to undertake complete synthesis of steroids, do possess *some* of the enzymes involved in the interconversion of steroids. Thus, they are able to take a circulating hormone, transform it enzymically and then utilize it locally. This property of some of the target tissues will be illustrated with two examples here, and will be referred to again in subsequent chapters.

The enzyme, 5$\alpha$-reductase, is present in many of the androgen target tissues, for example the male accessory sex glands, skin and tissues of the external genitalia. This enzyme converts testosterone, the main circulating androgen, to 5$\alpha$-

dihydrotestosterone (Fig. 2.2), which has a much *higher affinity* for the androgen receptor in these target cells, and thus is a much *more potent* androgen than testosterone. In the prenatal and immature male, in which both the testicular and blood levels of testosterone are relatively low, such testosterone as is secreted is converted to 5 α-dihydrotestosterone in the tissues possessing 5 α-reductase. The low levels of this highly active androgen then exert local effects on the external genitalia, increasing phallic and scrotal size and rendering them clearly distinguishable from the clitoris and labia in females. The consequences of this 5 α-reductase activity are made evident by examining cases where there is a genetic deficiency of this enzyme. Affected male infants have poorly developed male external genitalia, and at birth they are classed as females. At puberty, the external genitalia are suddenly exposed to much higher levels of circulating testosterone, and they respond with a sudden growth to normal size. This so-called *'penis-at-twelve' syndrome* vividly illustrates the role of local 5 α-reductase activity in mediating many of the actions of testosterone (see also Chapters 1 and 7).

A second example of local steroid interconversion appears perhaps even more dramatic. The hypothalamus of male and female rodents, and indeed primates, is able to take circulating testosterone and, by local aromatizing activity, convert it to oestradiol 17β. The high local levels of oestradiol 17β then interact with a local oestrogen receptor to stimulate activity. Thus, paradoxically, the central actions of testosterone are accomplished only after aromatization to oestradiol. This is most vividly seen when examining sexual differentiation of the brain, and maintenance of adult sexual behaviour in the *male* rat, as both depend critically on oestradiol—although this oestradiol is normally provided by conversion from testosterone (see Chapters 1 and 7 for details). This phenomenon emphasizes that the terms 'male' and 'female' hormones should be used with care.

## 3 Synthetic steroids

Thus far we have discussed only those steroid molecules synthesized, used and degraded naturally within the body. However, in recent years it has proved possible to synthesize artificial steroids that have been useful for therapeutic or contraceptive practice (Table 2.8). The synthetic steroids have a similar overall shape to that of their natural models, permitting binding to the various classes of steroid receptors. However, the synthetic steroids differ from natural steroids in two main respects. First, they tend to be much less rapidly degraded metabolically and thus are active over a longer period. Second, some synthetic analogues, after binding to receptors, may fail to activate them properly. Such analogues will compete with natural steroids and may therefore function

**Table 2.8.** Some synthetic steroids in clinical or experimental use.

Progestagens

1 Derivatives of 19-nortestosterone (testosterone lacking the C19-methyl group attached to C10 and with an ethynyl group —C≡CH at C17).
Norethisterone
  Norethynodrel ⎫
  Ethynodiol diacetate ⎬ converted to norethisterone by body before active
  Lynoestrenol ⎭
Norgestrel

2 Derivatives of 17 α-hydroxyprogesterone (esterification of 17-hydroxyl group).

Medroxyprogesterone acetate
Chlormadinone acetate
Magestrol acetate

3 Anti-progestins
RU 38486

Oestrogens
1 Derivatives of oestradiol 17β
Ethynyl oestradiol—ethynyl group at C17 α
Mestranol—ethynyl group at C17 α + —OCH$_3$ group at C3.

2 Clomiphene citrate

3 Diethylstilboestrol

4 Anti-oestrogens
Nafoxidene
Tamoxifen

Androgens
1 Anti-androgens
Cyproterone
Cyproterone acetate

---

as 'anti-steroids'. For example, clomiphene citrate binds to oestrogen receptors and intially causes stimulation. However, after exerting this stimulatory action the drug persists in its combination with the receptor, thereby preventing natural oestrogens from binding. Thus, a brief period of oestrogenic stimulation is followed by 24 hours or so of anti-oestrogenic action. Tamoxifen and nafoxidene function purely as anti-oestrogens without an initial stimulation. Nafoxidene binds to the receptor but the complex fails to bind to chromatin. The tamoxifen–receptor complex, in contrast, binds to the chromatin but is ineffective in stimulating the acceptor site. Anti-steroids are useful in controlling steroid-dependent tumours as well as for manipulating fertility (see Chapter 14).

**4 Summary**

In considering the activities of the sex steroids it is of fundamental importance to understand two general points. First, sex steroid activity in the body *may* be regulated by varying blood levels, which can reflect changing primary secretion, systemic secondary interconversions, metabolic clearance or the levels of binding proteins. Second, sex steroid activity may *also* be regulated at the level of the target tissues, either by altering the level or activity of endogenous steroid receptors, or by local enzymic interconversions of steroids from less active to more active forms within a class of steroids, or even by interconversion between different classes. These important conclusions applicable to steroids can also have more general applications to some of the other non-steroidal hormones of reproduction that we will encounter in Chapters 3, 4 and 5.

**Further reading**

American Physiological Society *Handbook of Physiology*. Section 7, Vol. II, Female Reproductive System. American Physiological Society, 1973.

Cole HM, Cupps PT (Eds). *Reproduction in Domestic Animals* 3rd Edn. Academic Press, 1977. (Chapters by JH Clark, EJ Peck Jr and SR Glaser.)

Fuchs E, Klopper A (Eds). *Endocrinology of Pregnancy* 2nd Edn. Harper and Row, 1977.

Gower, DB. *Steroid Hormones*. Croom Helm, 1979.

# Chapter 3
# Testicular
# Function

In Chapter 1 we described the morphological events involved in the formation of the fetal testis from the indifferent genital ridge. We saw that the development of endocrine function in the fetal testis was critical to the establishment of a normal masculine phenotype and behaviour. Further growth of the testis in the prepubertal period was slow, and output of androgens was low but detectably higher than in females. At puberty we saw how changes in androgen output were accompanied by a rapid growth in testis size, maturation of its organization to give an adult structure (Fig. 1.12), and maturation of the somatic tissues of the body. In this chapter, we consider the functioning of the adult testis.

## 1 The compartments of the testis

The testis has two major products: the spermatozoa that transmit the male's genes to the embryo; and the androgens required for completion of masculinization. The synthesis of androgens and spermatozoa occurs in two separate *compartments* within the testis. Spermatozoa develop *within* the tubules while androgens are synthesized *between* the tubules (Fig. 1.12). These two compartments are not only distinct structurally but are also separated physiologically in the adult by cellular barriers that affect the free exchange of water-soluble materials. Thus, fluid collected directly from the lumen within the seminiferous tubule differs markedly in composition from the extratubular fluids—the blood and the lymph draining the interstitial tissue (Fig. 3.1). This *blood–testis barrier* separating the two compartments may be studied by the injection into the blood of various 'marker

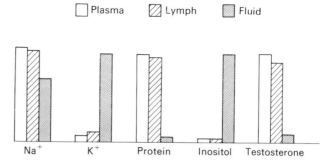

Fig. 3.1. Relative concentrations of various substances in the venous plasma (vascular compartment), lymph (from interstitial compartment) and fluid leaving the seminiferous tubules. Note that seminiferous tubule fluid differs markedly from plasma and lymph.

molecules', with subsequent measurement of both the rate at which the molecules enter the tubular fluid and the level at which the tubular fluid achieves equilibrium with the blood and lymph. Studies of this sort show that ions, proteins and charged sugars enter the interstitial fluid and lymph rapidly, indicating that little if any barrier exists at the capillary level. In contrast, these molecules do not gain free access to the tubular lumen, arriving there only as a result of selective transport and secretion.

The precise cellular location of the intercompartmental barrier has also been analysed by use of various dyes or electron-opaque materials. These are injected into the blood, and then visualized in the testis at the light- or electron-microscope level. Such markers equilibrate freely with interstitial tissue and lymph (marked I in Fig. 3.2) as would be predicted from dynamic studies, but penetrate more slowly through the peritubular wall into the *basal compartment* of the tubule (marked B in Fig. 3.2). However, the major barrier to diffusion lies not at the peritubular boundary itself, but between a basal compartment of the tubule and an *adluminal compartment*. The barrier comprises multiple layers of junctional complexes completely encircling each Sertoli cell, and linking it firmly to its neighbours (Fig. 3.2, upper insert). Marker molecules are never seen penetrating between the Sertoli cells past these junctional barriers and into the central or adluminal compartment of the tubule (A in Fig. 3.2). The presence of the blood–testis barrier ensures that the meiotic stages of spermatogenesis occur in a distinct and controlled environment (see later). It also precludes leakage of spermatozoal proteins into the systemic circulation where they can elicit an immune response, since the immune system is not tolerized to them. Moreover, should an immune reponse be elicited, the blood–testis barrier impairs access of antibody to the germ cells, and thus reduces the prospect of autoimmune

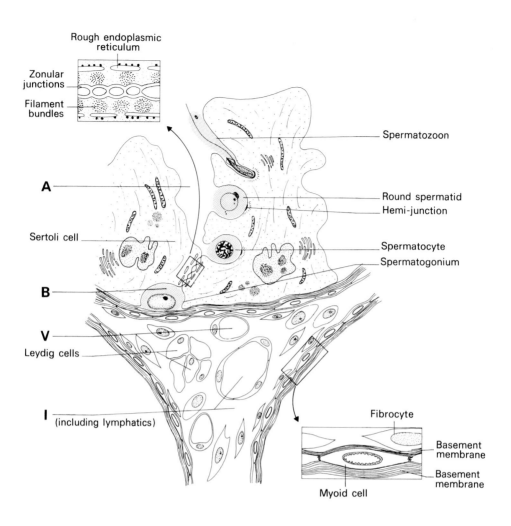

**Fig. 3.2.** Cross-section through part of an adult testis. There are four compartments in the testis: V = vascular; I = interstitial including the lymphatic vessels and containing the Leydig cells; the basal (B) and adluminal (A) compartments lie within the seminiferous tubules. An acellular basement membrane containing myoid cells and invested with a loose coat of interstitial fibrocytes separates the interstitial and basal compartments (see lower insert box). Myoid cells are linked to each other by punctate junctions. No blood vessels, lymphatic vessels or nerves traverse this boundary into the seminiferous tubule. Within the boundary, the basal and adluminal compartments are separated by rows of zonular junctional complexes (see upper insert box) linking together adjacent Sertoli cells round their complete circumference. Beneath the region of junctional complexes are bundles of filaments running parallel to the surface around the 'waist' of the Sertoli cells and beneath the filaments are cisternae of rough endoplasmic reticulum. Within the basal compartment are the spermatogonia, whilst spermatocytes, round spermatids and spermatozoa are in the adluminal compartment, and in intimate contact with the Sertoli cells which form anchoring 'hemi-junctions' with the elongating spermatids.

damage to the testis. However, on occasion the immunologic protection provided by the blood–testis barrier breaks down, and *autoimmune allergic orchitis* results.

The highly characteristic fluid present within the seminiferous tubule is a secretion to which the Sertoli cell itself is probably the major, if not the sole, contributor. The active secretion of fluid can occur against considerable hydrostatic and diffusional gradients. For example, if the outflow of fluid from the tubules towards the epididymis is blocked, secretion continues nonetheless and the tubules dilate with fluid and spermatozoa, generating a hydrostatic pressure that will lead eventually to pressure necrosis and an atrophy of tubular cells.

In summary, the testis may be divided into four major compartments: the *intravascular* compartment, which is in free communication with the *interstitial* compartment containing the lymphatics and in which androgen synthesis occurs. The interstitial compartment is in restricted communication with the *basal* intratubular compartment. A unique feature of the seminiferous tubule is that it also has a distinct *adluminal* intratubular compartment, which is effectively isolated from the other three, and in which most of the spermatogenic events occur.

Although androgens and spermatozoa are synthesized in discrete compartments, their production is interrelated functionally. Thus, as will be seen later, sperm production in most species is only possible when androgen synthesis occurs, and this ensures that mature spermatozoa are always delivered into an extragonadal environment suitably prepared for their efficient transfer to the female tract. In many adult mammals androgen and sperm production are continuous processes throughout the year, but in some mammals, for example the roe-deer, ram, voles, marine mammals and possibly some primates, the behaviour and morphology of the males show seasonal variations, which reflect the changing levels of androgen output. In these *seasonal breeders* the spermatogenic output also varies in parallel with the endocrine pattern. As the androgens appear to orchestrate all these changes, we will examine their nature and production first, and then proceed to discuss spermatogenesis and its endocrine control.

## 2 Testicular androgens

As we saw in Chapter 2, the androgens comprise a class of steroid with distinct structural and functional features. The principal testicular androgen is testosterone which is synthesized from acetate and cholesterol (Fig. 3.3) by the Leydig cells of the interstitial tissue. $3\beta$-Hydroxysteroid dehydrogenase, the enzyme involved in $\Delta 5$ to $\Delta 4$ conversions (Fig. 3.3) can be localized in the Leydig cells, and in a number of pathological conditions and in seasonal breeders, it is possible

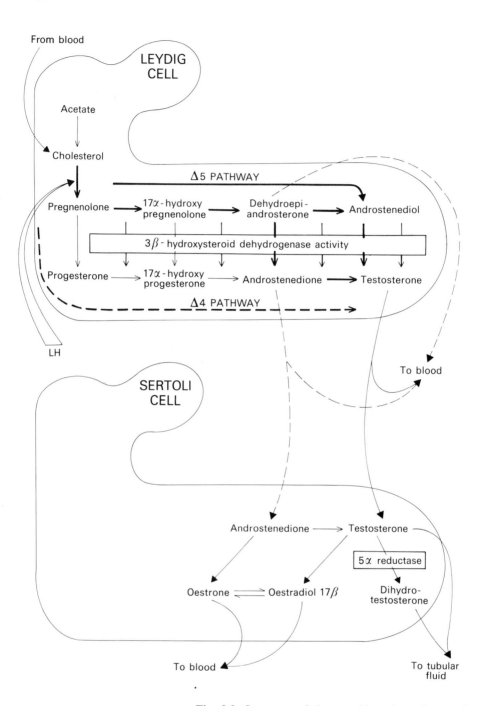

**Fig. 3.3.** Summary of the steroidogenic pathway relevant to the testis. The principal(Δ5) path for testosterone synthesis in the Leydig cells of the human is indicated by heavy lines, although the Δ4 pathway is also used and may be more important in other species. In addition to testosterone some of the intermediates in the pathway are released into the blood—androstenedione at 10% and DHA at about 6% of testosterone levels in man. Some testosterone and androstenedione is taken up by Sertoli cells which convert the androgens to oestrogens and to dihydrotestosterone.

to correlate testosterone output with the changes in the morphology of the smooth endoplasmic reticulum of the Leydig cells.

In man, 4–10 mg of testosterone are secreted daily. The hormone is released from the Leydig cells and rapidly enters both the blood (80ng/ml of testicular venous blood in the ram) and the lymph (50ng/ml). Although the concentrations in blood and lymph are similar, the major quantitative output of testosterone passes into the blood, because the capillary blood flow is much greater (17ml of blood/min in the ram compared to 0.2ml/min for lymph). Not all the testosterone goes into the blood and lymph however. Some enters the intratubular compartments of the testis. Testosterone is a relatively fat-soluble molecule (see Chapter 2), and can pass through the cellular barriers separating the major testicular compartments. Once in the tubule the androgens are bound within the Sertoli cells to an *androgen receptor* and within the tubular fluid of the adluminal compartment to an *androgen-binding protein* (ABP). This ABP is secreted by the Sertoli cells, and is similar to the testosterone-binding protein of plasma. The observations that androgens enter, and are bound, within the tubules make it reasonable to assume that they may have a function there. That this is the case will become evident when we consider the effect of androgens on the events of spermatogenesis. First, however, we will describe the process of spermatogenesis itself.

# 3 Cytodifferentiation of spermatozoa

The mature spermatozoon is an elaborate, highly specialized cell (Fig. 3.4). Whilst the basic elements of the spermatozoon are common to all cells, their organization is highly characteristic and shows evidence of an extensive and elaborate modelling during the terminal phases of cytodifferentiation. The generation of this complex structure involves the packaging of the genetically reshuffled haploid set of chromosomes in a parcel suitable for effective delivery to the egg. Large numbers of these complex cells are produced, between 300 and 600 sperm per gram of testis per second! Spermatogenesis is a complex process of cytodifferentiation that involves *mitotic proliferation* to produce large numbers of cells, *meiotic division* to generate genetic diversity and halve the chromosome number, and *extensive cell modelling* to package the chromosomes for transport.

*a Mitotic proliferation*

The interphase germ cells of the immature testis are reactivated at puberty to enter rounds of mitosis. Henceforth they are known as stem cell or A0 spermatogonia. This population of A0 cells proliferates slowly in the basal compartment of the tubule, and serves as a reservoir of stem cells from which, at intervals, spermatogonia with a distinct

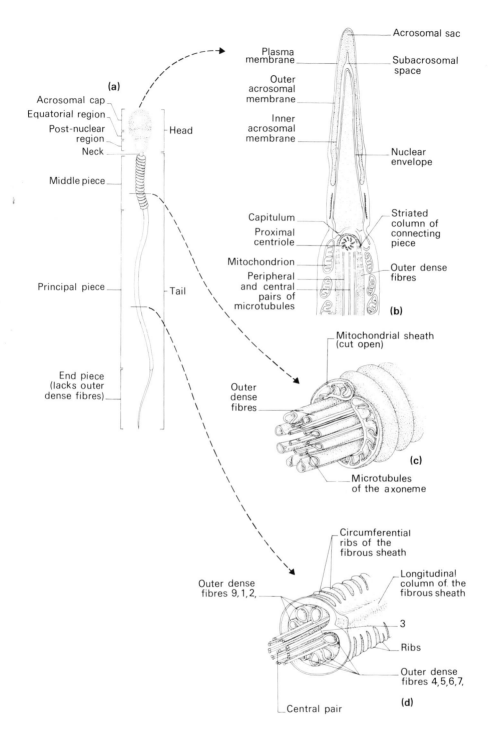

**(a)**

Acrosomal cap
Equatorial region
Post-nuclear region
Neck
Head
Middle piece
Principal piece — Tail
End piece (lacks outer dense fibres)

Plasma membrane
Outer acrosomal membrane
Inner acrosomal membrane

Acrosomal sac
Subacrosomal space
Nuclear envelope

Capitulum
Proximal centriole
Mitochondrion
Peripheral and central pairs of microtubules

Striated column of connecting piece
Outer dense fibres

**(b)**

Mitochondrial sheath (cut open)
Outer dense fibres
Microtubules of the axoneme

**(c)**

Circumferential ribs of the fibrous sheath
Outer dense fibres 9,1,2,
Longitudinal column of the fibrous sheath
3
Ribs
Outer dense fibres 4,5,6,7,
Central pair

**(d)**

**Fig. 3.4.** (a) Diagram of a primate spermatozoon (50 $\mu$m long) showing principal structural regions. (b) Sagittal section of head, neck and top of midpiece. Note elongated nucleus with highly compact chromatin, and acrosomal sac. (c) Sketch of midpiece (surface membrane removed). Note sheath of 'winding' mitochondria, axoneme of tail comprising nine circumferential doublets of

56    *Chapter 3*

morphology appear. These are called A1 spermatogonia, and their emergence marks the beginning of spermatogenesis. Each of these A1 spermatogonia undergoes a limited number of mitotic divisions at about 42-hour intervals, thus producing a *'clone'* of daughter cells. The number of divisions is characteristic for the species, and clearly will determine the number of daughter cells in the clone. Thus, in the rat there are six divisions leading to a maximum clone size of 64 cells although, since appreciable numbers of cells die during mitosis, the full-sized clone is not actually achieved. The morphology of the daughter cells produced at each mitotic division tends to differ slightly from the parent cell, and so one may identify roughly at what stage any individual spermatogonium is in the sequence of mitotic divisions by its morphology. In this way, spermatogonia in the rat are subclassified as being *'type A1–4'* during the first three mitoses, *'type intermediate'* after the fourth mitosis and *'type B'* after the subsequent fifth division (Fig. 3.5). All the spermatogonia type B of the daughter clone then divide to form *resting primary spermatocytes*.

There is some evidence to suggest that during the mitotic divisions of the type A spermatogonia, one of the A4 daughter cells of the clone reverts to a type A1 spermatogonium and thus serves as a second source of stem cells in addition to the A0 cells. Indeed, it is possible that in the mature adult this source is more important than the A0 cells which may be used primarily at puberty, after damage to spermatogenesis by, for example, X-irradiation, or on reactivation of spermatogenesis in seasonal breeders.

*b Meiosis*

During the proliferative phase of spermatogenesis the mitotically dividing cells are present in the basal intratubular compartment of the testis. Within the compartment the clone of resting primary spermatocytes duplicate their DNA content and then push their way into the adluminal intratubular compartment by disrupting transiently the zonular tight junctions between adjacent Sertoli cells. Within the new and distinctive microenvironment of the adluminal compartment, they then enter the first and prolonged meiotic prophase (Fig. 3.6; see Fig. 1.1 for details of meiosis). During prophase the sister chromatid strands on the paired homologous chromosomes come together, form points of contact at which the

---

microtubules and two central microtubules; peripheral to each outer doublet is a dense fibre. (d) Section and sketch of principal piece (surface membrane removed). Mitochondria replaced by a fibrous sheath comprising two longitudinal columns interconnected by ribs. The two fibrous sheath columns connect to underlying outer dense fibres 3 and 8.

57    *Testicular Function*

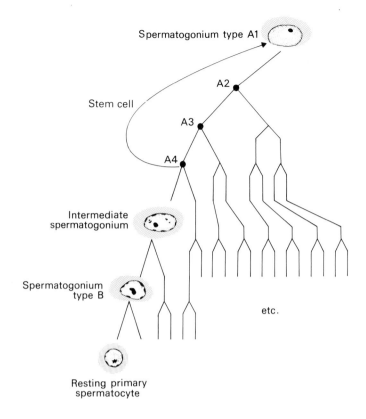

Spermatogonium type A1

A2

Stem cell

A3

A4

Intermediate
spermatogonium

Spermatogonium
type B

etc.

Resting primary
spermatocyte

**Fig. 3.5.** Cells of the mitotic phase of the spermatogenic lineage in the rat; present in the *basal* intratubular compartment. Type A spermatogonia have large, ovoid, pale nuclei with a dusty, homogeneous chromatin. The intermediate spermatogonia have a more scalloped chromatin pattern on their membranes, and this feature is heavily emphasized in Type B spermatogonia in which the nuclei are also smaller and rounded.

chromatids break, exchange segments of chromosomes and then rejoin, thus shuffling their genetic information (Fig. 1.1). At different steps in this sequence of chromosome pairing, interacting and separating the nuclei of the primary spermatocytes show characteristic morphologies, largely due to the differences in the state of their chromatin. This feature makes it possible to identify the position of any individual primary spermatocyte in meiosis (Fig. 3.6). During the prolonged meiotic prophase, and particularly during pachytene, the spermatocytes are especially sensitive to damage, and widespread degeneration can occur at this stage.

The first meiotic division ends with the separation of homologous chromosomes to opposite ends of the cell on the meiotic spindle, after which cytokinesis yields, from each primary spermatocyte, two daughter *secondary spermatocytes*, each containing a single set of chromosomes. Each chromo-

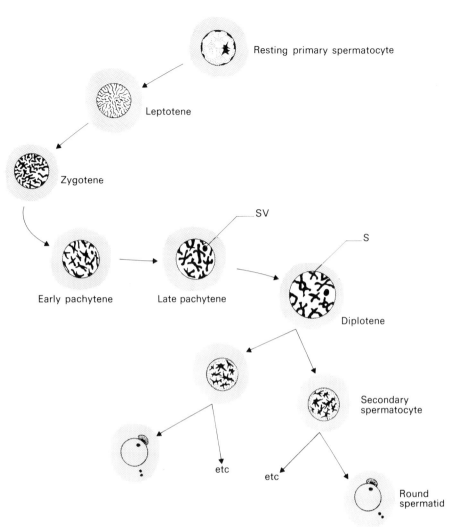

Resting primary spermatocyte

Leptotene

Zygotene

SV

S

Early pachytene

Late pachytene

Diplotene

Secondary
spermatocyte

etc

etc

Round
spermatid

**Fig. 3.6.** Progress of a resting primary spermatocyte through the meiotic phase of the spermatogenic lineage which occurs in the *adluminal* intratubular compartment. DNA synthesis is largely completed in the resting spermatocyte although 'repair DNA' associated with crossing over occurs in late zygotene and early pachytene. In leptotene the chromatin becomes filamentous as it condenses. In zygotene homologous chromosomes come together in pairs attached to the nuclear membrane by their extremities, thus forming loops or 'bouquets'. In pachytene the pairs of chromosomes (bivalents) shorten and condense, and nuclear and cytoplasmic volume increases. It is at this stage that autosomal crossing over takes place at 'synapsis' (sex chromosomes are paired in the 'sex vesicle': SV). The synapses (S) can be seen at light-microscopic level as chiasmata during diplotene and diakinesis as the chromosomes start to pull apart and condense further. The nuclear membrane then breaks down, followed by spindle formation, and the first meiotic division is completed with breakage of centromeric and chiasmatic contacts to yield two secondary spermatocytes each containing one set of chromosomes. These rapidly enter the second meiotic division, and the chromatids separate at the centromere to yield four haploid spermatids.

some is comprised of two chromatids joined at the centromere. The chromatids then separate at their centromere and move to opposite ends of the second meiotic spindle, and the short-lived secondary spermatocytes divide to yield haploid *early spermatids* (Fig. 3.6). Thus, from the maximum of 64 primary spermatocytes that entered meiosis (in the rat), 256 early spermatids could result. Again, the actual number is much less than this, since in addition to any losses at early stages of mitosis, the complexities of the meiotic process result in the further loss of a number of cells.

With the formation of the early round spermatids the important genetic events of spermatogenesis are substantially, but not fully, complete. Although the spermatid nuclei contain haploid sets of chromosomes, the autosomes (but *not* the sex chromosomes which are inactivated during meiotic prophase) continue to synthesize low levels of ribosomal and messenger RNA during the first phase of spermatid development. Moreover, protein synthesis continues throughout spermatid development. Therefore, the possibility exists for so-called '*haploid gene expression*', in which the haploid set of genes might code for mRNA which might then be translated into a protein for use in the third (packaging) phase of spermatogenesis. If such a process occurred, then it might be possible to separate spermatids and spermatozoa into two populations each carrying, and expressing, a distinctive genetic allele coding for a spermatozoal protein. Such a separation might also occur in the female genital tract which could thereby exert 'natural selection' on a population of spermatozoa which were genetically, and via haploid expression phenotypically, heterogeneous. A great deal of work has been done in an attempt to demonstrate haploid expression. It has been given impetus by the thought that X- and Y- bearing spermatozoa might be separated and thereby sex 'selected'. However, it must be admitted that haploid expression, both in autosomes and sex chromosomes, has been very difficult to prove decisively, and successful selection for sex by this approach has not been achieved. Recently however, successful separation of X- and Y-bearing spermatozoa has been accomplished not on the basis of the differential expression of chromosomes, but as a result of their differing masses (see Fig. 1.2b), which enables the spermatozoa carrying them to be segregated by buoyant density centrifugation.

*c Packaging (spermiogenesis or spermeteliosis)*

Although at least some autosomal genes are active in RNA synthesis immediately after the completion of meiosis, all synthetic activity ceases soon thereafter. The spermatid DNA becomes highly condensed or *heterochromatic*, and packed first with a set of nuclear basic proteins unique to the testis, which are then replaced during spermatid elongation with protamines, small basic proteins comprising 50% arginine.

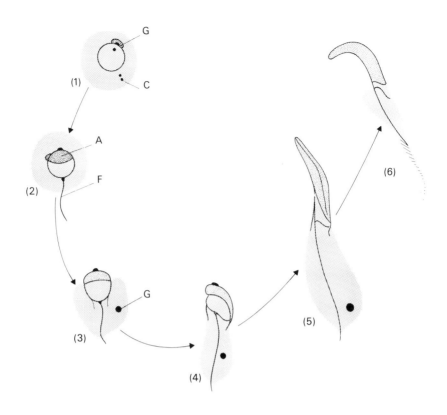

**Fig. 3.7.** Progress of a round spermatid through the packaging phase of the spermatogenic lineage in the rat. The Golgi apparatus (G) of the newly-formed round spermatid (1) gives rise to glycoprotein-rich granules, which coalesce to a single *acrosomal granule* (A) that apposes and grows over the nuclear surface to form a cap-like structure (2). The nuclear membrane at this site loses its nuclear pores. The two centrioles (C) lie against the opposite pole of the nuclear membrane, and a typical flagellum (F) (9 + 2 microtubules) grows outwards from the more distal centriole (2) whilst from the proximal centriole the neck or connecting piece forms, linking the tail to the nucleus. The nucleus moves with its attached acrosomal cap towards the cytoplasmic membrane and elongation begins (3 and 4). Chromatin condensation commences beneath the acrosomal cap, generating a nuclear shape which is characteristic for the species (3 to 6) and superfluous nuclear membrane and nucleoplasm is lost. The Golgi apparatus detaches from the now completed acrosomal cap and moves posteriorly, the acrosome starts to change its shape. Nine coarse fibres form along the axis of the developing tail, each aligned with an outer microtubule doublet of the flagellum (see Fig. 3.4 for details). In the final phase, the mitochondria migrate to the anterior part of the flagellum, and condense around it as a series of rods forming a spiral (Fig. 3.4). The superfluous cytoplasm appears 'squeezed' down the spermatid and is shed as the spermatozoa are released (6). The mature spermatozoon has remarkably little cytoplasm left.

61    *Testicular Function*

Spermatozoa have tight, inactive units of chromatin in which genetic expression is absent.

The major visible changes during spermiogenesis arise from the remarkable cytoplasmic remodelling of the spermatid. This occurs with the generation of: the *tail* for forward propulsion; the *midpiece* containing the mitochondria (energy generators for the cell); the *acrosome*, which functions like an 'enzymic knife' for penetration into the egg; and the *residual body* which acts as a dustbin for the residue of superfluous cytoplasm and is phagocytosed by the Sertoli cell after the spermatozoon departs (Figs 3.4 and 3.7). The process of complex, and, not suprisingly, not all spermatids complete it successfully.

The whole process of spermiogenesis occurs in close association with the Sertoli cells. Indeed, the spermatids and their spermatocyte progenitors in the adluminal compartment indent the Sertoli cell cytoplasm deeply and form unique, specialized, hemi-junctional attachments with the Sertoli cell membrane (Fig. 3.2). In addition, organelles of the Sertoli cell, such as the endoplasmic reticulum, cluster in the cytoplasmic regions adjacent to the heads of the developing spermatozoa. Consequently, it is tempting to ascribe an active role in sperm modelling to the Sertoli cells.

As meiosis and spermiogenesis proceed the spermatogenic cells are moved slowly towards the lumen of the tubule until, with the completion of spermatid elongation, the Sertoli cell cytoplasm around the cells retracts. The elongated spermatids are released into the lumen of the seminiferous tubule bathed in the distinctive tubular fluid. These newly-formed, immature spermatozoa will then be carried away by the fluid through the excurrent ducts of the testis at the start of their journey towards the egg.

The mature spermatozoon is one brother in a large clonal family, derived from one paternal spermatogonium type A. The family is large because of the number of premeiotic mitoses, and the spermatozoa are only brothers and not identical 'twins' because meiotic chiasmata formation ensures that each is genetically unique having a common ancestor. Within each testis tubule many such clonal families develop side by side, and there are 30 or so tubules within each rat testis.

**4 Organization of spermatogenesis**
*a How long does spermatogenesis take?*

The sequence of mitotic, meiotic and packaging events that constitute spermatogenesis is complex, and understandably requires a period of several weeks for its completion. One way to measure the length of time it takes to complete parts of the spermatogenic process is to 'label' cells at different points during the process, and then to measure the rate of progress of the labelled cells through subsequent cytodifferentiation. For

**Table 3.1.** Kinetics of spermatogenesis.

| Species | Time for completion of spermatogenesis (days) | Duration of cycle of the seminiferous epithelium (days) |
|---|---|---|
| Man | 64 | 16 |
| Cow | 56 | 14 |
| Sheep | 40 | 10 |
| Pig | 32 | 8 |
| Rat | 48 | 12 |

example, if radioactive thymidine is supplied to the resting primary spermatocytes as they engage in the final round of DNA synthesis prior to entry into meiosis, the cell nuclei will be labelled and their progress through meiosis and spermiogenesis can be followed. In this way the amount of time required for each step in spermatogenesis can be measured. In Fig. 3.8 (see folder, facing p. 36) the times required for each step in the rat are represented visually as blocks, the length of each being a measure of relative time. The absolute time for the whole process, from entry into first mitosis to release of spermatozoa, is recorded for several species in Table 3.1 (column 1).

Variation *between* species occurs both in total time and in the time required for individual components of the spermatogenic process. However, one dramatic observation that has come from this sort of study is that *within* a species or genetic strain the rate of progression of cells through the spermatogenic process is remarkably constant. Thus, all the spermatogonia type A within any testis of a given species seem to advance through spermatogenesis at the same rate, and take the same total time for completion of the process. Hormones, or other externally applied agents, do not seem to be able to either speed up or slow down the spermatogenic process. They may, as we will see, affect whether or not the process *occurs at all*, but they do not appear to affect *the rate at which it occurs*. This remarkable constancy of the rate of progress through spermatogenesis, and its apparent independence from external disturbing factors, suggests a high level of intrinsic organization.

*b The spermatogenic cycle*

So far we have considered the process of spermatogenesis from the viewpoint of a single spermatogonium type A generating a clone of spermatozoa at a highly characteristic rate, and also regenerating a new stem cell spermatogonium type A from which subsequent clones may be derived. This new stem cell will enter into generation of its own clone only after a period of several days quiescence. Remarkably, it has been found that this interval of quiescence, between successive entries into spermatogenesis of any given spermatogonial type A lineage, is absolutely constant for all spermatogonia in all testes of the same species, and is also characteristic for each species (Table 3.1, column 2). Somehow, the regenerated stem cell measures or is told the time and 'knows' when the interval of quiescence should end.

In the case of the rat this cyclic re-entry into spermatogenesis occurs about every 12 days. Now this 'spermatogenic cycle' is one quarter of the 48 days required for completion of spermatogenesis, so it follows that four successive spermatogenic processes must be occurring at the same time (Fig. 3.8). The advanced cells, in those spermatogenic clones which were initiated earliest, are displaced progressively towards a luminal position by subsequent clones. Thus, a transverse section through the tubule will reveal spermatogenic cells at four distinctive stages in the progression towards spermatozoa, each cell type representing a point in separate, successive cycles (Fig. 3.8).

Since the interval between successive entries of spermatogonia type-A stem cells into spermatogenesis is constant, and since the rate of progress of cells through spermatogenesis is constant, it must follow that the cells in successive cycles will always develop in parallel. Therefore, the sets of cell associations in any radial cross-section through a segment of tubule taken at different times will always be characteristic (Fig. 3.8 and 3.9). For example, since the cycle interval is 12 days, and it also takes 12 days for the six mitotic divisions, entry into meiosis will always be occurring just as a new cycle is initiated by the first division of a spermatogonium type A (Fig. 3.9, column 8). Similarly, it takes 24 days (i.e. $\times 2$ cycles) for the premeiotic spermatocyte to complete meiosis and the early phase of spermatid modelling. So not only will entry into mitosis and entry into meiosis coincide, but so will the beginning of spermatid elongation (Fig. 3.9, column 8). These events will also coincide with release of spermatozoa, since it takes a further 12 days for the completion of spermatid elongation.

These observations on spermatogenesis imply a remarkable degree of temporal organization amongst the spermatogenic cells, but the basis for this is not understood. It is tempting to invoke a role for the Sertoli cell, which has an intimate association with all the cells of successive spermatogenic clones. It also provides a radial axis of communication through the tubule which could ensure that all the cells remain adherent to the same rate of progress. The Sertoli cell itself shows a cyclicity in its structure and biochemistry that parallels the spermatogenic cycle, but whether this *reflects* or *directs* the spermatogenic cycle remains to be determined.

Up to this point we have considered the organization *in time* of one spermatogonium type A and its descendants.

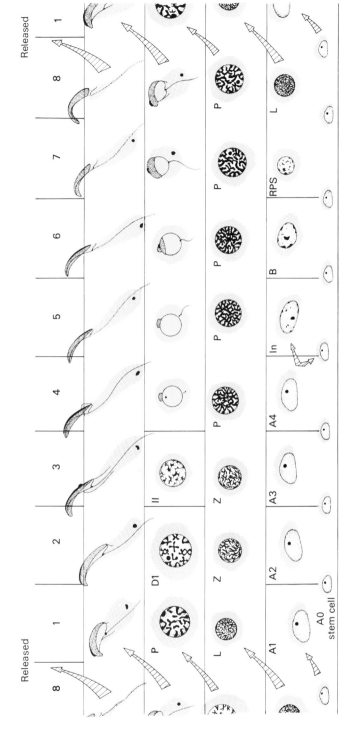

**Fig. 3.9.** The sections indicated in Fig. 3.8 are summarized here. Read from left to right. Bars between cells indicate cell divisions; otherwise the changes are not quantal but by progressive differentiation. Abbreviations as for Fig. 3.8.

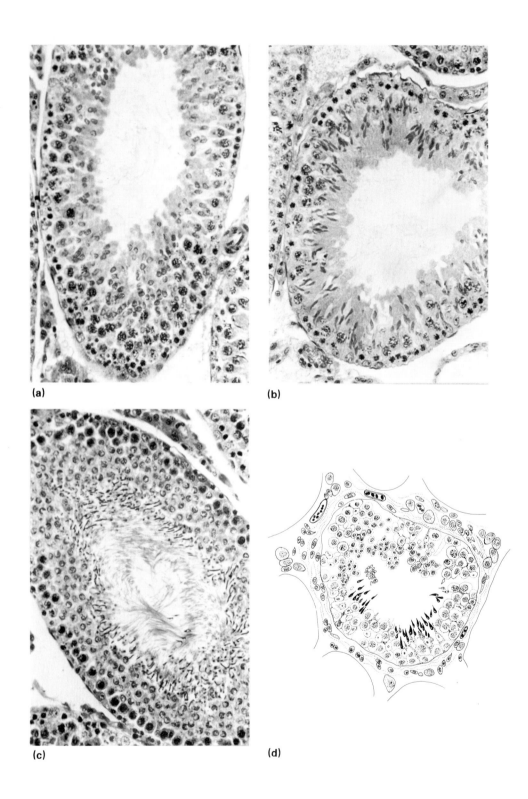

(a)

(b)

(c)

(d)

However, there are thousands of spermatogonia type A in any one testis at any one time—can we observe any organization *in space* between them?

*c Cycle of the seminiferous epithelium*

If all of the spermatogonia type A in the pubertal testis were to resume mitotic activity at exactly the same time then, since the time to complete spermatogenesis is constant, all the individual developing clones would release their spermatozoa at the same time. Moreover, as the entry of the regenerated stem cells into subsequent spermatogenic sequences occurs with a constant cycle of 12 days (in the rat), periodic pulses of spermatozoa would be released at 12-day intervals, which could result in an episodic pattern of male fertility. This problem could be overcome by making random the times at which pubertal spermatogonia type A initiated mitotic activity. Then, although different clones would all develop at both the same rate and the same cyclic interval, their relative times of entry into spermatogenesis would be staggered, thus eliminating the pulsatile release of spermatozoa and making it one of continuous flow. The testis functions in a manner somewhat between these two extremes, although the end result, as we will see, is continuous sperm production.

Close examination of the testes of most mammals, *excluding* man, shows that in a cross-section through a tubule the same *set of cell associations* is observed, regardless of the point on the circumference studied (Fig. 3.10a–c). Now many discrete clones are present around the basement membrane in any cross-section of tubule, and the fact that each of the clones is at the same stage of development with the same set of cell associations means that all the clones in that section of tubule must be synchronized in absolute time. It is as though at puberty a message passed circumferentially around a segment of tubule reactivating all the spermatogonia type A in that segment at the same moment. Once activated at the same time the constancy of both the spermatogenic cycle and the spermatogenic rate would keep all clones locked together. The fact that all the clones within one segment of tubule are synchronized is fortuitous experimentally because it means the whole cross-section of several clones 'cycles' together, and therefore the spermatogenic cycle of individual spermatogonia type A is amplified spatially to give the '*cycle of the*

---

**Fig. 3.10.** (a)–(c) Cross-sections through three adjacent rat seminiferous tubules. Note that within each tubule, the sets of cell associations along all radial axes are the same. However, each tubule has a different set of cell associations from its neighbour. Thus, tubule (a) has the Type 1 set of cell associations (Figs 3.8 and 3.9), tubule (b) has the Type 3 set and tubule (c) has the Type 8 set. Tubule (d) is a drawing of a human testis. Note that *within* each cross-section of tubule different sets of cell associations are evident.

*seminiferous epithelium'*. This makes for easier clinical and experimental analysis.

The human testis is atypical in this regard since a cross-section through an individual tubule reveals a degree of spatial organization which is more limited to 'wedges' (Fig. 3.10d). It is as if the activator message at puberty did not get all the way round the tubule and so the coordinated clonal development of different spermatogonia type A was initiated over a smaller area. This does not mean, of course, that the control of either the spermatogenic cycle or the rate of spermatogenesis in man differs fundamentally from control mechanisms in other species. It means merely that the extent of spatial coordination between individual clones is not so great.

One idea of the route by which adjacent clones might achieve this spatial coordination invokes the Sertoli cell. Adjacent Sertoli cells are linked to each other, not just by tight junctions separating basal and adluminal compartments, but also by gap junctions which permit intercellular communication. Thus, the Sertoli cells form an extensive network within the tubule through which information can pass radially, longitudinally or circumferentially.

*d Spermatogenic wave*

There is one further feature of the organization of spermatogenesis—the spermatogenic wave. If a rat seminiferous tubule is dissected and laid out longitudinally, and then cross-sections are taken at intervals along it and classified according to the set of cell associations in it, a pattern like that in Fig. 3.11 will often result. Quite extraordinarily, it seems that the tubule segments containing synchronized clones were activated serially at puberty. Thus, in Fig. 3.11 the most advanced segment (7) is at the centre, and as one moves along the tubule in either direction, sets of cell associations characteristic of progressively earlier stages of the cycle of the seminiferous epithelium are observed. It is as though the central segment was activated first at puberty, and then the hypothetical 'activator message' spread slowly along the tubule in both directions, progressively initiating mitosis and, thus, the first cycle of clonal growth. This phenomenon, which is often only incompletely observed, is referred to in the adult as the *spermatogenic wave*.

It is important not to confuse the *wave* with the *cycle of the seminiferous epithelium* although both phenomena appear very similar. Imagine that whereas the sequence of cell associations forming the *wave* could be recorded by travelling along the tubule with a movie-camera running, the same sequence of cell associations would only be captured in the *cycle* by setting up the movie-camera on time-lapse at a fixed point in the tubule. Thus the *wave* occurs in *space*, the *cycle* in *time*.

68    *Chapter 3*

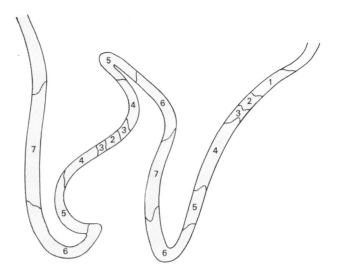

**Fig. 3.11** Dissected seminiferous tubule from a rat testis. Note that whole segments of the tubule are at the same stage (numbered) of the cycle of the seminiferous epithelium, and that adjacent segments tend to be either just advanced or just retarded.

The production of spermatozoa is a complex and highly organized process. Although it is now relatively straightforward to describe the sequence of events that creates a clone of spermatozoa from a single spermatogonium type A, at present it is not easy to understand how the *rate* at which these events occur is so constant, or how the interval between successive *cycles* of clonal growth is so invariant, and what the nature of the spatial interactions between adjacent clones is. As was pointed out earlier, the Sertoli cell is well placed to play a coordinating role; it is intimately associated with the spermatogenic cells, it provides a radial cytoplasmic axis to the tubule, and also, via gap junctional contacts with its neighbours, forms a potential network for information transfer around and along the tubule. Whether or not the Sertoli cell plays a role in control of rates, cycles and waves remains to be established. However, it is established that the Sertoli cell *does* play a role in mediating the actions of hormones on spermatogenesis.

## 5 Endocrine control of spermatogenesis

We saw earlier that testosterone was produced by the Leydig cells and that some of it was transferred selectively to the tubular compartments of the testis. Within the tubules, intracellular receptors and an extracellular binding protein take up the androgens. What happens to the testosterone within the tubules and what effects does it have there?

If tubules are isolated from the testis and incubated with radioactive testosterone, the radioactivity may be detected in a

69    *Testicular Function*

number of other steroids. This result suggests that some of the testosterone is metabolized by the tubules. Considerable amounts of the testosterone are metabolized to oestrogens, especially in the pubertal testis, and oestrogen release into the blood can be detected (Fig. 3.3). The significance of this oestrogen production is not clear. Perhaps more important is the conversion within the tubules of testosterone to $5\alpha$-dihydrotestosterone (DHT) (Fig. 3.3). DHT is a more powerful androgen than testosterone and does not appear to escape in appreciable quantities from the tubules.

The capacity of the tubules to bind, and indeed to generate, potent androgens suggests a role for the androgens within the tubule, and such a role has been demonstrated, albeit indirectly, mainly in studies on the rat. It has been known for many years that if the *pituitary* of the adult rat is removed (*hypophysectomy*) the testes shrink in size, sperm output declines, spermatogenesis arrests at the primary spermatocyte stage, the Leydig cells become involuted, testosterone output falls and the testosterone-dependent secondary sex glands, such as the prostate and seminal vesicle, regress. If large doses of testosterone are given at the time of hypophysectomy, spermatogenesis continues, albeit at a slightly reduced level, and the secondary sex characters show little sign of any regression, although the Leydig cells *do* still involute. These experiments established two important points: first, that testosterone was involved in the maintenance of spermatogenesis and, thus, does have an intratubular role; second, that the secretion of testosterone by the Leydig cells was under the control of the pituitary.

The nature of this control is shown clearly in two ways. If, after hypophysectomy, an extract of the anterior lobe of the pituitary is administered instead of testosterone, then not only are secondary sex characters and spermatogenesis maintained, but the Leydig cells do not involute and testosterone output is maintained. Fractionation and analysis of the pituitary extract has shown that a glycoprotein hormone called *luteinizing hormone* (LH, or sometimes interstitial cell stimulating hormone ICSH) is primarily responsible. The structure and properties of LH are summarized in Table 3.2. Further confirmation of the role for LH in stimulating testosterone production comes from the administration of an antiserum to LH to an intact adult male. Blood levels of free LH fall as it is bound by the antibody, and subsequently the level of plasma testosterone falls. Regression of the androgen-dependent secondary sex characters follow. These two experiments strongly suggest that pituitary-derived LH stimulates the Leydig cells to produce testosterone. LH has since been shown to bind specifically to Leydig cells and stimulate them, whereas it neither binds to, nor stimulates, isolated tubules. LH appears to function via high-

**Table 3.2.** Properties of human LH, FSH and prolactin.

| | Luteinizing hormone (interstitial cell stimulating hormone) LH (= ICSH) | Follicle stimulating hormone FSH | Prolactin |
|---|---|---|---|
| Secreted from | Anterior pituitary gonadotrophs | Anterior pituitary gonadotrophs | Anterior pituitary lactotrophs |
| Act upon (via surface receptors) | Leydig cells<br>Thecal cells—antral follicles (Ch. 4)<br>Granulosa cells—pre-ovulatory follicles (Ch. 4)<br>Luteal cells—corpus luteum (Ch. 4)<br>Interstitial glands of ovary (Ch. 4) | Sertoli cells<br>Granulosa cells— follicles (Ch. 4) | Leydig cells<br>Seminal vesicle and prostate (Ch. 7)<br>Ovarian follicles (Ch. 4)<br>Corpus luteum (Ch. 4)<br>Mammary gland (Ch. 13) |
| Molecular weight | 34 000 | 32 600 | 22 500 |
| Composition | Glycoprotein 16.4% carbohydrate<br><br>$a$ chain†<br>89 amino acids, 2 carbohydrate chains<br>$\beta$ chain‡<br>115 amino acids 1 carbohydrate chain | Glycoprotein 25.9% carbohydrate<br><br>$a$ chain†<br>identical to LH 2 carbohydrate chains<br>$\beta$ chain<br>115 amino acids, 2 carbohydrate chains (identical at only 7 amino acids to LH) | Single polypeptide chain of 198 amino acids |
| Biological half-life in blood | 30 minutes | 150 minutes | 10–20 minutes |
| Mean levels in human blood* | | | |
| Male | 5–10 mu/ml | 4–8 mu/ml | 60–370 mu/l |
| Female  early cycle | 5–10 mu/ml | 4–8 mu/ml | |
| mid-cycle | 50–95 mu/ml | 10–20 mu/ml | |
| late cycle | 2–8 mu/ml | 1–4 mu/ml | |

\* Note some variability between laboratories depending on reference standards used.
† The common or backbone subunit chain.
‡ The subunit chain conferring specificity.

affinity receptors in the surface membrane of the Leydig cell, and within 60 seconds of binding an elevation of intracellular cAMP levels occurs. A rise in testosterone output follows after 20 or 30 minutes, and this can be mimicked by adding cAMP directly. It is, therefore, proposed that the LH stimulates adenyl cyclase activity. The rise in cAMP that results leads to phosphorylation of intracellular proteins by activation of a protein kinase, and thereby to mobilization of steroid precursors and in particular the activation of pregnenolone synthesis from cholesterol. Recently a *second* anterior pituitary hormone has been found to facilitate the stimulatory action of LH on Leydig cells—*prolactin* (Table 3.2). Prolactin binds to receptors on the Leydig cells and in some way enhances the LH effect. However, prolactin *alone* will not itself stimulate testosterone production.

It seems clear, then, that LH stimulates Leydig cells to produce testosterone. The testosterone is transported into the tubules where it binds, both as testosterone and after conversion to DHT, to androgen receptors within the Sertoli cells and thereby aspermatogenesis is prevented. However, it is important to stress that although in the rat testosterone (or LH) administration after hypophysectomy prevents complete aspermatogenesis, some reduction in testis size and a reduction of about 20% in sperm output does occur. Moreover, in the past few years it has become clear that in many other species, including several primates, the decline in sperm output after hypophysectomy may be even more severe, despite administration of high doses of testosterone or LH. For *complete* and *continuing* maintenance of spermatogenesis, LH stimulation of Leydig cells is not enough. Is there yet *another* pituitary hormone required? Again experiments on the rat have proved to be informative.

It has been found that if adult male rats are hypophysectomized, and both Leydig cell function and spermatogenesis are allowed to *regress before* any therapeutic LH injections are commenced then, although testosterone levels are restored by LH, spermatogenesis is *not*. Under such conditions the spermatogenic cells advance to meiosis, but then they die. In order to complete spermatogenesis, it has been found that a *third* anterior pituitary glycoprotein hormone is required—*follicle stimulating hormone* (FSH, Table 3.2).

FSH binds to high-affinity, specific receptors detected only on the Sertoli cell membrane, and, like the effect of LH on the Leydig cells, stimulates cAMP production and, thus, activity of a protein kinase, which results in the phosphorylation of several intracellular proteins. Thirty to 60 minutes later mRNA and rRNA synthesis rises, and the general protein synthetic activity of the cells increases, *including synthesis of androgen-binding protein*. With this burst of synthetic activity the Sertoli cell increases in size and in its secretion of testicular fluid. This increased activity of the Sertoli cell, when coupled with the presence of testosterone, also allows spermatogenesis to go to completion. Thus, here is direct evidence of a role for the Sertoli cell in the events of spermatogenesis.

These experiments on the *restoration* of spermatogenesis in the rat indicate clearly a role for FSH. Furthermore, the drop of 20% in sperm output observed after hypophysectomy plus immediate *maintenance* doses of LH or testosterone is also prevented by adding FSH to the 'cocktail' of exogenous hormones. In those species, including man, where the drop in sperm output is considerably more than 20%, FSH is essential not only for restoration but also for maintenance of male fertility. Thus, we can conclude that male fertility involves *three* hormones from the anterior pituitary. LH and prolactin stimulate Leydig cells to generate androgens, and the androgens act with FSH to stimulate Sertoli cells. The Sertoli cells then permit spermatogenesis to proceed to completion.

## 6 Summary

In this chapter we have considered the two major secretory activities of the testis and their relationship. The *endocrine* secretion of testosterone by the Leydig cells is dependent upon pituitary LH. Some of the testosterone is transferred into the seminiferous tubule, and there it acts on the Sertoli cells together with FSH in order to help maintain production of the *cellular* secretion of the testis—the spermatozoa. Without testosterone spermatogenesis ceases. As we will see in more detail in subsequent chapters, testosterone also acts elsewhere in the body to affect anatomy, physiology and behaviour in such a way as to facilitate the efficient transfer of spermatozoa to the female tract. Thus it 'makes sense' for the body to link the production of spermatozoa to the production of the orchestrating hormone testosterone.

Although both FSH and testosterone have roles in the spermatogenic process, the actions of these hormones appear to be purely *permissive*. That is to say they determine whether or not spermatogenesis will take place. They do not appear, however, to regulate the *rate* at which cells develop along the spermatogenic lineage, the *frequency* with which the stem spermatogonia type A cells enter cycles of spermatogenic activity or the *spatial coordination* between adjacent clones of spermatogenic cells. These processes appear to be regulated internally in some way, possibly via the activity of Sertoli cells.

## Further reading

Bellve AR. The molecular biology of mammalian spermatogenesis, In *Oxford Reviews of Reproductive Biology* Vol. 1, pp. 159–261. Oxford University Press, 1979.

Davies AG. *Effects of Hormones, Drugs and Chemicals on Testicular Function*. Eden Press Annual Research Reviews Vol. 1, 1980.

Fawcett DW & Bedford JM. *The Spermatozoon*. Urban & Schwarzenberg, 1979.

Fritz IB. Sites of action of androgens and follicle stimulating hormone on cells of the seminiferous tubule. In: *Biochemical Actions of Hormones* 1978; **5**: 249–281.

Roosen-Runge EC. *Biol. Rev.* 1962; **37**: 343–377.

Roosen-Runge EC. *The Process of Spermatogenesis in Animals*. Cambridge University Press, 1977.

Setchel BP. *The Mammalian Testis.* Paul Elek, 1978.
Steinberger H & Steinberger E. *Testicular Development, Structure and Function.* Raven Press, 1980.

# Chapter 4
# Ovarian Function

In Chapter 1 we described the formation of the fetal ovary from the indifferent genital ridge. It was observed that the ovary differentiated later in development than the fetal testis, and that, unlike the testis, its endocrinological activity during fetal life was not essential for normal female development. The feminine phenotype develops spontaneously in the absence of gonadal hormones, and constitutes almost a 'neutral pattern' of sexual differentiation when compared to that of the male. However, we also observed that ovarian endocrine activity must occur postnatally if full sexual maturation is to occur at puberty. In addition, from puberty onwards, the ovarian germ cells must generate haploid ova for fertilization by spermatozoa. As in the testis, the production of gametes by the ovary is coordinated with its endocrine activity. However, adult ovarian function shows a major difference to that of the testis, for relatively few oocytes are re-leased, and their release is not a continuous stream as for spermatozoa, but occurs episodically at *ovulation*. The release of the two major steroid secretions of the ovary, the *oestrogens* and *progestagens*, reflects this episodic release of oocytes. Thus, the period prior to ovulation is characterized by *oestrogen dominance*, and the period *following* ovulation is characterized by *progestagen dominance*. Once this sequence of oestrogen–ovulation–progestagen has been completed, it is repeated. Therefore we speak of a *cycle of ovarian activity*.

The cyclic release of steroids imposes a corresponding cyclicity on the whole body and, in most species, on the behaviour of the adult female. These cycles are called the *oestrous cycle* in animals and the *menstrual cycle* in higher primates. The reason for this cyclicity of female reproductive activity lies in the fact that the genital tract of the female mammal, unlike that of the male, must serve two distinct reproductive functions, each with different demands. It must act to transport gametes to the site of fertilization, and it also provides the site of implantation of the fertilized egg and its subsequent development. Each cycle of the female reflects these two roles. During the first, oestrogenic part of the cycle, the ovary prepares the female for receipt of the spermatozoon and fertilization of the egg; during the second, progestagenic part of the cycle, the ovary prepares the female to receive and nurture the conceptus should successful fertilization have occurred. Sandwiched between these two endocrine activities of the ovary, the egg is released at ovulation. In this chapter, we will consider the sequence of changes *within the ovary itself* by which a coordinated and cyclic pattern of production of the egg and ovarian steroids is achieved. Later we will consider how this ovarian cyclicity leads to the oestrous and menstrual cycles.

## 1 The adult ovary

The adult ovary is organized on a pattern comparable to the testis (Fig. 4.1) with *stromal tissue* containing the *primordial follicles* (homologous to tubules) and also glandular tissue, the so-called *interstitial glands* (homologous to the Leydig cells). The primordial follicle, comprised of flattened mesenchymal cells condensed around a primordial germ cell (Fig. 4.7a), constitutes the fundamental functional unit of the ovary. In the first part of this chapter we will pursue the formation, growth and fate of a single follicle in some detail. In the second part of the chapter we will relate the activity of a single follicle to the activity of the ovary as a whole.

The pattern of gamete production in the female, like that in the male, shows processes of cell proliferation by *mitosis*, genetic reshuffling and reduction by *meiosis*, and packaging of the haploid chromosomes during *oocyte maturation*. In the female, however, the need for proliferation by mitosis is not as great because only one or a few eggs are shed during each cycle, unlike the massive sperm output of the testis. The female therefore uses a different approach to the mitotic phase of gametogenesis. The primordial germ cells that entered the gonad continue their mitotic proliferation well after ovarian morphology is established. At this stage they are known as *oogonia*, cf. spermatogonia in the proliferative phase of the male. However, unlike the situation in the male, the mitotic phase of the oogonia *terminates finally* before birth

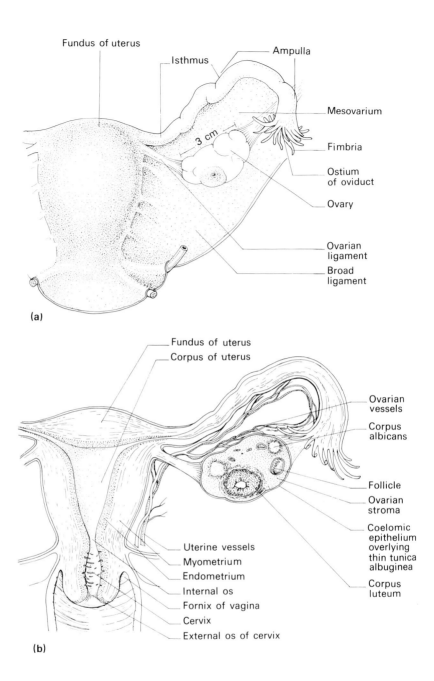

**(a)**

**(b)**

**Fig. 4.1.** Posterior views of human uterus and one oviduct and ovary: (a) intact; (b) sectioned. The ovaries have been pulled upwards and laterally, and would normally have their long axes almost vertical. Note that all structures are covered in peritoneum except the surface of the ovary and oviducal ostium. The ovary has a stromal matrix of cells, smooth-muscle fibres and connective tissue. Anteriorly at the hilus the ovarian vessels and nerves enter the medullary stroma via the mesovarium. Within the stromal tissue are follicles, corpora lutea and albicans and interstitial glands. (After Netter.)

77   *Ovarian Function*

(man, cow, sheep, goat, mouse), or shortly thereafter (rat, pig, cat, rabbit, hamster), by which time *all* oogonia have entered into their first meiotic division, thereby becoming *primary oocytes*.

This termination of mitosis and early entry into meiosis is evidently evoked by a *meiosis initiation factor* derived from cells of the ingrowing rete ovarii tissue, since if the rete is removed meiosis does not occur. The consequence of this early meiosis is that by the time of birth, a *woman has all the oocytes within her ovary that she will ever have* (Fig. 4.2). If these oocytes are lost, for example, by exposure to X-irradiation, they cannot be replaced from stem cells and the woman will be infertile. This situation is distinctly different from that in the male in which the fetal mitotic proliferation is not so marked, but is resumed in the adult when spermatogonial stem cells exist throughout reproductive life.

During their progress through the first meiotic prophase the oocytes become surrounded by ovarian mesenchymal

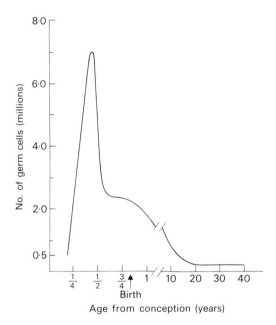

**Fig. 4.2.** Numbers of ovarian germ cells during the life of a human female. After initial period of migration, mitotic proliferation commences at around 25–30 days, and continues until around 270 days (birth). The first meiotic prophases can be detected at around 50–60 days, and the first diplotene stage chromosomes at around 100 days. All germ cells are at the dictyate stage by birth. Atresia of oocytes in meiotic prophase is first overt by about 100 days, and continues throughout fetal and neonatal life. The period from entry of first germ cells into meiosis to full attainment of dictyate stage by all germ cells varies with species (expressed in days post-conception—date of birth in brackets): sow 40–150 (114); ewe 52–110 (150); cow 80–170 (280); rat 17.5–27 (22). (After Baker.)

cells to form the primordial follicles. The follicular cells secrete a basement membrane, *membrana propria*, around the outside of this cellular unit. With the formation of the primordial follicles the oocytes *arrest in diplotene* of the first meiotic prophase, with the chromosomes still enclosed by a nuclear membrane within a nucleus generally known as the *germinal vesicle* (see Fig. 1.1 for details of meiosis). The oocyte halted at this point in meiosis is said to be at the *dictyate stage* (also called dictyotene).

It is not clear how or why meiosis is halted so soon after its initiation. Possibly the condensation of follicle cells on the oocyte generates a *meiosis inhibitory factor*. The primordial follicle containing the arrested dictyate oocyte may stay in this state for up to 50 years in women, the oocyte metabolically ticking over and waiting for a signal to resume development. The reason for storing eggs in this extraordinary protracted meiotic prophase is unknown. Although a few follicles may resume development sporadically and incompletely during fetal and neonatal life, regular recruitment of primordial follicles into a pool of growing follicles occurs first at puberty. Thereafter a few follicles recommence growth every day, so that a continuous trickle of developing follicles is formed. We will trace the development of one such follicle through its full history.

## 2 The phases of follicular development
### a The preantral phase

The earliest phase of follicular growth is characterized by an increase in follicular diameter from 20 $\mu$m to between 200 and 400 $\mu$m, depending on the species. The major part of this growth occurs in the primary oocyte which increases its diameter to between 60 and 120 $\mu$m. This growth phase is accompanied by massive synthetic activity and marked morphological changes (Fig. 4.3), but not by a reactivation of meiosis. Instead, the dictyate chromosomes are actively synthesizing large amounts of RNA, and the prominent dense nucleoli within the oocyte germinal vesicle show that some of this is ribosomal. This activity loads the oocyte cytoplasm with materials that are essential for later stages of egg maturation and for the first day or so after ovulation (see Chapter 9). It therefore constitutes part of the packaging process of gametogenesis.

Whilst the oocyte is enlarging, the surrounding *granulosa cells* divide to become several layers thick and secrete a glycoprotein material that forms an acellular layer, the *zona pellucida*, between themselves and the oocyte (Fig. 4.3). Contact with the oocyte is maintained via cytoplasmic processes that penetrate the zona and form gap junctions at the oocyte surface. Gap junctions also form in increasing numbers between adjacent granulosa cells thus providing the basis for an extensive network of intercellular communi-

79    *Ovarian Function*

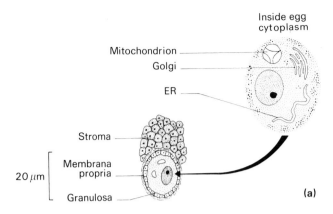

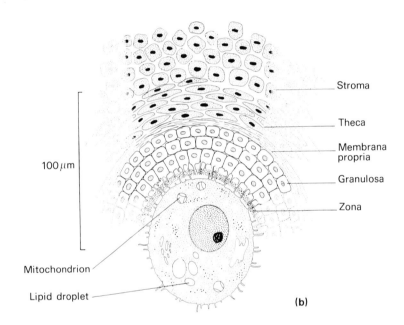

**Fig. 4.3.** Preantral phase of follicular development. (a) Stromal cells surrounding primordial follicle which has an outer membrana propria, flat granulosa cells enclosing primary oocyte in dictyate stage with germinal vesicle (equivalent to the nucleus) containing nucleolus, and ooplasm with mitochondria, Golgi complex and endoplasmic reticulum. (b) Stromal cells are condensed to form thecal layer, granulosa cells have divided, become cuboidal and secreted zona pellucida. Contact with primary oocyte is maintained by cytoplasmic processes which interdigitate with the many microvilli which are present on the oocyte. Within the oocyte, mitochondria increase in number and are small and spherical with columnar cristae, the smooth endoplasmic reticulum breaks up into numerous small vesicles, the Golgi complex breaks into small vesicular units often associated with lipid droplets. The chromosomes are still active synthetically. The oocyte has grown in size.

cation. Through this network low molecular weight bio-synthetic substrates such as amino acids and nucleotides are passed to the growing oocyte for incorporation into macro-molecules. This network is important since the granulosa layer is completely *avascular*. Cells in the ovarian stroma become condensed on to the membrana propria to form a loose matrix of spindle-shaped cells called the *theca* which, unlike the region of granulosa cells, is vascularized (Fig. 4.3). Thus, over the first phase of development the organization of the follicle becomes increasingly complex.

The progress through this first phase of follicular development occurs independently of any direct external control. However, progress on to the next stage of follicular development does require external support, and, as in the male, this support is provided by the pituitary. Hypophysectomy prevents the effective transition to the *antral* phase of follicular development.

*b The antral phase*

The continuous trickle of follicles through the first hormone-independent phase of follicular growth means that at any point in time there are a few follicles that have completed their growth and are termed advanced *preantral follicles*. Their subsequent fate depends upon the endocrine milieu in which they find themselves. Many of the follicles undergo the process of *atresia,* in which the outer layers of granulosa cells show reduced protein synthetic activity, accumulate lipid droplets and develop pyknotic nuclei. Death of the oocyte follows. The follicle is invaded by leucocytes and macro-phages and becomes fibrous scar tissue.

Atresia is prevented only if adequate *tonic* levels of FSH and also some LH are present in the circulation. These hormones bind to follicular *FSH and LH receptors* that develop towards the end of the preantral phase of follicular growth. In hypophysectomized animals, it is possible to save preantral follicles from becoming atretic by administering gonadotrophins. The effect of the gonadotrophins is to convert the preantral follicles to *antral follicles* (also called vesicular or Graffian follicles). In this process granulosa and thecal cells again proliferate, resulting in a further increase in follicular size (Fig. 4.4). However, there is little additional increase in the size of the oocyte itself, and its chromosomes remain in the dictyate stage, although synthesis of RNA and protein continues. The proliferating thecal cells divide into two distinct layers (Fig. 4.4b), a glandular, highly vascular *theca interna*, surrounded by a fibrous capsule, the *theca externa*. Fluid then starts to appear between the dividing granulosa cells, and the drops coalesce to form *follicular fluid* within the *follicular antrum.* The primary oocyte surrounded by a dense mass of granulosa cells, called the *cumulus oophorus,* is suspended in this follicular fluid connected only

81     *Ovarian Function*

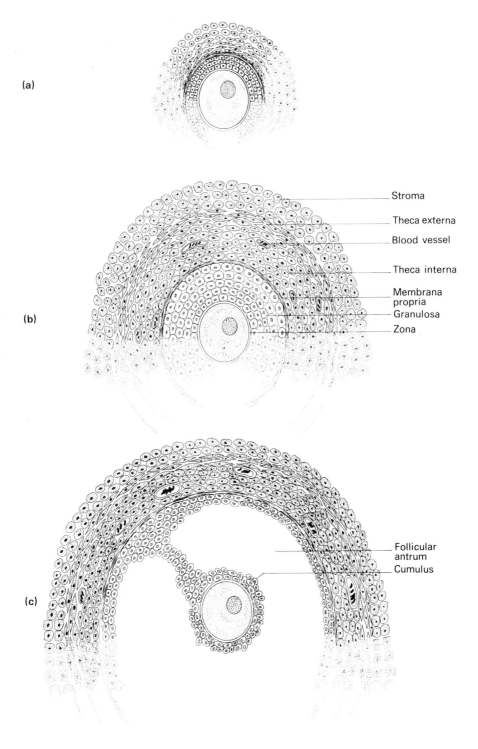

(a)

Stroma

Theca externa

Blood vessel

Theca interna

Membrana
propria
Granulosa

Zona

(b)

Follicular
antrum
Cumulus

(c)

**Fig. 4.4.** Antral phase of follicular development. The granulosa and thecal cells of the preantral follicle divide (a→b). The granulosa cells become polyhedral. The thecal cells comprise two layers, an outer fibrous theca externa and an inner theca interna, rich in blood

by a thin 'stalk' of cells to the rim of peripheral granulosa cells (Fig. 4.4c). The viscous fluid is comprised partly of muco-polysaccharides secreted by the granulosa cells, and partly of a serum transudate.

During this second phase of growth the follicles show a steady increase in the synthesis of androgens and oestrogens. Androstenedione and testosterone are the principal androgens while oestradiol 17$\beta$ is the principal oestrogen. Oestrone is also secreted in most species. As the follicles increase in size, the synthesis of oestrogen rises, and the largest, more advanced follicles release their steroids into the circulation culminating, towards the end of this phase of growth, in a *surge of circulating oestrogen*. This surge can be detected by daily monitoring of urinary oestrogen levels and provides a good guide to the state of maturity of the follicles.

The production of steroids is under the control of the gonadotrophins, hypophysectomy resulting in cessation of steroid output. Each gonadotrophin appears to exert its effect at different locations within the follicle. Only the *granulosa cells* of these follicles *bind FSH* whereas only the cells of the *theca interna bind LH*. If these two cell populations are carefully dissected apart from the antral follicles and grown separately *in vitro*, it is found that the cells of the *theca interna synthesize androgens* from acetate and cholesterol, and this conversion is greatly stimulated by LH (Fig. 4.5). Only limited oestrogen synthesis is possible by these cells, particularly in the early stages of antral growth. The granulosa cells, in contrast, are incapable of forming androgens. If, however, the *granulosa cells are supplied with androgens, they possess enzymes which will readily aromatize them to oestrogens*. This aromatization is stimulated by FSH (Fig. 4.5).

Thus, the androgens produced by developing follicles are derived exclusively from thecal cells, whereas the oestrogens arise via two routes. One involves *cell cooperation* in which *thecal androgens* are *aromatized* by the *granulosa cells*. The other route is by *de novo* synthesis from acetate in thecal cells. The balance between these two potential sources of oestrogen varies with different species, but it seems probable that in the antral follicle *all* of the oestrogen *within* the follicular fluid and *most* of the oestrogen released from the follicle to the blood results from cell cooperation.

---

vessels and comprising large, foamy cells with abundant smooth endoplasmic reticulum. With further growth (b—►c), the follicular antrum develops leaving the oocyte surrounded by a distinct and denser layer of granulosa cells, the cumulus oophorus. The oocyte itself shows only a small increase in size.

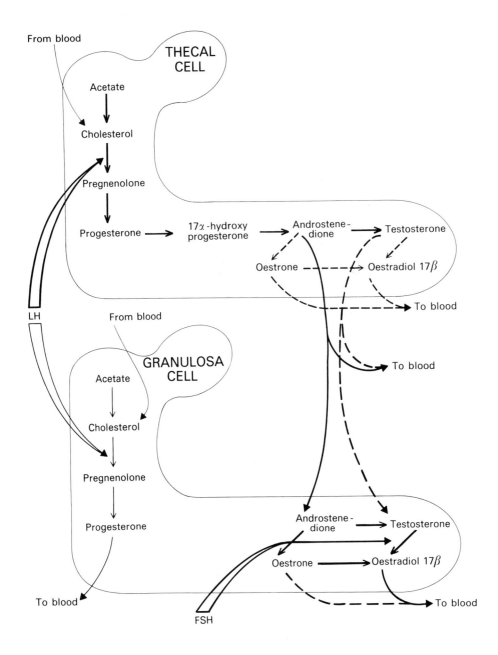

**Fig. 4.5.** Scheme outlining the principal steroidogenic pathways in developing follicular cells. Heavy arrows represent the main activities of later preantral and antral follicles. Light arrows represent activities developing in preovulatory follicles and persisting into the luteal phase. In the corpora lutea the oestrogenic synthetic capacity of thecal cells only persists in those species in which the thecal cells become incorporated into the corpus luteum. Where alternative pathways exist, the minor pathway is indicated by a dashed line.

Both production of steroids and the increase in follicular size are intimately interlinked, and it has become clear recently that the steroids, in addition to their systemic effects via secretion into the blood, may also have a local intra-follicular role. Oestrogens, progestagens and androgens are all detectable in follicular fluid. The androgens, as well as serving as substrates for conversion to oestrogens, also stimulate aromatase activity. The oestrogens can bind to receptors in the granulosa cells, which are then stimulated to proliferate and also to synthesize yet more oestrogen recep-tors. As the granulosa cells are the major site of conversion of androgens to oestrogens a system of *positive feedback* is operating in which oestrogen stimulates further oestrogen output. The surge in oestrogen observed towards the end of the antral phase may be partly explained by this positive feedback (see also Chapter 5).

Oestrogens, in conjunction with FSH, have a crucial role within the follicle towards the end of this second phase of growth. Together these hormones stimulate the appearance of *LH-binding sites* on the *outer layers of granulosa cells* which hitherto lacked them. These LH-binding sites are critical for successful entry of the antral follicle into the third (pre-ovulatory) phase of follicular growth.

*c The preovulatory phase*

Just as preantral follicles trickling through the first, hormone-independent phase of follicular growth will become atretic unless tonic FSH and LH levels coincide with the develop-ment of their receptors on the follicular cells, so the antral fol-licles that arise from them in the second phase of growth will also die unless a brief *surge of high levels of* gonadotrophin coincides with the appearance of LH receptors on the outer granulosa cells. If a surge of LH coincides with the terminal development of antral follicles, in which both the granulosa and thecal cells can bind LH, then entry into the preovulatory phase of growth occurs. The effects of the surge of LH on these advanced follicles are two-fold: first, it causes terminal growth changes in both the follicle cells and the oocyte that result in the oocyte's expulsion from the follicle at *ovulation*; second, it changes the whole endocrinology of the follicle which becomes a *corpus luteum* at ovulation.

Within 3–12 hours, depending on the species, of the beginning of a surge of LH dramatic changes occur in the oocyte. The nuclear membrane surrounding the dictyate chromosomes breaks down, and the arrested meiotic prophase is ended. The chromosomes progress through the remainder of the first meiotic division (see Fig. 1.1 for details of meiosis), culminating in an extraordinary cell division in which *half* the chromosomes, but almost *all* the cytoplasm, goes to one cell—*the secondary oocyte* (Fig. 4.6). The remaining chromo-somes are discarded in a small bag of cytoplasm, the *first polar*

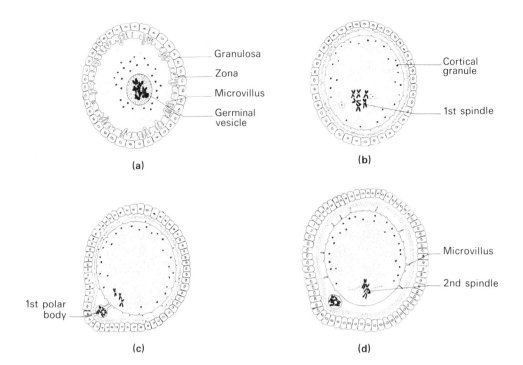

**Fig. 4.6.** Reactivation of meiosis in the preovulatory oocyte. A few hours after LH stimulation of the follicle (a→b) the germinal vesicle breaks down, and the chromosomes complete prophase and arrange themselves on the first meiotic spindle. Meanwhile, cytoplasmic contact between oocyte and granulosa cells ceases, and cortical granules made by the Golgi apparatus migrate to the surface. Subsequently, the first meiotic division is completed with expulsion of the first polar body (b→c). The chromosomes immediately enter the second meiotic division but stop at second metaphase. The cytoplasm around the eccentrically placed spindle is devoid of cortical granules and the overlying membrane lacks microvilli (c→d). The oocyte is ovulated in this arrested state (except in dogs and foxes in which the oocyte is ovulated at metaphase I, and the 1st polar body is extruded after ovulation).

*body,* which dies subsequently. This unequal division of cytoplasm conserves for the oocyte the bulk of the materials synthesized during earlier phases. The chromosomes within the secondary oocyte come to lie on the second metaphase spindle. Then, suddenly, meiosis arrests yet again, and the oocyte is ovulated in this arrested metaphase state. We do not know why this second meiotic arrest occurs.

The termination of the dictyate stage and the progress of *meiotic maturation* through to second metaphase and ovulation takes a matter of hours (see later), and is accompanied by *cytoplasmic maturation* in which changes occur in the cytoplasmic organization of the oocyte (Fig. 4.6). The intimate contact between the oocyte and the granulosa cells of the cumulus is broken by withdrawal of the cytoplasmic processes. The Golgi apparatus of the oocyte synthesizes lysosomal-like granules which migrate towards the surface of the oocyte to assume a subcortical position (*cortical granules*). Protein synthetic activity continues at the same rate, but new and distinctive proteins are synthesized. This activity prepares the oocyte for fertilization (see Chapter 8).

If an oocyte is shed from its follicle prematurely, or removed laparoscopically prior to completion of the maturational events, then its fertilizability is much reduced. For this reason, in clinical *in vitro* fertilization programmes, human oocytes aspirated from preovulatory follicles are cultured for a further 6–8 hours before addition of spermatozoa. This procedure improves the chance of complete maturation of the oocyte.

Meiotic and cytoplasmic maturation of the oocyte are stimulated by the surge of LH, yet it is clear that LH cannot, and does not, bind to the oocyte itself. Therefore its effect must be mediated via the cells of the follicle, although the nature of the intermediary steps is unknown.

In addition to acting on follicle cells in order to generate signals to the oocyte, LH also directly affects the growth and endocrinological activity of the follicle cells themselves. A final and considerable increase in follicular size occurs (reaching 25 mm diameter or more in the human) almost exclusively due to a rapid expansion of the volume of follicular fluid. This is accompanied by an increase in total blood flow to the follicle. Recent advances in ultrasound technology now make it possible to monitor these final stages of follicular growth in the conscious subject, and thereby ascertain how near to ovulation are the follicles.

The growth in the follicle is matched by changes in the pattern of steroid secretion. Within 2 hours or so of the beginning of the LH surge there is a transient rise in the output of follicular oestrogens and androgens, which then decline to very low levels. This rise coincides with distinctive changes in the thecal layer, which appears to be transiently stimulated and hyperaemic, but then becomes less prominent. The outer cells of the granulosa layer also show a marked change in their properties a few hours after the LH peak. If they are isolated from preovulatory follicles and placed in culture they show distinctive differences from granulosa cells isolated from the earlier antral follicles. First, they no longer convert androgen to oestrogen, but instead *synthesize progesterone*. Second, *LH stimulates this synthesis of progesterone* via the newly acquired LH receptors. Third, the cells have lost, or reduced, their capacity to bind oestrogen and FSH. This acquisition of the capacity to respond to LH by synthesizing progesterone results in a release of progesterone from the follicle which becomes significant in man several hours prior to ovulation, although in most species only just before or immediately after ovulation.

Although the third phase of follicular growth is the shortest (see Section 2g), it is also the most dramatic. It culminates in the remarkable process of *ovulation*.

*d Ovulation*

By the end of the preovulatory phase of follicular growth, the rapid expansion of follicular fluid has resulted in a relatively thin peripheral rim of granulosa cells and regressing thecal cells, to which the oocyte, with its associated granulosa cells, is attached only by a tenuous and thinning stalk of cells. The increasing size of the follicle and its position in the cortex of the ovarian stroma causes it to bulge out from the ovarian surface leaving only a thin layer of epithelial cells between the follicular wall and the peritoneal cavity. At one point in this exposed periphery the wall becomes even thinner and avascular, the cells in this area dissociate and then appear to degenerate, and the wall balloons outwards. The follicle then ruptures at this point, *the stigma,* causing the fluid to flow out on to the surface of the ovary carrying with it the oocyte and its surrounding mass of cumulus cells. In many species, including man, the ovarian surface is directly exposed to the peritoneal cavity but in some (for example the sheep, horse, rat) a peritoneal capsule or bursa encloses the ovary to varying degrees and acts to retain the egg mass(es) close to the ovary. There they are collected by cilia on the *fimbria* of the oviduct which sweep the egg mass into the oviducal *ostium* (Fig. 4.1). The residual parts of the follicle within the ovary collapse into the space left by the fluid, the oocyte and the cumulus mass, and within this cavity a clot forms. Thus, the post-ovulatory follicle is comprised of a fibrin core, surrounded by several collapsed layers of *granulosa cells,* enclosed within a *fibrous outer thecal capsule.*

The precise biochemistry of the ovulatory events has attracted a great deal of attention, but as yet there is no agreement on the details. It seems probable that enzymes, such as plasmin (a proteolytic enzyme) and collagenase, are activated in the preovulatory period and digest the follicular wall. The intrafollicular activity of prostaglandins and histamine has also been implicated as necessary for ovulation. It does seem reasonably clear that increased hydrostatic pressure within the follicle is *not* responsible for rupture at ovulation.

*e The corpus luteum*

The collapsed follicle now transforms into a *corpus luteum* (Fig. 4.7c). The fate of thecal cells is uncertain, and probably varies with different species. In many the thecal cells disperse to the stromal tissue, but in others, for example the great apes, the pig and man, some thecal cells become incorporated within the developing *corpus luteum.* Within the granulosa

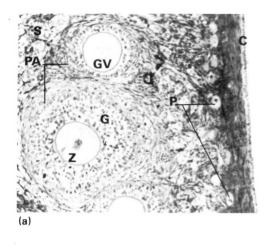

(a)

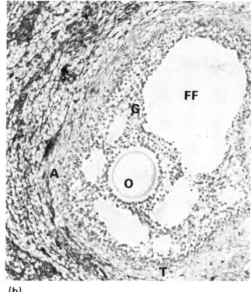

(b)

**Fig. 4.7.** (a) Primordial follicles (P) lying adjacent to the coelomic epithelium (C) in the stroma (S) grow in size to give preantral follicles (PA) with enlarged oocytes (O) containing a 'nucleus' or germinal vesicle (GV), condensed thecal cells (T), proliferated granulosa (G) and a zona pellucida (Z). (b) Further growth leads to antral follicles (A) secreting follicular fluid (FF). (c) After a preovulatory growth spurt and ovulation, granulosa cells luteinize (L) within a fibrous thecal capsule (T).

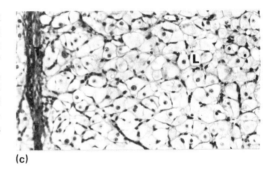

(c)

cell layer the fibrin core undergoes fibrosis over a period of several days, and this process is assisted by the breakdown of the membrana propria between the granulosa and thecal layers and the vascularization of the granulosa that occurs at ovulation. The granulosa cells cease dividing and hypertrophy instead, becoming rich in mitochondria, smooth endoplasmic reticulum, lipid droplets, Golgi bodies and, in many species, a carotenoid pigment, lutein, which may give the corpora lutea a yellowish or orange tinge. This transformation is referred to as *luteinization* and is associated with a steadily increasing secretion of progestagens from the corpus luteum. In most species, the principal progestagen secreted is progesterone, but secretion of significant quantities of $17\alpha$-hydroxyprogesterone in primates and of $20\alpha$-hydroxyprogesterone in the rat and hamster also occurs. In a few species, notably the great apes and man and to a lesser extent the pig, it is clear that the corpus luteum also secretes oestrogens, particularly oestradiol $17\beta$. The source of oestrogens may be

89    *Ovarian Function*

**Table 4.1.** Hormones luteotrophic for the non-pregnant corpus luteum of different species.

| Species | LH* | Prolactin | Oestrogen | FSH** |
|---------|-----|-----------|-----------|-------|
| Man | + + | +? | Luteolytic | — |
| Cow | + + | +? | Luteolytic | — |
| Sheep | + | +? | — | — |
| Pig | + | + | + | — |
| Rabbit | + | + | + + + | + |
| Rat/hamster | + | + + + | + | + |
| Dog | + | ? | ? | ? |

*LH may be directly luteotrophic, but in species in which follicles are growing through the luteal phase, the LH may also stimulate oestrogen synthesis (see Section 3b for details).
**FSH is probably only luteotrophic indirectly by stimulating oestrogen production from developing follicles (see Section 3b for details).

the thecal cells incorporated after the collapse of the follicle, but this is not proven. In most species (monkey, sheep, cow, rabbit, rat and horse) however, the corpus luteum secretes only trivial amounts of oestrogen.

The endocrine support of the corpus luteum shows considerable species variation. The conversion of a follicle to a corpus luteum requires that high surge levels of LH provoke ovulation. This gonadotrophin is then also required, albeit at much lower levels, for the maintenance of the corpus luteum. However, in some species, prolactin is also an important component of the so-called *luteotrophic complex*. In those species in which prolactin is luteotrophic (Table 4.1), prolactin receptors can be detected on granulosa cells from the preovulatory stage onwards.

*f Luteolysis*

The life of the corpus luteum in the non-pregnant female varies from 2 to 14 days (see later and Chapter 10). Luteal regression or *luteolysis* involves a collapse of the lutein cells, ischaemia and progressive cell death with a consequent fall in the output of progestagens. The whitish scar tissue remaining, the *corpus albicans,* is absorbed into the stromal tissue of the ovary over a period which varies from weeks to months, depending on the species. It is clear that luteolysis can be caused by withdrawal or inadequacy of the luteotrophic complex. However, in many species it is not primarily a failure of luteotrophic support, but active production of a *luteolytic factor* which brings about normal luteal regression.

In many mammals studied, *with the notable exception of primates,* it is found that luteal life can be prolonged considerably by removing the uterus. A similar effect can be achieved by ligating the tissues, including the blood vessels, between the uterus and the ovary. If the endometrium of the

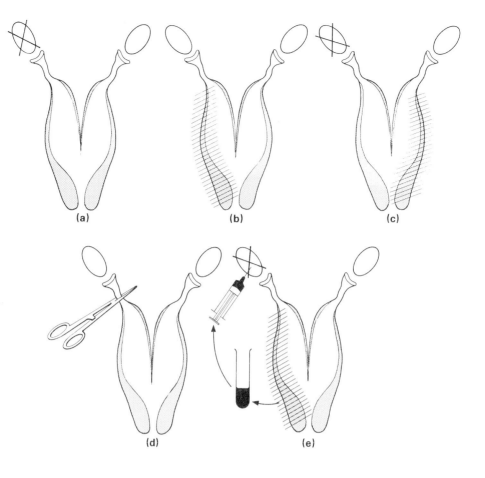

**Fig. 4.8.** (a) Non-pregnant sheep uterus and ovaries—a single corpus luteum (CL) present in left ovary regresses (x); (b) remove ipsilateral uterine horn (hatched)—CL survives; (c) remove contra-lateral horn—CL regresses; (d) clamp blood supply between horn and ovary—CL survives; (e) homogenize and reinject endometrium of ipsilateral horn—CL dies.

excised uterus is homogenized and injected, then luteolysis does occur. These results led to the suggestion that a humoral factor passes from the endometrium to the ovary, and causes luteolysis. Two observations on ewes suggested that this humoral factor was highly labile. First, if instead of removing the whole uterus, which in the ewe has two separate horns, only one horn was removed, then only the corpus luteum on the opposite side regressed (Fig. 4.8). Second, if the whole uterus was transplanted elsewhere in the body, luteolysis was prevented *unless* the ovaries had also been transplanted with the uterus as a unit. Thus, the luteolytic 'factor' does not appear to travel well!

The identity of the endometrial substance responsible has now been established for the sheep, cow, guinea-pig and

**Table 4.2.** Duration of phases of follicular development in the non-pregnant animal.

| Species | Preantral phase (days) | Antral phase (days) | Preovulatory phase (hours) | Luteal phase (days) |
|---|---|---|---|---|
| Mouse | 14 | 4 | 11 | 2 |
| Man | NK | 8–12 | 37 | 12–15 |
| Sheep | NK | 4–5 | 22 | 14–15 |
| Cow | NK | c. 10 | 40 | 18–19 |
| Pig | NK | c. 10 | 41 | 15–17 |
| Horse | NK | c. 10 | 40 | 15–16 |

NK = not known.

horse as *prostaglandin $F_{2\alpha}$* (PGF$_{2\alpha}$). (See Chapter 12 for details of the biochemistry of PG.) PGF$_{2\alpha}$, which is largely destroyed on one passage through the systemic circulation, is secreted in a series of pulses from the endometrium some 10–15 days, depending on the species, after corpus luteum formation, and corpus luteum regression follows shortly thereafter. If this PGF$_{2\alpha}$ is neutralized experimentally by specific antibodies, then luteolysis is prevented. Conversely, premature injections of exogenous PGF$_{2\alpha}$ lead to rapid luteolysis; indeed, PGF$_{2\alpha}$ will cause regression of luteinized cells *in vitro*.

The control of luteolysis in *primates*, unlike that for other species, does *not* involve uterine prostaglandins. The levels of prostaglandins secreted *do* rise in the late luteal phase, but neither hysterectomy nor antibodies to prostaglandins prolong luteal life. Moreover, injections of prostaglandins are without effect on the corpus luteum unless very high doses are used when a transient drop in progesterone output occurs. What then controls luteolysis in the primate? One possibility is that the very low levels of LH during the luteal phase of primates (see Section 3b this chapter) are insufficiently luteotrophic and the corpus luteum simply regresses slowly.

*g How long does follicular development take?*

For many species we do not have precise measurements of the duration of each of the phases of follicular development. However, the approximate values, recorded for some species in Table 4.2, give an idea of the relative duration of each phase. The briefest is always the preovulatory phase, which lasts only hours compared with other phases.

*h How many follicles ovulate?*

The number of follicles ovulating together will be determined by the number that are successfully rescued from atresia at late preantral and late antral stages. The number tends to be characteristic for each species and ranges from one to several hundred. Thus, in man it has been estimated that about 15–20 follicles start the process of antrum formation. This number can be increased if tonic FSH levels are elevated by exogenous hormone. Of the 15–20 follicles, usually only one is sufficiently primed with LH receptors on granulosa cells for the LH surge to induce preovulatory changes. Increasing the LH by exogenous hormone can increase the number of follicles ovulating. *Superovulation* can be achieved with extra FSH for antral growth and extra LH for preovulatory growth, and it is used clinically and in veterinary practice to improve fertility.

*i Summary of events*

Follicular growth and differentiation through to a functional corpus luteum are complex processes that are only completed successfully by under 0.1% of follicles. Most become atretic at some point during the process. The number of follicles which normally complete the process successfully varies with the species. For successful follicular growth there must be a delicate interplay between the levels of circulating gonadotrophins and the acquisition, by the follicle, of hormonal responsiveness due to differentiation of hormone receptors. *If appropriate hormone levels and acquisition of receptors coincide, then follicular development continues. If hormone levels are inappropriate when the receptors develop, then follicular atresia ensues.*

The first *preantral phase* of development is hormone-independent, and involves the resumption of development by primordial follicles. The stimulus for preantral growth, which occurs continuously throughout adult life, is obscure at present. During this phase the egg undergoes its major growth, and the follicle acquires receptors for FSH and oestrogen (granulosa cells) and for LH (thecal cells). During the shorter *antral phase,* growth of follicular cells and formation of an antrum occur with a rising output of oestrogens and androgens. Adequate levels of LH and FSH are required for this phase to be completed successfully, and the higher the levels the more follicles will be saved. Towards the end of this phase oestrogen output increases from the largest antral follicles, and the granulosa cells acquire LH receptors. During the short *preovulatory phase* a large surge of LH gives a sudden stimulus to those follicles with LH receptors on both granulosa and thecal cells. This surge of LH first stimulates, and then stops, the endocrine activity of the thecal cells, and also switches off the aromatizing activity of granulosa cells and diverts them to the production of progestagens. This dramatic switch in endocrine activity is accompanied by the renewed meiotic and cytoplasmic maturation of the oocyte and culminates in *ovulation.* During this phase the granulosa cells of many species acquire prolactin receptors. The final *luteal phase* of follicular development is characterized by a rising output of progestagens, on which may be superimposed, in primates and pigs, a rise in oestrogen output. A luteotrophic complex of some, or all, of the three hormones, prolactin, LH and oestrogen, supports the luteal phase. During this phase prostaglandin receptors develop within the corpus luteum, at least in non-primates. The whole sequence terminates with luteolysis, due either to the production of the uterine luteolytic hormone $PGF_{2\alpha}$ or due to failure of luteotrophic support.

## 3 Follicular development and the ovarian cycle

In the previous section, we described the development of an *individual follicle* through to either ovulation and luteinization, or to atresia en route. But each of the two ovaries has *many* primordial follicles and we must now consider the relationships between the various follicles developing in the two ovaries in order to obtain an overall picture of ovarian function and cyclicity.

### a The ovarian cycle

One complete *ovarian cycle* is the interval between successive ovulations, where each ovulation is preceded by a period of oestrogen dominance. Since the oestrogens are derived from the follicles, the period prior to ovulation is often called the *follicular phase of the cycle*. Correspondingly, the post-ovulatory period is often called the *luteal phase*, because progesterone is derived from the corpus luteum. The duration of the ovarian cycle and its constituent follicular and luteal phases in various species are summarized in Table 4.3. It is clear immediately that there are major differences between species in both the absolute length of the ovarian cycle and in the relative duration of its follicular and luteal components. In fact, these apparent major differences mask a fundamentally similar organization, and result only from minor but significant modifications to a basic pattern. First

**Table 4.3.** Ovarian cyclicity in different species.

| Species | Length of cycle (days) | Follicular phase (days) | Luteal phase (days) |
|---|---|---|---|
| Man | 24–32 | 10–14 | 12–15 |
| Cow | 20–21 | 2–3 | 18–19 |
| Pig | 19–21 | 5–6 | 15–17 |
| Sheep | 16–17 | 1–2 | 14–15 |
| Horse | 20–22 | 5–6 | 15–16 |
| Mouse and rat* (+infertile male) | 13–14 | 2 | 11–12 |
| Rabbit* (+infertile male) | 14–15 | 1–2 | 13 |
| Mouse and rat | 4–5 | 2 | 2–3 |
| Rabbit | 1–2 | 1–2 | 0 |

*See text for discussion.

we will discuss the human ovarian cycle, as it is the easiest to understand. We will then relate that pattern to those of other species.

*b The ovarian cycles of man in relation to those of the cow, pig, sheep and horse*

Figure 4.9a shows a basic outline of two sequential human ovarian cycles. The pattern of measured blood steroids and gonadotrophins is indicated below, and the activities of the four follicular stages are indicated above. A continuous trickle of developing preantral follicles occurs throughout the cycle; growth of these follicles does not require gonadotrophic support, and the preantral follicles do not secrete significant levels of steroids and thus do not affect blood levels of steroids. Mature preantral follicles are doomed to atresia, however, unless rescued during the first 8–12 days of an ovarian cycle. During this period tonic LH and FSH levels are higher than during the luteal phase and therefore adequately support the

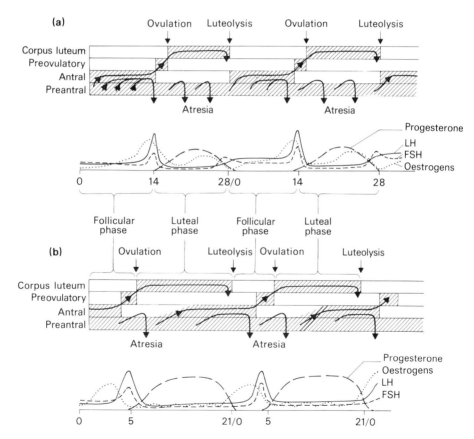

Fig. 4.9. Summary of follicular activity in two sequential ovarian cycles: (a) man; (b) pig. Activity of the different follicular stages is indicated by cross-hatching in the bars. Relative blood hormone levels are as indicated. In order to follow the history of an ovulatory follicle from preantral to luteolytic stages follow the continuous line marked with closed arrows.

95    *Ovarian Function*

growth of new antral follicles. The earliest of the rescued preantral follicles will become the most advanced antral follicle, secreting high levels of oestrogens at maturity 8–12 days later—witness the rising blood oestrogen levels. This advanced follicle is then converted to a preovulatory follicle by the transient high levels of LH that may be measured in the blood at this time. The less mature antral follicles lack the capacity to respond to high LH levels, and when they mature subsequently there is no elevated LH to rescue them and they become atretic. The successful ovulatory follicle forms a corpus luteum which secretes progesterone, and some oestrogen, until luteolysis 14 or so days later. A new cycle then begins as tonic gonadotrophin levels are elevated.

Two important features emerge from this description of the human ovarian cycle: first, we need to understand what controls the fluctuating tonic and surge levels of gonadotrophin output—this will be discussed in detail in the next chapter; second, *the complete antral phase, ovulation and the complete luteal phase occupy one complete ovarian cycle.* This second feature distinguishes man from the cow, pig, horse and sheep. In these species, as indicated for the pig in Fig. 4.9b, all or most of the antral growth occurs whilst the luteal phase of the previous crop of follicles is still proceeding. Thus, in these species the complete antral phase, ovulation and the complete luteal phase *occupy longer than one complete ovarian cycle* (follow and compare the lines with solid arrows in Fig. 4.9a and b). It is almost as though in the cow, etc., the human cycle has been 'telescoped' by 'pushing' the follicular part of one cycle backwards into the luteal half of the previous cycle. In the large domestic animals, but not man, the tonic activity of gonadotrophins in the luteal phase is adequate to maintain antral growth of follicles such that conversion to a preovulatory state can occur sooner after luteolysis. This follicular growth during the luteal phase also means that in these species, some follicular oestrogen secretion will occur in the luteal half of the cycle. The absence of this follicular oestrogen in the higher primate is made good by the oestrogenic secretion of the corpus luteum.

The ovarian cycles of most other species can be understood quite simply in terms of the above discussion. We will examine those of the rat, mouse and rabbit because they exhibit a certain distinctive feature of great importance in understanding the ovarian cycles of man and the large farm animals.

*c The ovarian cycles of the rat and mouse*

The ovarian cycles of the rat and mouse are basically of the 'telescoped' sort seen in large farm animals. They show, however, a curious and distinctive feature; the cycle differs in length, depending upon whether the females mate or not. If the female has an infertile mating at the time of ovulation, for example with a vasectomized male, her luteal phase is 11–12 days in duration (often called *pseudopregnancy*), and the ovarian cycle is generally similar to that of the pig. However, if she fails to mate at the time of ovulation the luteal phase is only 2–3 days long. In the latter case the corpora lutea only become transiently functional in producing progestagens, secreting a small amount of progesterone, but mainly 20 $\alpha$-hydroxyprogesterone.

The explanation for this phenomenon lies in the mechanical stimulus to the cervix provided naturally by the penis at coitus. The presence of such stimulation is relayed via sensory nerves from the cervix to the central nervous system, and activates the release of *prolactin* from the pituitary. This hormone, as was seen earlier, is an essential part of the luteotrophic complex in the rat and mouse, and without it luteal life is abbreviated from the 'normal' extended pattern characteristic of the large farm animals to only 2 or 3 days. The rat and mouse will derive increased reproductive efficiency from this evolutionary modification, since without the abbreviating device they would only be fertile every 13 or 14 days instead of every 4 or 5. As their pregnancy only lasts 20–21 days, this is a highly significant economy. Evolutionary pressures for such a truncation of the luteal phase would not apply to the pig, cow, sheep and horse where pregnancy is a very extended event compared with the luteal phase.

This neat neural device which shortens the luteal phase introduces an important new concept into our discussion. It illustrates how the central nervous system (CNS) can influence ovarian function. Our next two chapters will discuss this problem in detail. Before starting that discussion we will look at one other type of modification to the ovarian cycle which re-emphasizes the role that the CNS can play.

*d The ovarian cycle of the rabbit*

A female rabbit caged alone shows little evidence of a cycle. Blood levels of oestrogens are high, progestagens are low, ovulation cannot be detected, and yet her ovary contains waves of developing antral follicles. It is as though she is in a continuous follicular phase. If she is mated with a vasectomized buck, or if her cervices are stimulated mechanically, she ovulates 10–12 hours later and has a luteal phase, or 'pseudopregnancy' of 12 or so days. If the vasectomized buck is left in with her she will show a 14-day cycle with a 2 + 12 day follicular:luteal pattern, much like the porcine pattern.

As in the rat, cervical stimulation in the rabbit is the source of a sensory input to the CNS which, in this case, induces a surge of LH, high levels of which rescue any advanced ovarian antral follicles from atresia and ovulate them. In essence, the rabbit has abbreviated her cycle even more than the rat or mouse by eliminating the luteal phase completely. This phenomenon is known as *induced ovulation*. It is a

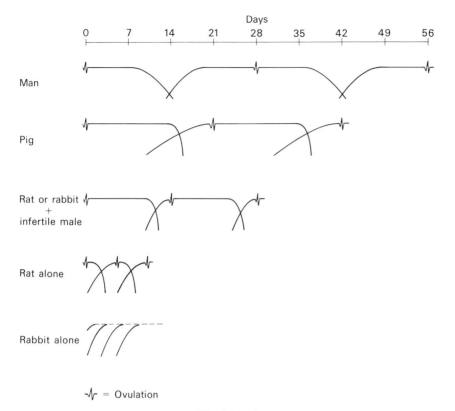

**Fig. 4.10.** Comparison of two ovarian cycles of man, pig and rat or rabbit caged with infertile male, and rat or rabbit caged alone. Day 0 = first ovulation. Each continuous line represents a complete and successful sequence of growth of one crop of preantral follicle(s) through preovulatory follicle(s) to corpora lutea and ultimately luteolysis.

subject to which we will return in subsequent chapters on the CNS, because not only is induced ovulation observed in several species, e.g. the rabbit, cat, ferret, camel and llama, but it is often suggested that it may also occur in man. Our discussion on different ovarian cycles is summarized in Fig. 4.10.

## 4 Interstitial glands

We have discussed at length the activities of the maturing follicles composed of primordial follicles and accreted stromal cells—the thecal layers. However, within the stroma, and between the developing follicles, lie the interstitial glands. Do they have a role to play in ovarian function and cyclicity?

The interstitial glands show considerable variation amongst different species, and little is known of their activity. Interstitial glands are composed of aggregates of steroidogenic-like cells which contain extensive smooth endoplasmic reticulum and lipid droplets. In the rabbit, in which these cells are well developed, they synthesize progesterone and 20α-hydroxyprogesterone, and are sensitive to the LH activity observed after coitus. In the rat the interstitial glands are also stimulated by the surge of LH to form progestagens.

Interstitial cells have also been implicated in the synthesis of androgens in the human, rat and rabbit ovaries, and it is possible that this tissue serves as a source of androgens for both secretion and aromatization in the follicles.

## 5 Conclusions

The ovary, like the testis, produces both gametes and steroids. However, whereas in the male the gametes and the androgens are produced continuously and concurrently, in the female the ovarian output is cyclic and shows two distinct phases separated by the ovulatory release of an oocyte. The follicular phase, prior to ovulation, is dominated by a rising output of oestrogen and some androgens, whereas in the luteal phase, after ovulation, progestagens predominate. In many species the follicular phase is brief, and the major part of follicular growth occurs during the luteal phase of the previous cycle. In man, however, the follicular phase is more extended. We have considered the cyclicity of reproduction in the female from the viewpoint of the ovary. However, the cyclic output of steroids imparts cyclicity to the physiology and behaviour of the female as we will see in more detail in Chapter 7. This *external* manifestation of the *internal* ovarian cycle can be recognised in many mammals by a behavioural characteristic. Female mammals of most species are only receptive to males (*on heat*), and therefore willing to mate, around the time of ovulation. This period of heat is conditioned by the hormonal milieu of the preovulatory phase, and can be accompanied by marked changes in behaviour patterns. The frisky state observed in some species, particularly in horses, was given the name *oestrus* (or 'attacked by gadflies') and thus the internal ovarian cycle is manifested externally in the so-called *oestrous cycle*. Day 1 of each oestrous cycle is generally considered to be the day of first appearance of oestrous behaviour. Higher primates show little evidence of oestrus and thus it is inappropriate to speak of their oestrous cycle. However, in higher primates, another external manifestation of ovarian cyclicity is observed—the shedding of bloody endometrial tissue via the vagina at the end of the luteal phase. This hormonally conditioned event is called *menstruation* (=monthly event), and is used as the external basis for the measurement of the *menstrual cycle* in which the first day of menstruation is considered Day 1 of the cycle. As may be observed in Fig. 4.11, although both the oestrous and menstrual cycles reflect ovarian cyclicity, starting days for each cycle occur at different points in the underlying ovarian cycle.

99      *Ovarian Function*

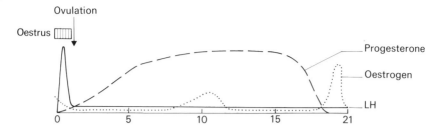

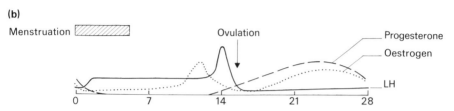

**Fig. 4.11.** The ovarian cycles: (a) cow; (b) man. Plotted as oestrous and menstrual cycles. The first day of the cycle coincides with the beginning of either oestrus or menstruation (plasma concentrations not to scale).

We have seen in this chapter that the levels of gonadotrophins affect critically the successful completion of the ovarian (and thus of the oestrous/menstrual) cycle, and that the circulating levels of gonadotrophins vary naturally during the cycle. Moreover, we have observed that the CNS can affect the levels of circulating gonadotrophins. In the next chapter we will examine how gonadotrophin levels are regulated, and how the CNS exerts its effects.

**Further reading**

Channing CP, Marsh J, Sadler WA. Ovarian follicular and corpus luteum function. *Adv Exp Med Bio* 1979; **112**.

Cole HH, Cupps PT (Eds) *Reproduction in Domestic Animals.* Academic Press, 1977.

Hillier SG. Sex steroid metabolism and follicular development in the ovary. In *Oxford Reviews of Reproductive Biology* 1985 Vol. 7. Oxford University Press, 1985.

Perry J. *The ovarian cycle of mammals.* University Reviews in Biology 13. Oliver and Boyd, 1970.

Peters H, McNatty KP. *The Ovary.* Paul Elek, 1980.

Richardson MC. Hormonal control of ovarian luteal cells. *Oxford Reviews of Reproductive Biology* Vol 8. Oxford University Press, 1986.

Van Blerkom J, Motta P. *The Cellular Basis of Mammalian Reproduction.* Urban and Schwarzenberg, 1979.

Zuckermann Lord, Weir BJ. *The Ovary* (3 Volumes). 2nd Edn. Academic Press, 1979.

# Chapter 5
# The Regulation of Gonadal Function

In Chapters 3 and 4 we saw the ways in which the anterior pituitary gonadotrophins, FSH and LH, together with prolactin, regulate the cellular and endocrine functions of the

101    *The Regulation of Gonadal Function*

ovary and testis. In this chapter we will discuss the mechanisms by which secretion of FSH, LH and prolactin are regulated. In order to do this a new variable must be introduced into our discussion, namely the central nervous system (CNS), and, in particular, the part of the brain most intimately connected to the pituitary, the *hypothalamus*.

## 1 Hypothalamic–pituitary axis
### a The pituitary

The pituitary gland weighs about 500 mg in the adult human; it is ovoid in shape and lies in the *hypophyseal fossa* of the *sphenoid bone*. Here it is overlapped by a circular fold of dura mater, the *diaphragma sellae*, which has a small central opening through which the pituitary stalk, or *infundibulum*, passes (Fig. 5.1). The gland has an extremely rich blood supply derived from the internal carotid artery via its superior and inferior hypophyseal branches. Venous drainage is by short vessels which emerge over the surface of the gland and enter neighbouring dural venous sinuses.

There are three *lobes* of the pituitary gland. The *anterior* lobe *(adenohypophysis)* is derived embryologically from a small diverticulum (Rathke's pouch) that becomes pinched off from the dorsal pharynx. It is closely apposed to a distinct

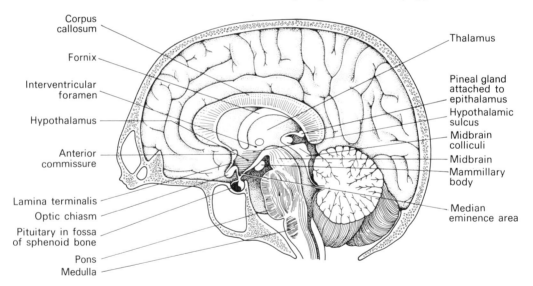

Corpus callosum · Fornix · Interventricular foramen · Hypothalamus · Anterior commissure · Lamina terminalis · Optic chiasm · Pituitary in fossa of sphenoid bone · Pons · Medulla · Thalamus · Pineal gland attached to epithalamus · Hypothalamic sulcus · Midbrain colliculi · Midbrain · Mammillary body · Median eminence area

**Fig. 5.1.** A sagittal section of the human brain with the pituitary and pineal glands attached. Note the comparatively small size of the hypothalamus, which is shaded, and its rather compressed dimensions ventrally. The pineal is attached by its stalk to the epithalamus (habenula region) and lies above the midbrain colliculi. The interventricular foramen is the communication between lateral and third ventricles, the latter being a midline, slit-like structure which has been opened up by the midline cut exposing this medial surface of the brain. The thalamus (above) and the hypothalamus (below) form one wall (the right in this view) of the third ventricle, which is best viewed in Fig. 5.3.

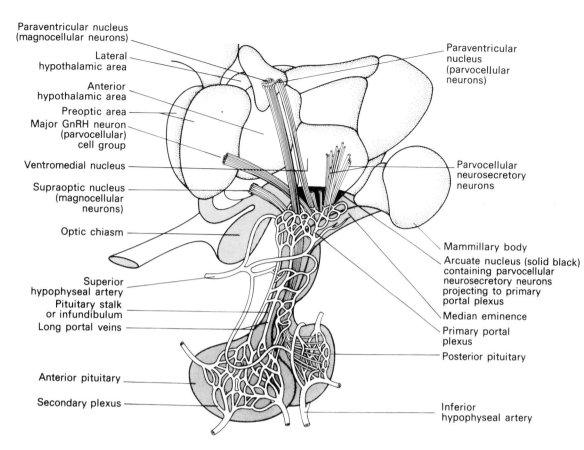

Paraventricular nucleus
(magnocellular neurons)

Lateral
hypothalamic area

Anterior
hypothalamic area

Preoptic area

Major GnRH neuron
(parvocellular)
cell group

Ventromedial nucleus

Supraoptic nucleus
(magnocellular
neurons)

Optic chiasm

Superior
hypophyseal artery

Pituitary stalk
or infundibulum

Long portal veins

Anterior pituitary

Secondary plexus

Paraventricular
nucleus
(parvocellular
neurons)

Parvocellular
neurosecretory
neurons

Mammillary body

Arcuate nucleus (solid black)
containing parvocellular
neurosecretory neurons
projecting to primary
portal plexus

Median eminence

Primary portal
plexus

Posterior pituitary

Inferior
hypophyseal artery

**Fig. 5.2.** A highly schematic and enlarged view of the human hypothalamus showing structures mentioned in the text. Note the portal capillary system derived from the superior hypophyseal artery and running between the median eminence/arcuate region of the hypothalamus and the pituitary (anterior lobe). The ventromedial, paraventricular and arcuate nuclei are well defined and contain neurons which terminate in close association with the portal capillaries. The anterior hypothalamic and preoptic areas should be viewed as a continuum in functional and anatomical terms. Note the latter is in front of the lamina terminalis and strictly (embryologically) not within the hypothalamus proper. The major group of GnRH-containing neurons is indicated in the preoptic area.

and smaller *posterior* lobe (*neurohypophysis*) which is derived embryologically from neurectoderm. This origin is reflected in its connection by a stalk of nervous tissue, the infundibulum, to the overlying hypothalamus in the region of the *median eminence* (Figs 5.1 and 5.2). An *intermediate* lobe of the pituitary, a small division of the adenohypophysis, lies between the two.

The anterior lobe of the pituitary contains a variety of cell types, amongst them the so-called *gonadotrophs*, basophilic cells containing granules of FSH or LH, and *lactotrophs*, acidophilic cells containing prolactin. In addition to FSH,

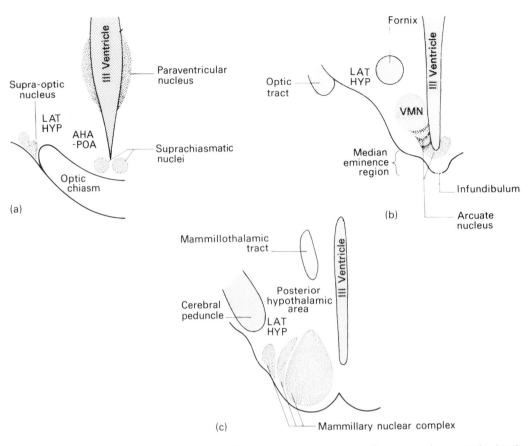

**Fig. 5.3.** Three coronal sections at different anterior–posterior levels
of the right half of the human hypothalamus (consult Fig. 5.2 to
construct planes of sections). (a) Through the optic chiasm, note the
third ventricle in the midline flanked by the paraventricular (magnocellular) nuclei, which, together with the supraoptic nuclei, synthesize oxytocin and vasopressin, which are transported along the
hypothalamo-hypophyseal tract (axons of neurons in these nuclei) to
the posterior pituitary. The region of the suprachiasmatic nuclei and
the anterior hypothalamic–preoptic area (AHA–POA) are shown.
(b) Through the infundibulum and showing the relationship between the arcuate and ventromedial nuclei (VMN). The capillary
loops of the portal plexus are found in this median eminence region.
(c) Through the level of the mammillary bodies and showing the
mammillary nuclear complex. The area labelled LAT HYP in all
three sections is the lateral hypothalamus, and is composed of many
ascending (largely aminergic neurons from the brain stem) and
descending (from the rostral limbic and olfactory areas) nerve fibres.
This pathway represents a major input/output system for the more
medially placed hypothalamic nuclei.

LH and prolactin the anterior pituitary secretes growth
hormone (GH) from the *somatotrophs*, ACTH from the
*corticotrophs*, and TSH from the *thyrotrophs*. The posterior
lobe of the pituitary secretes two nonapeptide hormones,
*arginine vasopressin (AVP or antidiuretic hormone: ADH)* and

*oxytocin*. The importance of oxytocin in reproductive events is discussed in Chapters 12 and 13. Both lobes of the pituitary are connected *anatomically* and *functionally* to the overlying hypothalamus.

*b The hypothalamus*

The hypothalamus is a relatively small area at the base of the brain. It lies between the *midbrain* behind and below, and the *forebrain* in front and above (Fig. 5.1). The boundaries of the hypothalamus are conventionally described as the *hypothalamic sulcus* separating it from the *thalamus* (above), the *lamina terminalis* (in front) and a vertical plane immediately behind the *mammillary bodies* (Fig. 5.1). The hypothalamus is split symmetrically into left and right halves by a cavity, *the third ventricle*, containing cerebrospinal fluid and lying in the midline so that the hypothalamus forms its floor and lateral walls (Figs 5.1 and 5.3).

Despite its small size, the hypothalamus is an extremely complicated structure with many diverse functions which include the regulation of sexual and ingestive behaviours, the control of body temperature and the integration of autonomic activity. Each function is associated with one or more small areas of the hypothalamus consisting of aggregations of neurons, called *hypothalamic nuclei* (Figs 5.2 and 5.3). The functions of the hypothalamus associated with reproduction involve the *supraoptic, paraventricular, arcuate, ventromedial* and *suprachiasmatic* nuclei, and also two less easily defined areas, the *anterior hypothalamic* and *preoptic* areas. These nuclei are in intimate contact with the underlying pituitary gland by *neural* and/or *vascular* connections.

The hypothalamus has rich interconnections with many parts of the brain, in particular the brain stem *autonomic* and *reticular* structures and areas of the limbic forebrain such as the *amygdala* and *septum*. Also of interest is the photic input from the retina to the suprachiasmatic nuclei. By this route, the hypothalamus is made aware of the cycles of light and dark which in turn affect reproductive function (see later— Section 7a).

*c Connections between hypothalamus and pituitary*
*i The posterior pituitary*

The axons of neurons in the supraoptic and paraventricular nuclei project directly to the posterior lobe of the pituitary via the hypothalamo-hypophyseal tract (Fig. 5.2). The hormones of the posterior pituitary, vasopressin and oxytocin, are synthesized in the cell bodies of the two hypothalamic nuclei, packaged by combination with binding proteins (*neurophysins*) which are specifically associated with each peptide and transported along the axons by a process of axoplasmic flow to be stored in the posterior lobe of the pituitary. From here, release of the hormone into the bloodstream occurs. This system of neurons in the hypothalamus is called the *magnocellular* (i.e. large-celled) *neurosecretory* system (Fig. 5.4a).

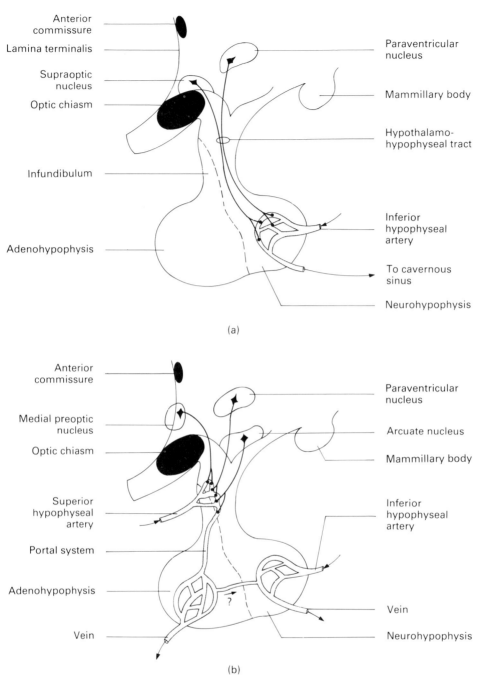

Anterior
commissure

Lamina terminalis

Supraoptic
nucleus

Optic chiasm

Infundibulum

Adenohypophysis

Paraventricular
nucleus

Mammillary body

Hypothalamo-
hypophyseal tract

Inferior
hypophyseal
artery

To cavernous
sinus

Neurohypophysis

(a)

Anterior
commissure

Medial preoptic
nucleus

Optic chiasm

Superior
hypophyseal
artery

Portal system

Adenohypophysis

Vein

Paraventricular
nucleus

Arcuate nucleus

Mammillary body

Inferior
hypophyseal
artery

Vein

Neurohypophysis

(b)

**Fig. 5.4.** (a). Schematic representation of the magnocellular neurosecretory system. Neurons in the supraoptic and paraventricular
nuclei send their axons via the hypothalamo-hypophyseal tract
through the infundibulum to the neurohypophysis where terminals
lie in association with capillary walls, the site of neurosecretion. (b).
Schematic representation of the parvocellular neurosecretory system. Neurons in the medial preoptic area (e.g. GnRH-containing),
anterior periventricular nucleus, medial parvocellular paraventricular nucleus (e.g. VIP-containing) and arcuate nucleus (e.g. GABA-

In contrast to the direct neural connections linking hypothalamus and posterior pituitary, the hypothalamus communicates with the anterior lobe of the pituitary by a *vascular* route (Fig. 5.2). A variety of neurohormones are synthesized in hypothalamic neurons designated the *parvocellular* (i.e. small-celled) *neurosecretory* system. The axons of these neurons terminate in the pericapillary space of the *primary portal plexus* of vessels in the external zone of the *median eminence*. These capillaries are derived from the superior and inferior hypophyseal arteries and pass from this area of the hypothalamus down to the anterior pituitary. By this route, the neurohormones manufactured by hypothalamic neurons reach and act upon the gonadotrophs, thyrotrophs, corticotrophs, somatotrophs and lactotrophs to regulate the synthesis and release of their various hormones. Three main hypothalamic areas contain neurons which synthesize releasing hormones and transport them to the capillary-rich lateral and medial palisade zones of the external layer of the median eminence. These are: the parvocellular paraventricular nucleus, the medial preoptic area and the arcuate nucleus (Fig. 5.4b).

iii Hypothalamic control of gonadotrophin secretion

The glycoprotein hormones LH and FSH are secreted by gonadotroph cells in the anterior pituitary. Immunocytochemical studies have revealed that each hormone is generally elaborated in different cells, but occasionally both may be found in the same cell.

The synthesis and secretion of FSH and LH are dependent on *gonadotrophin hormone. releasing hormone* or *GnRH* (a decapeptide with the structure (Pyro)-Glu-His-Trp-Ser-Tyr-Gly-Leu-Arg-Pro-Gly-NH2). GnRH is the neurosecretory product of neurons in the hypothalamus and is carried to the anterior pituitary through the portal vessels. GnRH is derived from a larger, precursor molecule called prepro-GnRH whose structure has been established using recombinant DNA technology. This molecule (molecular weight 10 000) comprises the decapeptide GnRH preceded by a signal sequence of 23 amino acids and followed by a Gly-Lys-Arg sequence necessary for enzymatic processing and C-terminal amidation of GnRH (Fig. 5.5). The C-terminal region of the precursor is occupied by a further 56 amino acids that constitute the so-called *GnRH-associated peptide* or *GAP* (Fig. 5.5). GAP has been shown both to release LH and FSH equipotently from the anterior pituitary and to inhibit the

---

and dopamine-containing) send their axons down to the portal vessels in the external layer (palisade zone) of the median eminence where neurosecretion occurs.

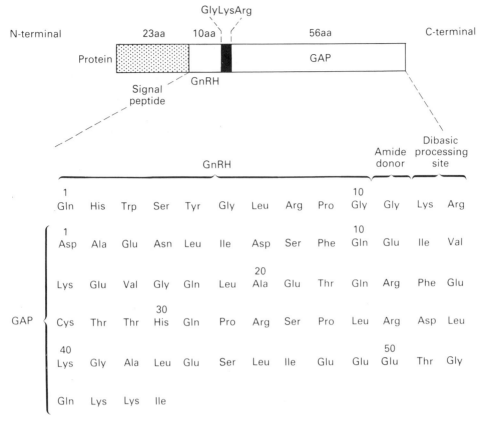

**Fig. 5.5.** Structure and encoded amino acid sequence of human cDNA for prepro-GnRH. The amino acid sequences of GnRH and GAP and the enzymatic processing site are indicated.

basal secretion of prolactin. In all species examined, immuno-cytochemical techniques have located the GnRH precursor, together with GAP and GnRH itself, in a major group of neurons within the medial preoptic area and adjacent anterior hypothalamus. However, in primates, but not rodents, numbers of GnRH-containing neurons have also been demonstrated in the arcuate nucleus. Nerve terminals containing GnRH are found in the external layer of the median eminence, particularly in association with portal capillaries in the lateral palisade zone (Fig. 5.6a), which is the site, therefore, of GnRH neurosecretion.

GnRH is presently assumed to be the most important final common mediator of all influences on reproduction conveyed through the CNS. Any abnormality in GnRH synthesis, storage, release or action will result in partial or complete failure of gonadal function. Destruction of GnRH-producing neurons in the hypothalamus, or immunization against the peptide, prevents gonadotrophin function and results in gonadal atrophy. In the former case, this can be reversed by

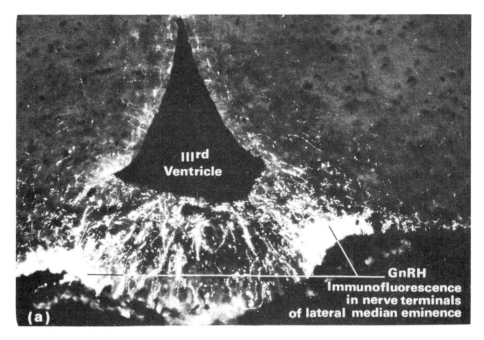

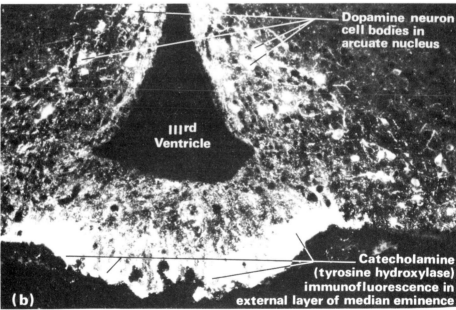

**Fig.5.6.** (a) GnRH detected by immunofluorescence in the median eminence (middle region). Note the very dense network of GnRH neuron *terminals* in the lateral region, but no fluorescent cell bodies are visible. Compare with Figs 5.22 and 5.23. (b) Tyrosine hydroxylase immunofluorescence in an adjacent section through the median eminence. The dense terminal fluorescence in the external layer largely represents dopamine, but noradrenaline-containing terminals will also be labelled by this procedure. Large dopamine-containing cell bodies are clearly visible in the arcuate nucleus. Compare with Figs 5.22 and 5.23.

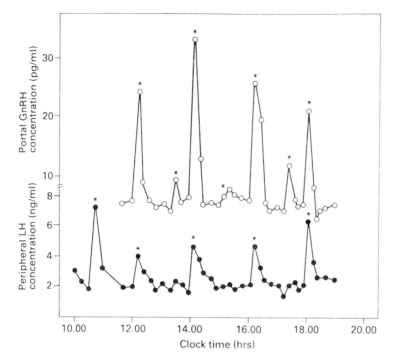

**Fig. 5.7.** Concentrations of GnRH in portal plasma (○—○) and LH in jugular venous plasma (●—●) of four ovariectomized ewes. Asterisks indicate secretory episodes (pulses) of GnRH and LH.

appropriate treatment with synthetic GnRH given intravenously. The precise importance of GAP in the physiological regulation of FSH and LH secretion has not yet been established.

Relatively recent technological advances have made it possible to sample portal blood in order to measure GnRH secretion. By concurrently sampling peripheral blood for gonadotrophins, the relationship between GnRH and gonadotrophin secretion has been defined. The secretion of both GnRH and LH occurs in a pulsatile manner, approximately one pulse being measured about every hour or so and therefore called *circhoral pulses*. The GnRH pulses in portal blood have been shown to be coincident with peripheral serum LH peaks (Fig. 5.7). However, the data in Fig. 5.7 not only demonstrate that each LH pulse is coincident with a GnRH pulse but also show that each GnRH pulse is not *necessarily* followed by a LH pulse. The location of the *circhoral oscillator* which controls GnRH secretion, and hence pulsatile LH and FSH secretion, is unclear, although it is assumed to reside in the hypothalamus or be a property of one of its neuronal inputs.

Is the fact that GnRH secretion occurs in a pulsatile mode important physiologically? The question can be answered by

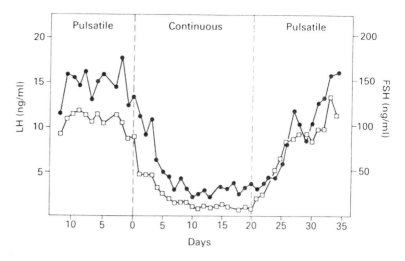

**Fig. 5.8.** Suppression of plasma LH and FSH concentrations after initiating (Day 0) a continuous GnRH infusion (1 $\mu$g/minute) in an ovariectomized, hypothalamically-lesioned rhesus monkey in whom pulsatile gonadotrophin secretion has been re-established by pulsatile GnRH infusions (1 $\mu$g/minute for 6 minutes once per hour). The inhibition of gonadotrophin secretion was reversed by reinstating the pulsatile pattern of GnRH administation on Day 20. ●—● = LH (ng/ml). □—□ = FSH (ng/ml).

destroying the mediobasal hypothalamus, whereupon a dramatic fall in the level of gonadotrophin secretion occurs, and then restoring GnRH using an intravenous infusion pump programmed to deliver either pulses or a continuous infusion. Only with pulses at a frequency approximating that seen naturally were normal patterns of LH and FSH secretion and apparently regular menstrual cycles restored (Fig. 5.8). Clearly, the pulsatile pattern of GnRH secretion is critical. The reason for this critical requirement can be understood by examining how GnRH interacts with the gonadotrophs.

GnRH binds to specific receptors on the cell membrane of the gonadotrophs, as a result of which the inositol trisphosphate second messenger system is activated. Three types of response are then seen. First, an initial release of stored LH and FSH occurs within minutes and lasts for 30–60 minutes. The secretory response to a second pulse of GnRH is larger than that to the first—the so-called 'self-priming' effect of GnRH. Using electron microscope immunocytochemistry, it has been demonstrated that a first pulse of GnRH, which primes the gonadotroph, induces movement of secretory granules into the marginal zone beneath the plasmalemma. The granules also become smaller, perhaps indicating maturation of their contents, but few are lost. A further exposure of GnRH results in 'primed' release of LH and a marked loss of small granules. Second, over a period of hours to days,

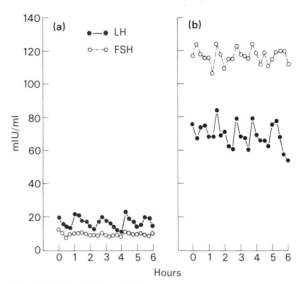

**Fig. 5.9.** Circulating FSH and LH levels in a postmenopausal woman (b) and a woman at Day 7 in the (early) follicular phase (a). The difference between the two represents the negative feedback effect of oestradiol. Note: (1) the reversal of LH and FSH levels between (b) and (a) which may reflect preferential inhibition of FSH secretion due to inhibin (see text); (2) the changes in pulsatility of FSH and LH between (b) and (a).

gonadotrophin biosynthesis is stimulated. Third, again over a period of days, further binding of GnRH is required to maintain the secretory morphology of the gonadotroph.

After binding, some GnRH–receptor complexes remain in the surface while others are internalized via coated pits to lysosomal structures where degradation or other processing of the peptide can occur. Continuous exposure of gonadotrophs to GnRH, or infusion of a long-acting GnRH analogue, results in maintained occupancy of the receptors and is followed eventually by a *reduction* in pituitary LH and FSH contents and secretion (as the middle panel in Fig. 5.8 shows). It appears that continued receptor occupancy uncouples the receptors from subsequent biochemical events involved in signal transduction.

In summary, we have discussed the widely accepted view that hypothalamic GnRH neurons regulate the synthesis and secretion of FSH and LH by the anterior pituitary. The GnRH is released as a series of pulses into the portal vessels, reaches and binds to receptors on the gonadotrophs and drives gonadotrophin secretion in a similar, pulsatile manner. Alterations in the output of GnRH, LH and FSH may be achieved, therefore, by increasing or decreasing the *amplitude* or *frequency* of these pulses of GnRH or by modulating the response of the gonadotrophs to the pulses. As we shall now see each of these mechanisms is employed.

## 2 Regulation of gonadotrophin secretion in the female

During the primate's menstrual cycle the regulation of gonadotrophin output is exercised mainly, if not exclusively, by the secretory products of the ovary. In order to understand the dynamic relationship between the ovary and gonado-trophin output two phenomena must be examined. The first is a *depressant* effect on gonadotrophin output induced by elevation of the plasma concentrations of oestrogens, pro-gestagens and also of a third ovarian hormone that we have not previously encountered and that is called *inhibin*. The second phenomenon concerns an *increase,* or *surge,* of LH and FSH secretion induced principally by oestradiol. Throughout our discussion we will use the data from the primate as a model, referring to other species where corrobor-ative evidence or species differences exist.

### a Oestradiol and feedback control of FSH and LH secretion

Plasma concentrations of circulating FSH and LH increase markedly after ovariectomy or the menopause (Fig. 5.9b). This rise is largely attributable to removal of oestradiol, as infusion of this hormone results in the rapid decline of FSH and LH to levels seen in the early follicular phase of the menstrual cycle (Fig. 5.9a). The important characteristics of this feedback relationship are (i) that only low circulating levels of oestradiol are required to exert this very marked effect, and (ii) the effects of oestradiol are very rapid in onset, detectable within 1 hour and maximal by 4–6 hours. Since oestradiol is here responsible for *suppressing* gonadotrophin levels, the process is termed *negative feedback*.

By contrast, if plasma concentrations of oestradiol increase greatly, say 200–400% greater than in the early follicular phase of the cycle, and remain at this high level for 48 hours or so, then LH and FSH secretion is *enhanced* not suppressed (Fig. 5.10). Under these conditions we speak of a *surge* of LH and FSH and the term *positive feedback* has been used to describe the relationship whereby *high* levels of oestradiol *increase* the secretion of gonadotrophin (although this term is strictly incorrect so far as the terminology of feedback control systems is concerned).

Thus, we see that oestradiol has a dual function in regulating gonadotrophin secretion. At low circulating levels, it exerts rapidly expressed, *negative feedback* control over FSH and LH. At high, maintained circulating levels *positive feedback* becomes the dominant force and induces an LH and FSH surge.

### b Progesterone and feedback control of FSH and LH secretion

The single most important effect of progesterone is that *high* plasma concentrations, 4–8 ng/ml in humans, *enhance the negative feedback effects of oestradiol* and hold FSH and LH secretion down to a very low level. At these high plasma concentrations progesterone may have another important action, which is to *block the positive feedback effect of oestra-*

113    *The Regulation of Gonadal Function*

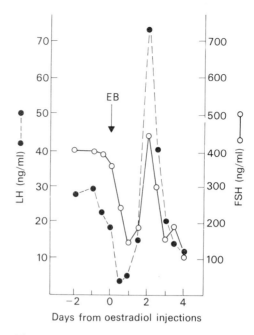

Days from oestradiol injections

**Fig. 5.10.** The negative and positive feedback effects of a sudden and large injection of oestradiol benzoate (EB) in a female rhesus monkey. Note the *early* action of the oestrogen is to decrease FSH and LH levels (negative feedback, maximal at 6 hours or so). But if oestradiol levels are of sufficient magnitude and duration a surge of the gonadotrophins occurs after 36–48 hours (positive feedback). Note that one effect naturally follows the other and each is critically dependent on time–dose relationships with the steroid.

*diol.* Thus, injection of oestradiol into women during the progesterone-dominated phase of the menstrual cycle (luteal phase) is *not* followed by an LH surge.

*c Inhibin and feedback control of FSH secretion*

The secretion of FSH by the anterior pituitary is also regulated by non-steroidal, high molecular weight proteins termed 'inhibins'. Two forms of inhibin (A and B) have been purified from porcine follicular fluid and characterized as heterodimers of molecular weight 32 000. Each type of inhibin molecule consists of an $\alpha$-subunit (molecular weight 18 000: the same in both inhibins) and a $\beta$-subunit (molecular weight 13 800–14 700: distinct for A and B inhibins). The subunits are linked by disulphide bonds.

In experiments in the ewe, intravenous administration of bovine follicular fluid (bFF) rich in inhibin significantly reduces pituitary and plasma FSH concentrations. Furthermore, treatment with bFF completely prevents the rise in FSH which usually follows ovariectomy and also blocks the FSH, but not the LH, response to a GnRH agonist (Fig. 5.11). Furthermore, infusion of a polyclonal antiserum to inhibin during the late antral phase causes an increase in FSH but not

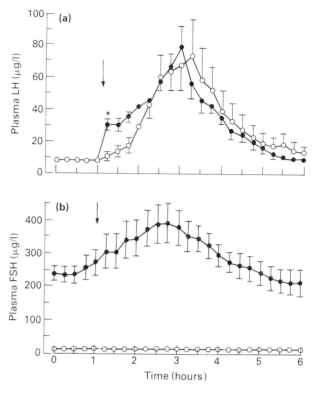

**Fig. 5.11.** Responses in plasma (a) LH and (b) FSH following an injection of a GnRH agonist (arrow), in control ovariectomized ewes (●) and ovariectomized ewes treated with bovine follicular fluid (○). Note the complete absence of an FSH response to GnRH in the inhibin-treated ewes in (b).

LH concentrations in plasma (Fig. 5.12). Clearly, these experimental data strongly indicate a role for inhibin in regulating FSH secretion.

In both the rhesus monkey and women, inhibin activity in plasma is low in the early follicular and luteal phases of the menstrual cycle, but is detected in ovarian vein blood at much higher concentrations in the late follicular phase, and is found to be 100–1000 times greater in follicular fluid than in ovarian vein blood. It has been found clinically that women in whom a rising inhibin output is *not* observed are likely to have inadequate follicles and to be subfertile.

The role of inhibin in regulating FSH secretion has become complicated by the observation that a heterodimeric protein composed of the two β-subunits of inhibins A and B linked by interchain disulphide bonds possesses marked *FSH-releasing* properties. This finding indicates that the ovary may be a major regulator of FSH secretion, with hypothalamic GnRH having a lesser role than in the control of LH secretion.

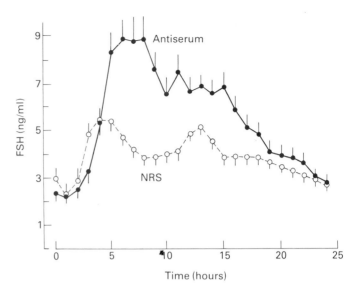

**Fig. 5.12.** The effect of intravenous infusion of normal rabbit serum (NRS) or inhibin antiserum on plasma FSH levels during the late antral phase. Note the marked rise in plasma FSH concentrations in the female rats immunized against inhibin.

*d Role of feedback in*
*the normal menstrual cycle*

Let us now re-examine the blood levels of steroids, inhibin and gonadotrophins during a normal human menstrual cycle (Fig. 5.13) and interpret them in the light of positive and negative feedback. The menstrual cycle begins with the first day of menstruation. Prior to this event luteal oestrogen and progesterone levels fall and FSH levels rise, followed, after a day or so, by an increase in LH levels. These events coincide with the initiation of the antral phase of follicular growth (Chapter 4). During the first half of the follicular phase the ovarian output of steroids changes little, but towards the middle of the phase plasma oestradiol levels have clearly increased. In the second half of the follicular phase oestradiol and oestrone levels rise steadily, culminating in an oestradiol surge. Concomitant with the rise in oestrogens, inhibin levels rise, FSH levels fall, LH levels rise slowly, testosterone and androstenedione levels rise steadily and peak with oestradiol. Plasma concentrations of 17$\alpha$-hydroxyprogesterone also rise slowly over this period. This leads to the preovulatory phase, during which LH and FSH levels rise steeply, or surge, leading to ovulation. Oestradiol levels then fall precipitously, and progesterone levels start to rise. The first half of the cycle is complete.

It is clear that the fall in oestrogen and progesterone prior to menstruation relaxes negative feedback inhibition of gonadotrophin secretion, allowing LH and FSH tonic levels to rise. This rise permits antral growth to proceed, resulting in a slowly rising output of oestrogens and androgens. Inhibin

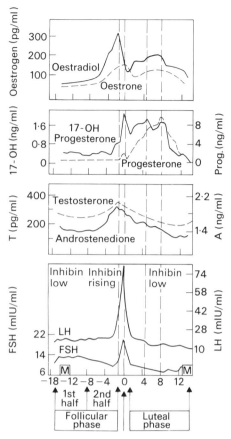

**Fig. 5.13.** Serum hormone levels during the human menstrual cycle. Note the units of measurements; FSH and LH are expressed in milli-international units/ml; testosterone and oestradiol in pg/ml; 17 α-hydroxyprogesterone, progesterone and androstenedione in ng/ml. M = menstruation.

output also rises as the follicles grow and it exerts its negative feedback effect on FSH, leading to its decline. Rapidly, however, oestradiol levels surge, and therefore move into the positive feedback range, triggering the gonadotrophin surge. The gonadotrophin surge triggers the preovulatory phase, in which oestrogen, androgen and inhibin output by the follicle falls and progesterone output begins to rise. The LH and FSH levels now fall equally precipitously because, at least in part, they lack a continuing positive feedback stimulus.

The crucial role of the oestrogen surge in triggering the LH surge is shown by actively immunizing female rhesus monkeys against oestradiol. Neither an LH surge nor ovulation occur. However, if the synthetic oestrogen stilboestrol, which is not neutralized by the antiserum, is given to the immunized animal the neutralization of endogenous oestradiol is bypassed and the synthetic oestrogen is able to reinstate an LH surge. Thus, the pattern of FSH, LH and steroid secretion during the first half of the cycle is explicable largely in terms of the positive and negative feedback effect of oestrogens (primarily oestradiol) and inhibin on gonadotrophin secretion. The length of the follicular phase appears to be determined by the rate at which the main source of these hormones, the principal preantral follicle, matures and, since it is a major arbiter of cycle length, it is often referred to as the *ovarian* or *pelvic clock*.

The luteal phase of the cycle is characterized by rising plasma progesterone concentrations which peak around 8 days after the LH surge, when luteal oestradiol, oestrone and $17\alpha$-OHP levels also peak for a second, somewhat lesser, time. By contrast, testosterone, androstenedione, inhibin, FSH and LH levels decline to their lowest levels in the cycle (Fig. 5.13). The nadir in LH and FSH plasma concentrations is clearly related to the negative feedback effects of oestrogens, which are further enhanced by the presence of high levels of progesterone. Growth of antral follicles is therefore suppressed and so androgens are at a low level. Although oestrogens from the corpus luteum may rise to levels which previously induced positive feedback, they fail to induce an LH surge, and this failure reflects the effects of the uniquely high levels of progesterone found during the luteal phase of the cycle which (a) are dominantly negative and (b) probably inhibit oestradiol positive feedback (see Section 2b above). At the end of the luteal phase, if conception has not occurred, both oestrogens and progesterone decline at luteolysis, the negative feedback effect of these hormones is relaxed, and LH and FSH levels thereby rise, again permitting the rescue of preantral follicles and the initiation of another cycle.

*e Sites of positive and negative feedback*

Having described the nature of the relationships between ovarian hormones and gonadotrophins during the menstrual cycle, we must now consider the sites at which these hormones exert their regulatory influences. There are two obvious candidates. First, the *anterior pituitary*: the hormones might regulate FSH and LH secretion by a (direct) action on the gonadotrophs, decreasing (negative feedback) or increasing (positive feedback) the sensitivity to GnRH pulses of invariant frequency and magnitude arriving from the hypothalamus. There are abundant receptors for oestrogens, progestagens and inhibin in the anterior pituitary, a fact which emphasizes the potential importance of this site. Second, the *hypothalamus*: the ovarian hormones might change the GnRH signal by either a direct effect on the GnRH neurons in the hypothalamus, or indirectly by changing the activity of other neural systems which exert a modulatory influence on GnRH release in the median eminence. Large regional concentrations of receptors for oestradiol and progesterone exist in the preoptic area, anterior hypothalamus and median eminence. However, there is no evidence to support a hypothalamic action of inhibin. Of course, these two potential sites of feedback are not mutually exclusive and a *third* possibility is, therefore, that ovarian hormones alter *both* the GnRH signal *and* the response of the anterior pituitary to it.

i The anterior pituitary

The most convincing demonstration that the pituitary, independent of the CNS, can respond to both the negative and positive feedback effects of oestradiol comes from experiments on the ovariectomized rhesus monkey. Large lesions of the mediobasal hypothalamus, destroying the arcuate and ventromedial nuclei and a large part of the median eminence (see Fig. 5.2), result, as would be expected, in a decrease in serum FSH and LH to undetectable levels. This emphasizes the dependence of gonadotrophin secretion on hypothalamic GnRH. Infusion, using a programmable pump, of hourly pulses of GnRH restored pulsatile LH and FSH secretion. The subsequent administration of surge levels of oestradiol results first in a fall, and then a dramatic rise (surge) in serum FSH and LH levels. In the presence, therefore, of invariant GnRH pulses of constant amplitude and frequency, oestradiol can exert both its negative and positive feedback effects on gonadotrophin secretion. This action can only have been exerted on the anterior pituitary. These experiments imply that *modulation of GnRH pulse amplitude or frequency is not an essential requirement* for effective positive or negative feedback by oestrogen. However, this does not mean that steroid-dependent alterations in the GnRH signal never normally occur (see below).

A number of clinical studies on postmenopausal, hypogonadal or ovariectomized women give less direct, but nonetheless convincing, evidence that the responsiveness of the anterior pituitary to exogenous GnRH changes in response to oestrogen injections. Perhaps the most decisive studies, however, relate to pituitary sensitivity during the normal cycle. These experiments are performed by measuring the change in plasma levels of LH and FSH induced by repeated injections, 'pulses', of GnRH on different days of the human cycle (similar results have been obtained for the rat oestrous cycle). Figure 5.14 shows results of an experiment, which clearly demonstrates that secretion of FSH and LH by the pituitary in response to constant GnRH pulses increases during the follicular phase, suggesting an increasing sensitivity related to rising plasma oestradiol levels. However, it is also important to notice that the responsiveness of the pituitary to GnRH is still very high in the luteal phase of the cycle, a time when FSH and LH levels are normally at their lowest, and when it is impossible to induce an LH surge with

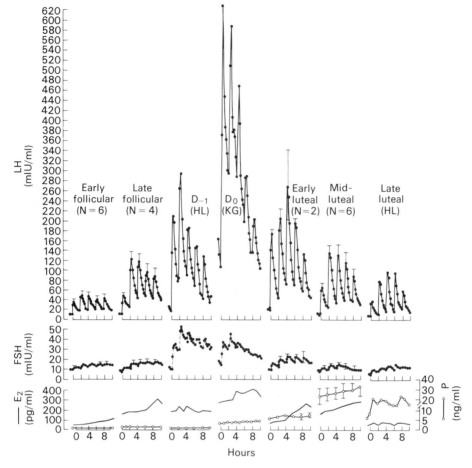

Fig. 5.14. The release of LH and FSH in response to 'pulse injections' of GnRH (10 µg at 5 × 2-hour intervals) during various phases of the menstrual cycle. The magnitude of the LH and FSH responses can be measured against the naturally changing circulating levels of oestradiol and progesterone. Note the correlation between oestrogen levels and *increasing* pituitary responsiveness to the injected GnRH. Note particularly that this remains high in the early and mid luteal phases, at times when further LH/FSH surges cannot be elicited by oestradiol (see text). $D_{-1}$ and $D_0$ = the day before, and the day of, the spontaneous LH surge.

an oestrogen surge (see above). This must mean that oestradiol and/or progesterone, which are both high at this time, exert an important component of their negative feedback action on FSH and LH secretion *somewhere other than the pituitary*. This site is almost certainly the hypothalamus as we will see in the next section. The demonstration that inhibin is able to reduce the FSH secretory response to GnRH strongly indicates that its effects are mediated on the anterior pituitary.

ii The hypothalamus

It has long been suspected, using the pattern of LH and FSH pulses as an indicator of GnRH secretory episodes, that

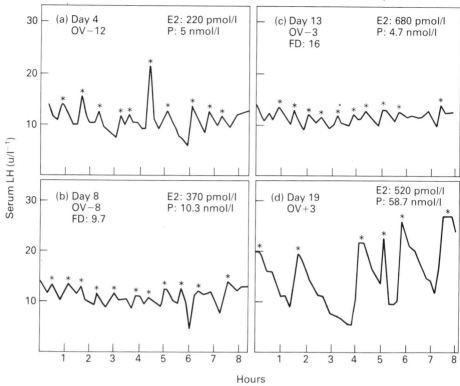

**Fig. 5.15.** LH pulsatility during an ovulatory cycle (ovulation occurred on Day 16) in 10 minute serum samples. (a) Early follicular phase. (b) Mid-follicular phase. (c) Late follicular phase. (d) Early luteal phase. ★ = LH peaks. E2 = 17β oestradiol. P = progesterone. FD = follicular diameter (mm). OV = +/− number of days after/ before ovulation. Note the unique high-amplitude, low-frequency pulses characteristic of the luteal phase when progesterone plasma concentrations are high.

GnRH secretion does vary during the menstrual cycle. Thus, during the follicular phase of the menstrual cycle, LH is secreted in a series of high-frequency, low-amplitude pulses occurring approximately once every hour. By contrast, the luteal phase of the cycle is characterized by a unique pattern of high-amplitude, low-frequency, irregular LH pulses, often with long intervals between them of up to 6 hours (Fig. 5.15). The conclusion to be drawn from these data is that the GnRH pulse generator, which drives LH secretion, is subject to feedback modulation by gonadal steroids and that its activity is increased during the follicular phase and decreased in the luteal phase.

Observing the consequences of manipulating steroids independently reveals that while progesterone acts primarily to reduce pulse *frequency*, oestrogen acts to reduce pulse *amplitude*, suggesting that progesterone acts mainly on the hypo-

121        *The Regulation of Gonadal Function*

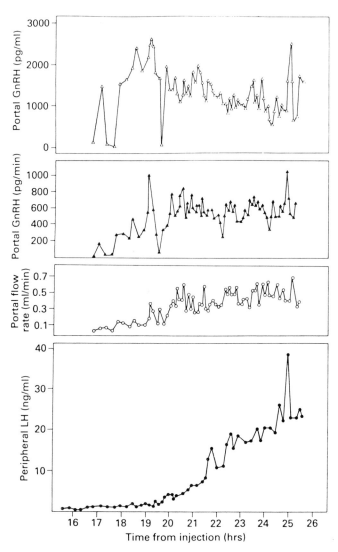

**Fig. 5.16.** GnRH secretion rate and concentration in portal blood, portal blood flow rate and peripheral plasma LH levels in an ovariectomized ewe give an injection of 50 μg oestradiol monobenzoate to induce the LH surge. Note the increased frequency of GnRH and LH pulses during the LH surge.

thalamus, while oestrogen exerts some effect on the pituitary and, as we will see, also on the hypothalamus.

Direct information on the modulation of GnRH secretory activity by steroids requires measurement of the peptide in portal blood. Obviously this is impossible in humans and, even in experimental animals, presents considerable methodological difficulties. However, it has been achieved in experiments on the rat, sheep and rhesus monkey and the data show that GnRH secretory activity *is* subject to modulation by steroid feedback during the menstrual and oestrous cycles, so confirming the indirect evidence described above. Thus, in rats, GnRH secretion is increased on the afternoon immediately before the LH surge. Similarly, the LH surge induced by exogenous oestradiol administration in rhesus monkeys and ewes is associated with elevated concentrations of stalk blood GnRH (Fig. 5.16), indicating a hypothalamic site of positive feedback. In contrast, the negative feedback actions of oestradiol can occur in the absence of changes in GnRH secretion, indicating a pituitary site of action. It would be premature, however, to rule out a hypothalamic site of oestradiol negative feedback on this evidence since the sensitivity of GnRH assays may not always be adequate to measure confidently decreases in the already low levels of the peptide.

The problem of defining hypothalamic sites of steroid feedback has also been approached using the classical techniques of neurobiology, namely the intracerebral implantation of sex steroids and other substances, or localized lesions of the suspected sites of steroid action. There are considerable species differences in the results of these studies. Part of the explanation for these differences may lie in the variable extent to which aspects of the environment influence reproduction in different species (see Section 7). In addition, there is considerable variability in the location of GnRH-containing neurons within the hypothalamus of different species. In the rat, the major group of GnRH neurons is located in the medial preoptic area and they project back to the median eminence. Knife cuts placed behind the optic chiasm, which will sever these axons, prevent cyclicity in the female and may disrupt gonadal function in the male. However, in the female rhesus monkey, similar cuts do not disrupt cyclicity and it has been suggested that this is at least in part due to the presence of a second, major group of GnRH neurons in the arcuate nucleus (and, indeed, in the mammillary region as well in the primate).

Experiments in which oestradiol has been placed into the hypothalamus have consistently implicated the arcuate nucleus as a site at which its negative feedback influence on GnRH secretion is expressed. Thus, in the rat, oestradiol infused into the arcuate nucleus suppressed gonadotrophin secretion without detectable amounts of the steroid reaching the anterior pituitary. Similar experiments in the female rhesus monkey showed that oestradiol, in amounts 1000-fold less than those administered systemically, exerted marked negative feedback actions when infused into the arcuate, premammillary and ventromedial nuclei (Fig. 5.17). These actions of oestradiol seem not to involve GnRH-containing neurons directly, but to be mediated indirectly through other neural systems with convergent effects on GnRH neurosecretion.

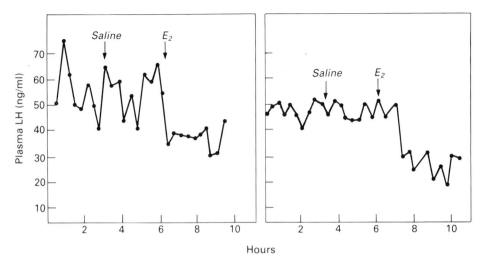

**Fig. 5.17.** The negative feedback effect of oestradiol (800 pg) on plasma LH when infused into the hypothalamic arcuate nucleus of ovariectomized rhesus monkeys ($E_2$=oestadiol). Note the lack of response to earlier, control injections of saline.

Attempts to establish a site of positive feedback effects of oestradiol have met with difficulty. However, in the rat, implantation of oestradiol in the anterior hypothalamic–preoptic area has been shown to induce an LH surge without evidence of diffusion of the steroid into the anterior pituitary. There are few data in the primate on this issue of positive feedback actions of oestradiol exerted in the brain.

The arcuate region, where oestradiol-induced progesterone receptors are found in considerable number, appears to be an important site for the negative feedback actions of progesterone during the luteal phase of the cycle.

The data summarized above clearly indicate that, while oestradiol and progesterone can act directly on the anterior pituitary to exert feedback effects on gonadotrophin secretion, the hypothalamus is also an important site for these effects via the modulation of GnRH secretion. Inhibin appears to exert its effects on FSH secretion solely in the anterior pituitary.

**3 Regulation of gonadotrophin secretion in the male**

The neuroendocrine mechanisms which govern testicular function are fundamentally similar to those which regulate ovarian activity as seen in the previous sections. The hypothalamo–pituitary unit of the male is responsible for the secretion of gonadotrophins which (see Chapter 3) regulate the endocrine and spermatogenic activities of the testis and are themselves subject to feedback regulation as a consequence of these actions. The major differences between the sexes in the control of gonadal activity is that gamete and attendant steroid hormone production in the male occur

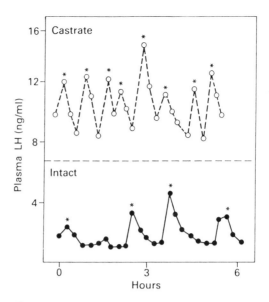

**Fig. 5.18.** The negative feedback effect of testosterone on plasma LH levels in male red deer. (a) Levels in the intact male (●---●). (b) Levels in the castrate (○---○). Note the increased frequency of pulses (*) after castration.

*continuously* after puberty and not cyclically, as is the case in the female. A reflection of this fact is seen in the *absence of a positive feedback influence* of testicular products on gonadotrophin release, since there is no abrupt cyclic phenomenon in gamete or hormone production to underly it.

The synthesis and secretion of FSH and LH by the gonadotrophs in the anterior pituitary depends, as in the female, on a GnRH signal from the hypothalamus which arrives via the portal blood. The pulsatile nature of GnRH secretion, which presumably reflects the coordinated activity of groups of hypothalamic neurosecretory GnRH neurons, is generally thought to drive the pulsatile pattern of FSH and LH secretion readily measured in plasma, particularly after castration when levels are high (Fig. 5.18). The last statement tells us immediately that gonadotrophin secretion in the male is under negative feedback control by the testis. The evidence to date suggests that at least two hormonal products of the testis participate in this control system, and that they are differentially concerned with the regulation of LH and FSH.

*a Testosterone and the regulation of the pituitary–Leydig cell axis*

In Chapter 3 we saw that testosterone secretion by the testis is the result of LH stimulation of the Leydig cells. It is now widely accepted that testosterone is, in turn, the principal hormone responsible for regulating LH secretion. Immunization against, and thus neutralization of, testosterone in the rhesus monkey results in increased circulating levels of LH. Conversely, administration of testosterone causes an abrupt decline in LH levels in castrate males of all species (Fig. 5.18). The negative feedback effect of testosterone is achieved largely by decreasing the frequency of episodic LH peaks, but there is also some change in pulse amplitude. This phenomenon is often taken to indicate a hypothalamic site of action of testosterone since the pulse frequency of LH and FSH discharge is generally held to be determined by similar pulses of GnRH secretion.

However, there is little doubt that testosterone can alter the responsiveness of the pituitary to GnRH and so, as in the female, both sites are believed to mediate its negative feedback effects. Androgen receptors are found in abundance in both places and implantation of testosterone in the medio-basal periarcuate hypothalamus of castrate male rats causes a significant fall in circulating LH levels. At least in rodents, $5\alpha$-dihydrotestosterone has some effect on LH secretion, whether given systemically or implanted in the hypothalamus. This suggests that this metabolite of testosterone may also participate in negative feedback mechanisms, in addition to its role in the periphery.

In man the anabolic steroids sometimes used by athletes to increase muscular bulk have been seen to induce infertility, principally by decreasing LH and, hence, endogenous testosterone secretion. These changes are reversible, but they emphasize that any androgenic steroid has the potential to modify, to lesser or greater extent, LH secretion.

*b Inhibin and the regulation of the pituitary–seminiferous tubule axis*

It was observed in the above experiments that testosterone, was also able to reduce circulating FSH levels when given to the castrate male, but that it exerted relatively lesser regulatory effects on the secretion of FSH than of LH. This led to the experimental manipulation of other testicular steroids, such as dihydrotestosterone and oestrogens, to see if they possess more selective FSH-suppressing activity. The essentially negative data so obtained led to the proposal of an additional, non-steroidal secretory product of the testis, indeed of its tubular compartment, which both reflects the actions of FSH on the Sertoli cell and exerts negative feedback effects on FSH secretion by the pituitary. Inhibin appears to be the product involved, but it has a rather more controversial role in the regulation of FSH secretion in the male than in the female (see earlier).

Preparations with inhibin-like (i.e. FSH-lowering) properties have been obtained from testicular extracts, testicular lymph, rete testis fluid, semen and the medium from Sertoli cell cultures, although at 100-fold lower levels than in follicular fluid. Some data suggest that the output of testicular inhibin from the Sertoli cells is related to the completion of spermiogenesis. Thus, GnRH treatment of infertile men results in increased gonadotrophin output from the pituitary, elevated serum levels of testosterone and the induction of spermatogenesis. As the output of spermatozoa rises, serum FSH levels decline while LH levels remain unaltered. Conversely, failure to complete spermatogenesis in man is correlated with elevated serum FSH levels, perhaps through reduced inhibin secretion by the testis. Such data are consistent with the concept of a testicular inhibin, but firm proof is lacking.

## 4 Is the hypothalamo-pituitary-gonadal system sexually dimorphic?

We have observed that males are distinguishable from females in failing to show a positive feedback effect naturally, although both show negative feedback effects. Is the absence of positive feedback in the intact male merely a secondary consequence of the lack of an ovary? Or is the male fundamentally different and incapable under any circumstances of manifesting a positive feedback response to oestradiol?

Experiments indicating that fetal hormones might influence the physiology of the brain were first carried out in 1936 by Pfeiffer. He demonstrated that ovaries transplanted to normal male rats castrated in adulthood failed to show any cyclic changes, whereas in recipient males castrated *at birth* the transplanted ovary underwent cyclic ovulation. Clearly, presence of the testis at birth prevented the subsequent support of ovarian cyclicity, and it was demonstrated soon afterwards that testicular androgens were responsible for this effect. Thus, testosterone injected during the first few days after birth in female rats suppresses the later ability of these animals to show oestrous cycles. Their ovaries contain follicles which secrete oestrogens, and hence their vaginal smears are constantly cornified, but since ovulation does not occur there are no corpora lutea.

How is the neonatal androgen acting to cause this acyclicity in rats? A number of experiments suggest that the effect is in some way mediated by suppression or modification of the oestradiol positive feedback mechanism. If castrated, adult male or female rats are injected or implanted with oestradiol, only the females give a positive feedback response with a surge of gonadotrophins (Fig. 5.19). If the same experiment is undertaken with female rats given testosterone during the first few (about 5) days of life, the capacity for a positive feedback response to oestradiol in adulthood is suppressed. There is a *critical period* when androgens exert this effect on the CNS, which varies from species to species, depending on the length of gestation and the relative immaturity of the newborn.

At what site does testosterone act to suppress the positive feedback response to oestradiol? Earlier we saw that the mid-cycle gonadotrophin surge was elicited by a surge of ovarian oestrogen acting probably on both the pituitary and the hypothalamus. Thus, the ovaries, pituitary *and* hypothalamus are

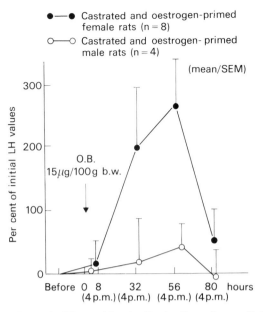

●——● Castrated and oestrogen-primed
female rats (n = 8)
○——○ Castrated and oestrogen- primed
male rats (n = 4)

**Fig. 5.19.** The positive feedback effect of oestradiol in castrate male and female adult rats. The LH response, measured as % change from resting values, to a single large injection of oestradiol benzoate (O.B.) is shown to be present only in the female and not in the male. This is taken to be evidence of sexual differentiation of the hypothalamus, neonatal androgens in the male preventing the ability to respond to an oestrogen surge with an LH surge in the adult.

all *potential* targets for the early effects of androgen on the development of acyclicity in the adult. However, the ovaries of androgenized females are quite capable of secreting surge levels of oestrogen if transplanted into normal females, and thus their functional capacity does not seem to be grossly impaired. Similarly, pituitaries from males or androgen-exposed females can function cyclically if transplanted adjacent to a normal female's hypothalamus. The 'masculinizing' effect of neonatal testosterone (or its active metabolites) would therefore appear to be exerted on the hypothalamus. We described in Chapter 1 a number of sexually dimorphic features in the rat hypothalamus that were conditional on neonatal exposure to androgens. However, it is not clear which, if any, of these is responsible for the capacity to yield a positive feedback response in the female rat.

It has been widely assumed that 'masculinization' of the brain occurs in all species. The phenomenon has been demonstrated in many rodents, sheep and some carnivores, but, in these species, the 'critical periods' of sensitivity to the effects of androgens vary considerably. For example, guinea-pigs have a gestation period of 68 days, compared to 21 days in the rat, and are born in a state of relative maturity. The critical period during which androgens exert their effects on

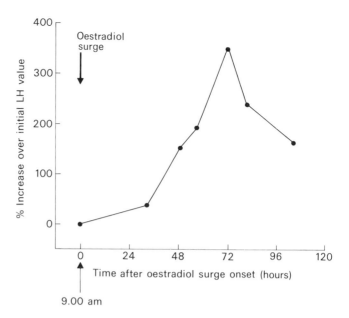

**Fig. 5.20.** Positive feedback effects of oestradiol in a castrate male talapoin monkey. Unlike the results in Fig. 5.11 the normal male monkey is able to respond to an oestradiol surge with an LH surge similar to that seen in the female. This suggests that exposure of the primate's brain to testosterone *in utero* does not 'masculinize' the hypothalamus as it is believed to do in non-primate mammals.

the brain is pre- and not post-natal. Similar considerations apply to the large domestic animals and carnivores.

In contrast to the results of experiments on the above animals, it has recently become clear that 'masculinization' of the hypothalamus does not occur in the same way in primates, and that the *capacity for positive feedback exists in normal male monkeys and men*. Thus, in male primates, if endogenous androgens are removed from the adult circulation by castration and an oestrogen surge administered, a gonadotrophin surge follows (Fig. 5.20). The same has been reported to be true of hypogonadal men. Furthermore, female rhesus monkeys exposed to high levels of testosterone during fetal life show menstrual cycles as adults after a puberty which occurs only slightly later than normal. Indeed, in some of these monkeys the external genitalia are so completely 'masculinized' that menstruation occurs through the penis! Similarly, human females exposed to high levels of androgens during gestation, for example in subjects suffering from the adrenogenital syndrome, will, when the increased androgen secretion is controlled after birth, show menstrual cycles as adults, often after a somewhat delayed puberty.

Clearly, 'masculinization' of the hypothalamic mechanism underlying positive feedback *does not occur in the normal male primate*, and, thus, by these criteria there are no apparent hypothalamic consequences in female primates exposed to high levels of androgens *in utero*, provided the androgen-producing disorder is controlled after birth. We must conclude, therefore, that sexual differentiation of the brain is not a global phenomenon so far as the mechanisms regulating ovarian and testicular function in the adult are concerned. This conclusion is reminiscent of the less overt effects that neonatal androgens appeared to have on sexual behaviour in the adult primate as compared to its more rigid effects on behaviour in the rat (Chapter 1).

## 5 Mechanisms of feedback regulation of gonadotrophin secretion

Little has been said during this account of the cellular mechanisms by which negative and positive feedback are exercised at either neural or pituitary loci. What does oestradiol do to gonadotrophs to increase or decrease their sensitivity to GnRH? How does oestradiol induce a GnRH surge, or decrease the frequency of GnRH pulses when given to ovariectomized subjects? The answers to some of these questions remain obscure and are the subject of intensive research.

Within the anterior pituitary, oestradiol appears to exert its positive feedback effects by inducing and maintaining GnRH receptors and by interacting with the self-priming process whereby GnRH induces its own receptors.

The small amplitude GnRH pulses which, as we saw earlier, do not by themselves cause an LH pulse (Fig. 5.7), are probably most important in the 'self-priming' effect, since they probably ensure a full LH response to the next adequate GnRH pulse. In Fig. 5.21 it may be seen that the presence of oestradiol enhances this interaction between GnRH and its receptor. These events therefore contribute to the magnitude of the oestradiol-induced LH surge.

There is less information on the ways in which oestradiol in the female, or testosterone in the male, causes a decrease in gonadotrophin secretion. It is not known whether the steroids contribute to a decrease in GnRH receptors or an uncoupling of the receptors from subsequent biochemical events in the gonadotrophs. Nor is there detailed information on the way that inhibin exerts its selective effects on FSH secretion in either sex.

If, as much of the evidence now indicates, oestradiol also induces a GnRH surge, then the hypothalamus is clearly a likely site of positive feedback. The cell bodies of GnRH neurons lie within the preoptic and anterior hypothalamic areas which are rich in oestradiol receptors, and it has been demonstrated immunocytochemically that the GnRH content of these neurons changes in response to oestradiol.

However, using a combined immunocytochemical/autoradiographic procedure, it has now been demonstrated that

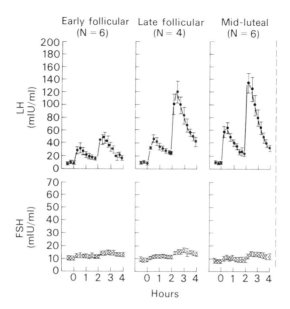

**Fig. 5.21.** The 'self-priming' effect of GnRH and its dependence on steroid hormone levels. The figure shows the enhanced response to a second injection of GnRH during three stages of the menstrual cycle. Note the large increase between the early and late follicular phases, which correlates well with the rise in oestradiol secretion and, further, how this effect remains pronounced in the mid-luteal phase when oestradiol, as well as progesterone, secretion is high.

oestradiol is not taken up directly by neurons containing GnRH, indicating that the effects of the steroid on GnRH *are mediated indirectly* by other neural systems which are oestrogen targets and which converge onto GnRH neuronal cell bodies or terminals.

Among the hypothalamic neural systems which have been investigated in this context, those containing the *catecholamines dopamine and noradrenaline*, the *amino acid GABA* and the *opioid peptide β-endorphin* have been particularly linked to the regulation of GnRH secretion (Fig. 5.22).

Within the arcuate nucleus is a group of neurons known as the *tuberoinfundibular dopamine, or TIDA*, neurons (Figs 5.3 and 5.6b). These neurons are known to take up radiolabelled oestradiol, and to project to the region of the portal capillaries where they lie in close association with GnRH terminals (Figs 5.6a,b and 5.22). The evidence for a functional relationship between these dopamine and GnRH terminals relies on pharmacological and correlative neurochemical data. Although there is not complete accord on this point, it is seen that dopamine infusions or treatment with dopamine agonists dramatically decrease plasma levels of LH, presumably by decreasing GnRH release. Similarly, a number of experiments have implicated adrenergic mechanisms in pulsatile

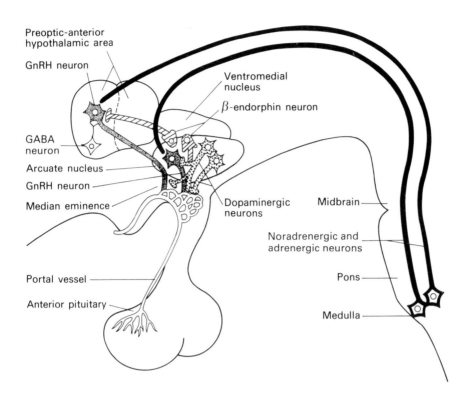

Preoptic-anterior hypothalamic area

GnRH neuron

Ventromedial nucleus

β-endorphin neuron

GABA neuron

Arcuate nucleus

GnRH neuron

Median eminence

Dopaminergic neurons

Midbrain

Noradrenergic and adrenergic neurons

Portal vessel

Pons

Anterior pituitary

Medulla

**Fig. 5.22.** Schematic diagram to show some of the postulated neurochemical interactions which may control GnRH secretion. GnRH neurons lie mainly in the medial preoptic area, but also in the arcuate nucleus. They project to the portal vessels in the median eminence. Dopamine neurons in the arcuate nucleus may modulate GnRH release by mechanisms described in text. Neurons within the hypothalamus which contain β-endorphin also modulate anterior pituitary secretion, perhaps by modulating GnRH neuron activity in the preoptic area, but perhaps also by altering dopamine neuron activity and hence GnRH (and dopamine) neurosecretion. Noradrenergic and adrenergic neurons in the medulla project to the hypothalamus and preoptic area and have been seen by pharmacological techniques to enhance GnRH secretion. GABA-containing neurons have been shown to accumulate oestradiol and may exert local control over GnRH neuron activity.

GnRH discharge and oestrogen-induced LH surges, but the precise physiological circumstances in which noradrenergic or adrenergic neurons in the brain stem, which innervate the hypothalamus, become activated and so influence GnRH secretion are unclear.

Neurons containing γ-aminobutyric acid (GABA) have also been implicated in the regulation of gonadotrophin secretion. Levels and turnover of GABA in the anterior hypothalamus are high when plasma LH concentrations are low. Interestingly, 70–80% of GABA-containing neurons in the preoptic and mediobasal hypothalamic areas have been shown to accumulate oestradiol, and systemic oestradiol treatment which causes a surge of LH is associated with a marked decrease in GABA release in the mediobasal hypothalamus. It has been suggested, therefore, that some of the effects of oestradiol on LH secretion are mediated by local, oestrogen-sensitive GABA neurons in the preoptic area and mediobasal hypothalamus whose axons converge onto GnRH-containing neurons (Fig. 5.22).

The opioid peptide β-endorphin, which is derived from the large precursor molecule, *pro-opiomelanocortin*, is found in a major group of neurons in the arcuate nucleus which richly innervate the medial preoptic area where the majority of GnRH-containing neurons are found (Fig. 5.22). Hypothalamic β-endorphin concentrations fluctuate during the cycle, with highest levels in the luteal phase and lowest levels in the follicular phase. β-endorphin given intraventricularly suppresses pulsatile and preovulatory surge gonadotrophin release whereas *naloxone*, an antagonist of β-endorphin at some opiate receptors, markedly elevates serum LH levels. The powerful effects of β-endorphin in suppressing both the amplitude and frequency of LH pulses have been taken to indicate that β-endorphin-containing neurons in the arcuate nucleus may mediate the negative feedback effects of gonadal steroids—particularly progesterone. Infusion of naloxone during the luteal phase of the menstrual cycle in the rhesus monkey accelerates LH pulse frequency to that more associated with the follicular phase. Thus naloxone is able to counteract the slowing of LH (hence GnRH) pulse frequency known to reflect the negative feedback actions of progesterone (see above). Clearly, these data are compatible with a role for hypothalamic β-endorphin neurons in the regulation of GnRH secretion. The puzzle which remains is to understand how these various neural mechanisms, which can affect GnRH secretion and are sensitive to steroid hormones, interact under physiological circumstances.

Finally, some interest has also been focussed on the fascinating structures called *ependymal tanycytes*, which, among other things, appear to form a direct link between the base of the third ventricle and the pituitary portal plexus (Fig. 5.23). Thus, it is conceivable that biologically active substances in the cerebrospinal fluid (amines, peptides and steroids) could exert effects on the anterior pituitary by being transported to the portal blood by the tanycytes. Indeed, there is little doubt that tanycytes are able to perform this function. However, it is far from clear why this system is required in addition to all the others discussed above, particularly since the transport mechanism in the tanycytes appears to be nonselective. The regulatory functions they might subserve are therefore obscure at present, but nonetheless this fascinating route of communication between the cerebral ventricles and

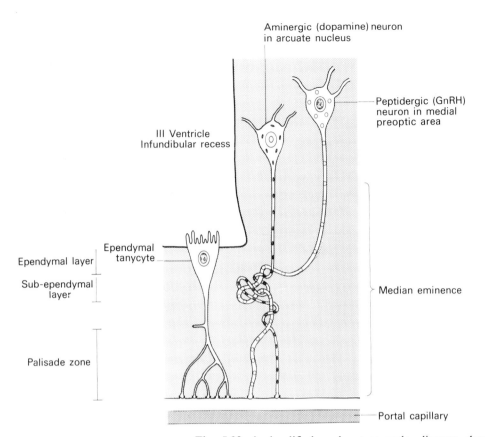

Aminergic (dopamine) neuron
in arcuate nucleus

Peptidergic (GnRH)
neuron in medial
preoptic area

III Ventricle
Infundibular recess

Ependymal
tanycyte

Ependymal layer

Sub-ependymal
layer

Palisade zone

Median eminence

Portal capillary

**Fig. 5.23.** A simplified, and not to scale, diagram showing the anatomical relationships of the ependymal tanycytes to the capillaries of the portal plexus and the juxtaposition of their 'terminals' to the terminals of amine (for example dopamine) and peptide (for example GnRH) neurons. It has been suggested that the tanycytes could convey biologically active substances from the CSF in the ventricle to the anterior pituitary via the portal capillaries. That they *can* do this seems clear, but the physiological importance of this route remains to be defined. (Note the GnRH cell body is not accurately placed.)

the pituitary exists, and may well prove to be of some importance in the future.

## 6 The reproductive functions of prolactin

Prolactin, as we have seen, is a polypeptide hormone of 190–200 amino acids (molecular weight 25 000). Although there is one major molecular form, several variants of different molecular size have also been demonstrated, both in the pituitary and in the blood. Since only a single gene for prolactin has been identified, these variations in the size of the prolactin molecule are probably the result of different forms of post-translational processing. The pituitary lactotrophs are distributed evenly throughout the anterior pituitary and four different prolactin-secreting cell types have been identified. Prolactin is stored in secretory granules in the cell, and the hormone is released in a pulsatile manner which probably reflects the pulsatile release of hypothalamic hormones controlling prolactin secretion.

*a Hypothalamic control of prolactin secretion*

Prolactin, unlike other pituitary hormones, is secreted in large amounts when the vascular links between the pituitary and hypothalamus are disconnected. This has led to the search for a hypothalamic factor(s) that inhibits release (*prolactin inhibitory factor or PIF*) as the primary controller of prolactin secretion. However, it is now clear that *prolactin releasing factors* also exist and that prolactin secretion is the result of complex interactions between these various influences.

i Prolactin inhibitory factors

Probably the most important PIF is the catecholamine, *dopamine* (Fig. 5.24), which is found in neurons of the arcuate nucleus whose axons project to the portal capillaries in the medial and lateral palisade zones of the external layer of the median eminence (Fig. 5.6). Dopamine secreted into the portal blood from the terminals of this tubero-infundibular dopamine (TIDA) system is carried to the lactotrophs which contain dopamine receptors. The amine and its agonists (such as bromocriptine) suppress prolactin secretion (Fig. 5.25) while dopamine antagonists (such as *haloperidol, metaclopramide and domperidone*) increase prolactin secretion by direct actions on the dopamine receptors of the lactotroph. Having bound to the receptors, dopamine is internalized and acts, at least in part, to increase lysosomal degradation of prolactin within the secretory granules, thereby making less of the hormone available for release.

While dopamine is widely acknowledged to be the most important PIF, it is not the only neurohormone to have this effect. For example, GABA also inhibits prolactin secretion, and arcuate neurons containing GABA also project to the capillaries of the median eminence—indeed, GABA and dopamine coexist in the same arcuate neurons in many instances. GABA is found in portal blood, although its concentrations there are generally said to be less than is

$$HO-\underset{HO}{\bigcirc}-CH_2-CH_2-NH_2$$

Dopamine (3, 4, dihydroxyphenylethylamine)

**Fig. 5.24.** Structure of the catecholamine *dopamine*. It is synthesized from the amino acid L-tyrosine and is also a precursor of noradrenaline and adrenaline.

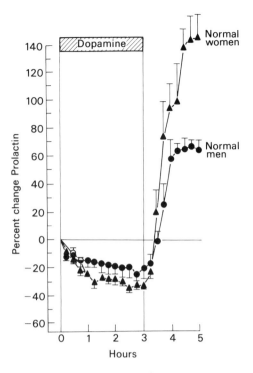

**Fig. 5.25.** The effects of dopamine infusion on prolactin levels in normal men and women. Note the rapid onset of effect of dopamine and the rebound increase in prolactin levels on stopping the infusion. In women with hyperprolactinaemia, bromocriptine, a dopamine-receptor agonist, will cause serum prolactin concentrations of 80–100 ng/ml (normally they are less than 18 ng/ml) to fall to below 10 ng/ml within 5 or 6 hours.

necessary to significantly affect prolactin secretion and thus its physiological significance in the regulation of prolactin secretion is open to question.

Finally, the peptide associated with gonadotrophin releasing hormone, GAP (see Fig. 5.5 and Section 1c), has also been shown to inhibit prolactin secretion, as well as being able to promote gonadotrophin secretion, and it is found in, and may be released from, GnRH-containing nerve terminals in the median eminence along with GnRH itself and the precursor molecule. However, like GABA, its importance in the physiological regulation of prolactin secretion has not been established.

**ii Prolactin releasing factors**

A variety of hormones has been shown to stimulate prolactin secretion. Hypothalamic thyrotrophin releasing hormone (TRH) was first shown to stimulate prolactin secretion markedly and TRH receptors have been localized to lactotrophs. However, the physiological role of TRH in the regulation of prolactin secretion has not been established.

The *vasoactive intestinal polypeptide* (VIP) is an extremely potent releaser of prolactin, and receptors for the peptide have been demonstrated on the lactotrophs. Furthermore, a decrease in dopamine secretion from the median eminence is associated with an increased prolactin secretory response to VIP. Only very low levels of VIP are present in neurons in the brains of normal males or non-lactating females, although VIP is measurable in portal blood. However, in the suckling female rat, VIP-containing neurons are readily seen in the parvocellular paraventricular nucleus and a rich terminal plexus becomes visible in the external layer of the median eminence. Moreover, increased amounts of VIP mRNA are also detectable in the paraventricular neurons at this time. These observations suggest that VIP may be a particularly important prolactin releasing factor during periods of high prolactin secretion such as occur during lactation. This will be discussed further in Chapter 13.

Oestrogens also induce hyperprolactinaemia, probably by decreasing the sensitivity of lactotrophs to dopamine and by increasing the number of TRH receptors. Chronic exposure to oestrogen results in increased pituitary DNA and RNA synthesis with consequent hyperplasia and increased prolactin secretion. The prolactin secretory response to oestrogens is, therefore, not acute and the steroid should not be regarded as a releasing factor in the way that VIP and TRH are.

In addition to these humoral controls, prolactin secretion is also subject to daily variation, the highest plasma concentrations occurring during the nocturnal sleep period. Reversal of the sleep–waking cycle results in reversal of the daily rhythm of prolactin secretion, demonstrating that it is sleep-entrained rather than light-entrained (Fig. 5.26). In a normal sleep–waking cycle, prolactin release begins to increase 1–1.5 hours after sleep onset, and this is achieved by progressive increases in pulse amplitude. Plasma concentrations are elevated during the remaining hours of sleep and fall in the early

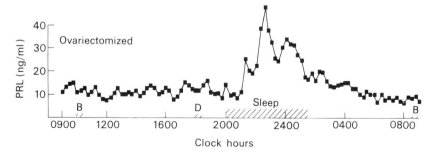

Fig. 5.26. Daily variation of prolactin secretion. Note the onset of sleep is associated with the rise in serum prolactin concentrations, which then begin to fall towards the period of wakening. B = Breakfast; D = Dinner.

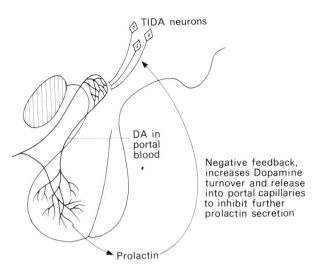

TIDA neurons

DA in
portal
blood

Negative feedback,
increases Dopamine
turnover and release
into portal capillaries
to inhibit further
prolactin secretion

Prolactin

**Fig. 5.27.** Diagrammatic summary of the proposed negative-feedback relationship between prolactin and dopamine (PIF). Prolactin is believed to accelerate dopamine (DA) turnover in the arcuate nucleus neurons (tubero-infundibular dopamine or TIDA neurons), and the amine is then released into the portal capillaries to gain access to the lactotrophs. Hyperprolactinaemia could be caused by a failure either of PIF activity at the dopamine receptor level in the anterior pituitary or a reduction of TIDA neuron activity in the hypothalamus.

morning, shortly before awakening. Lowest concentrations are found between about 10.00 a.m. and 12.00 noon. Interestingly, the bursts of prolactin secretion seem to occur during slow wave or 'non-rapid eye movement' (non-REM) sleep, while REM (or paradoxical) sleep is associated with the smallest episodic prolactin pulses.

An oestrous rhythm of prolactin secretion in animals such as the rat has been demonstrated, with a prolactin surge occurring coincident with the LH surge. However, in women it seems generally to be agreed that no clear *menstrual* rhythm in serum prolactin levels exists. Occasional mid-cycle and premenstrual peaks have been reported, and the latter has been suggested to be related to the premenstrual syndrome. Serum prolactin concentrations do not seem to alter significantly after the menopause. Changes in prolactin secretion during pregnancy and lactation will be discussed in Chapters 10 and 13.

Prolactin is also seen to be released by acute stressors, both in animals and man. The functions of prolactin as a 'stress hormone' when released in this way are unclear.

iii Feedback regulation of prolactin secretion

Dopamine released from TIDA neurons into the portal capillaries is generally regarded to be the main controller of prolactin secretion. What regulates the activity of TIDA neurons? It appears that the answer to this question is dopamine itself. Increases in circulating prolactin levels result in an increase in dopamine turnover within TIDA neuron terminals of the median eminence and a reduction therefore in prolactin secretion. The increase in dopamine turnover is related to an increase in *tyrosine hydroxylase*, the enzyme which is rate-limiting in the intraneuronal synthesis of dopamine. This so-called 'short-loop' feedback control of hypothalamic TIDA neuron activity, and hence dopamine release, by circulating prolactin is illustrated in Fig. 5.27.

*b Functions of prolactin*

Extramammary functions of prolactin in mammals are too numerous even to list, but include such diverse examples as regulation of kidney and adrenocorticotrophic activity (these tissues having higher prolactin-binding activity than mammary tissues) in addition to synergistic actions with ovarian and adrenal steroids and gonadotrophins. Few of these actions have been established in man.

In the context of reproductive functions in the human male and in the non-pregnant and non-lactating female the situation is far from clear. There is some evidence that prolactin may participate in the regulation of steroidogenesis in the follicle, particularly the inhibition of progesterone secretion in the early stages of follicular growth and its enhancement in the luteal phase. In addition, prolactin appears able to modulate the number of ovarian receptors for LH and so affect steroidogenesis indirectly. However, the exact importance of these findings to our understanding of anterior pituitary regulation of ovarian function remains to be established.

Plasma levels of prolactin do not change markedly during the follicular phase of the oestrous cycle, but they do increase around the time of the preovulatory LH surge. In the rat, the mid-cycle increase in prolactin is related to the increase in oestradiol, but it is less clear that this is so in women. Prolactin plasma levels are not generally increased in the luteal phase, even in species such as the rat, sheep and goat where prolactin is an essential part of the luteotrophic complex. Prolactin appears to act in these species by increasing the amount of oestradiol receptor in the corpus luteum.

In the male, most of the information concerning prolactin and testicular function comes from experiments on rodents. It has been shown that, while exerting little effect on its own, prolactin may increase the number of LH receptors and potentiate the steroidogenic effect of LH on Leydig cells; testicular prolactin receptors seem to be confined to the interstitial tissue of the testis. Similarly, prolactin increases the uptake of androgen and increases $5\alpha$–reductase activity in the prostate. Prolactin also potentiates the effects of testosterone on the seminal vesicle, while testosterone maintains the number of prolactin receptors in the prostate. In man, the functional importance of prolactin in testicular and reproductive tract activity is unclear.

*c Hyperprolactinaemia*

Pathological elevation of serum prolactin levels in men is associated with impaired fertility (see Chapter 14), decreased circulating levels of testosterone and loss of libido. In women, the syndrome is characterized by amenorrhoea, and hence infertility (see Chapter 14), with or without galactorrhoea (abnormal milk secretion, see Chapter 13) and loss of libido. Clearly, prolactin in such circumstances exerts profound effects on reproductive function.

The causes of hyperprolactinaemia are multiple and varied. They may of course be 'physiological', for example in pregnancy and during the first few months of breast feeding. It is commonly induced by dopamine receptor-blocking neuroleptic drugs used widely in psychiatry; by oestrogens, including oral contraceptives; or as the result of some underlying pathology, for example some pituitary tumours—prolactinomas (adenomas or microadenomas).

The high prolactin levels are largely due to absence of the fall in prolactin secretion normally seen in the morning associated with awakening. Thus, high-amplitude prolactin pulses are seen abnormally during the daytime and sleep augmentation is therefore masked. The syndrome is commoner in women than men. The endocrine profile of women with hyperprolactinaemia is characterized by an absence of pulsatile LH secretion, a failure of positive feedback and hence chronic anovulation and amenorrhoea. Contrary to earlier opinion, it is now clear that the ovarian responsiveness to FSH and LH is not necessarily impaired and does not underlie the amenorrhoea; indeed, normal cyclicity can be maintained in the presence of high prolactin levels if exogenous gonadotrophins are administered. However, the pituitary LH response to injected GnRH *is impaired*, suggesting that a decreased pituitary sensitivity to GnRH may, in part, underlie the failure of cyclicity in the female and also presumably infertility in the male.

Exactly *how* elevated prolactin levels cause these changes is unclear, but what evidence there is suggests it to be a major *causal* factor and not simply an *associated* one. Thus, increasing prolactin levels in normal women by a variety of means (for example TRF, dopamine antagonists) induces an identical pattern of endocrine changes and amenorrhoea. The loss of pulsatility in LH secretion, as we have argued before, strongly suggests an action on GnRH secretion. The decrease in pituitary sensitivity to the latter might, therefore, reflect an indirect effect of prolactin via decreased ovarian oestradiol secretion, rather than an effect exerted directly on the pituitary.

Treatment of hyperprolactinaemia has become effective and simple. Dopamine agonists such as *bromocriptine* are now used to lower serum prolactin concentrations immediately, and daily treatment results in the return of ovulation and cyclicity in the vast majority of women within 2 months.

In the case of prolactin-secreting tumours, dopamine agonists have an anti-mitotic action and, in addition to lowering plasma prolactin concentrations, reduce tumour size. However, in the majority of cases, cessation of treatment is followed by a resumption of prolactinoma activity and surgical removal is often necessary.

The loss of libido in men and some women with hyperprolactinaemia is unexplained. The fascinating idea that this represents an action of prolactin in the brain remains to be demonstrated.

## 7 Environmental influences on reproduction

Our foregoing discussion has established a major role for the CNS in providing a GnRH supply and acting as a target for steroid modulation. It would be quite wrong, however, to dismiss a more extensive role on present evidence. The CNS mediates the effects of environmental factors such as coital stimuli, olfactory cues and light on the regulation of reproductive activity. In addition, factors such as those arising from social interaction, including anxiety or other forms of emotional distress, which can have profound effects on cyclicity and fertility in men and women, are also mediated by the CNS. Although the effects of these social factors are readily observed in humans, studying causal relationships depends in large part upon careful analysis of animals living in social groups, where both behavioural and endocrine variables may be experimentally controlled. We will now go on to examine the way in which these various environmental influences on reproduction are mediated.

### a Effects of light on fertility

Reproduction is only one of a host of activities in which an individual may engage. To ensure that it is carried out effectively and with minimum interference from other processes, it is important that reproductive activity occurs only at the most appropriate time. This temporal control of reproductive activity is further complicated in the female because the production of a viable gamete, the oocyte, is itself a cyclical event which must be matched to other cyclical events occurring within the life of the animal. In a nocturnally active rodent, for example, potential encounters with mates will be restricted to the hours of darkness. This selection pressure has led to the development of an oestrous cycle which is tightly locked to the best indicator of external time, the daily light/dark cycle. As we saw in Chapter 4, the oestrous cycle of the rat is much shorter than that of the primate, but the

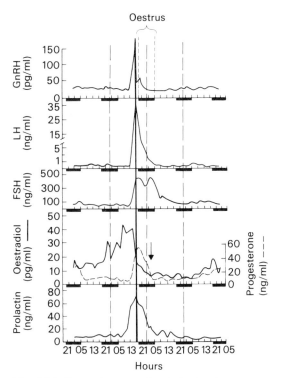

**Fig. 5.28.** Serum hormone levels during the oestrous cycle in the rat. The black bars represent dark periods (18.00–06.00 hours) centred around midnight while the arrow denotes the time of ovulation. Also shown are the levels of GnRH measured in portal vein blood which show a clear surge just prior to the FSH and LH surges. Note the surge in prolactin coincident with the gonadotrophin surges.

temporal relationships of oestradiol, LH and FSH secretion are remarkably similar (Fig. 5.28). The really dramatic differences are, first, the surge of progesterone secretion accompanying the LH surge which is very important in the cyclical control of sexual behaviour (see Chapter 7 for details). Second, the LH surge itself is precisely timed to occur between 5 and 7 hours before darkness. This ensures that ovulation occurs *during* the night when the female is active because of her nocturnal habits and is behaviourally receptive (see Chapter 7). The likelihood of conception is therefore maximized. In the intact animal, the LH surge occurs only every 4 or 5 days because it is dependent on the trigger of rising oestrogen production. However, the neural signal which determines the *time* of the LH surge is present every day. In an ovariectomized female given a constantly high level of oestrogen delivered from a subcutaneous capsule, an LH surge occurs every day at precisely the same time. By controlling oestrogen levels, *the ovary* therefore determines *the day of ovulation*, but a *neural timer*, controlling a critical period of sensitivity to oestrogen, determines *when, during that particular day*, the LH surge and thus ovulation should occur.

Reversal of the light/dark cycle causes a 12-hour shift in the timing of the critical period of sensitivity to oestrogen, the LH surge and ovulation, demonstrating the essential role of information about light in setting the neural timer. However, even in constant light or dark, daily LH surges will continue, given the appropriate oestrogen environment, indicating that the timing mechanism is a *self-sustaining biological clock* (or oscillator). Under these constant conditions, the oscillator and the rhythm it controls *free-run* with a period of approximately 24 hours, which is therefore termed *circadian* (*circa* = about, *diem* = day). It is important to realize that the circadian system controls a wide range of other behavioural and endocrine rhythms, both reproductive and non-reproductive, which are held in a very strict, temporal relationship to each other.

The majority of these rhythms are driven by the *suprachiasmatic nuclei* of the hypothalamus. This cluster of neurons located above the optic chiasm, adjacent to the third ventricle (see Fig. 5.3), has the ability to generate a circadian signal even when isolated from the rest of the brain. In life, this approximately 24-hour signal is converted to a precise period of 24 hours by the *entraining effect of photic stimuli* which reach the suprachiasmatic nuclei via a direct retinal input, the *retino-hypothalamic tract*. Lesions of the suprachiasmatic nuclei disrupt many circadian functions, including the LH surge, and thereby cause a condition of permanent anoestrous. However, the neuroanatomical route linking the circadian signal to the release of GnRH is poorly understood. Furthermore, it is unclear whether there is a circadian rhythm in GnRH secretion, normally peaking in the afternoon, which is amplified by the oestrogen surge, or whether the circadian input leads to a transient increase in the sensitivity of GnRH to the positive feedback effects of oestrogen.

In female primates, patterns of reproductive activity are much more flexible than in rodents and the social rather than the physical environment has a much more pronounced effect upon reproductive processes. The primate has a menstrual cycle in which ovulation may occur at any time of day, the period of sexual receptivity is far more extensive and the circadian system makes little contribution to the control of reproductive function. However, another form of temporal programming is apparent in some primates (although not modern human society) and in a wide range of other mammalian groups in which purely circadian influences are not initially evident.

In seasonal environments, where adverse climate and the availability of food will be major determinants of offspring survival and therefore of the reproductive success of the

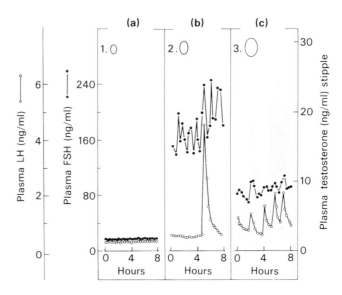

**Fig. 5.29.** The seasonal sexual cycle of a Soay ram. Changes in the plasma concentration of FSH, LH and testosterone are shown at three times of the year. Testis size is shown inset at each time: (a) in the non-breeding season with the testes fully regressed, LH, FSH and testosterone levels are all low; (b) towards the onset of the breeding season, the testes are redeveloping and, associated with this process, FSH concentrations are very high; (c) during the mating season, testosterone levels are very high and this reflects the marked increase in the frequency of pulsatile LH discharge.

parent, it is adaptive for individuals to ensure that young are born in the equable, productive conditions of spring or early summer. This tight control over birth season, apparent in many domestic and wild species, is achieved by a precise regulation of the month(s) of fertility and hence the timing of conception. In species with short gestation times, winter is a time of infertility with gonadal development suspended until spring. In species with longer gestation times, the anticipation of spring must begin much earlier and seasonal changes in autumn act as a stimulus to reproductive function. This leads to the dramatic spectacle of the rut when animals which have been reproductively quiescent for the entire year suddenly develop pronounced secondary sexual features such as antlers, become fertile, aggressive and territorial and spend their whole time engaged in an intense competition for access to mates. In a third group, which includes marsupials, mustelids and seals, the total length of the gestation period can be varied because of delayed implantation and embryonic diapause (see Chapter 9). These processes are sensitive to environmental influences and provide a second level of control over the timing of the birth season. For the reproductive physiologist, these seasonal phenomena offer an important opportunity to investigate the central mechanisms which regulate the fertility of an individual.

In some species, such as deer and ground squirrels, there is good evidence that seasonal cycles are under the control of an endogenous *circannual* oscillator, a biological clock with a period of approximately one year. In other species, there is no endogenous rhythmicity and the seasonal rhythms observed in the field are triggered by cyclical stimuli within the environment. Of these, photoperiod is by far the most important so that, in the laboratory, artificial manipulation of *daylengths* can be used to drive all of the components of the annual reproductive cycle. For example, exposure of Syrian hamsters to less than 12.5 hours of light per day (pseudo-winter) leads to gonadal regression. In contrast, these short photoperiods stimulate gonadal activity in species such as sheep which normally mate in the autumn. All of these effects are mediated by changes in the frequency of the GnRH pulse generator in the hypothalamus which then determines the level of secretion of gonadotrophins and steroids (Fig. 5.29). Photic information obviously has access to the GnRH neurons, but does it use the same pathways involved in the circadian control of reproduction? Certainly the suprachiasmatic nuclei have an important role to play because lesions of these structures completely block photoperiodic sensitivity. However, the pathways involved are not exclusively intrahypothalamic. It is now well recognized that the *pineal gland*, which sits attached to the epithalamus in the posterior portion of the third ventricle (Fig. 5.1), is the mediator of photoperiodic time-measurement. Removal of the gland or interruption of its sympathetic innervation leaves animals insensitive to changing daylength. The primary pineal hormone, *melatonin*, is synthesized and released into the bloodstream only in the hours of darkness, exhibiting a true circadian rhythmicity driven by the suprachiasmatic nuclei. At night, the circadian signal increases sympathetic activation of the gland, resulting in a dramatic rise in the activity of the enzyme, *N-acetyl transferase*, the rate-limiting step in melatonin biosynthesis. The crucially important feature of the circadian melatonin signal is that it provides a precise representation of the length of the night, so that as days shorten and nights lengthen in autumn, the duration of the nocturnal melatonin peak is increased. Conversely, after the winter solstice the photoperiod increases and the duration of the melatonin signal falls (Fig. 5.30 a and b). The changing shape of the rhythm of circulating melatonin is detected within the hypothalamus and somehow leads to alterations in GnRH secretion. The melatonin signal is such a powerful regulator of neuroendocrine state that in pinealectomized animals the entire reproductive axis can be turned on or off by repeated, nightly administration of programmed infusions of

145    *The Regulation of Gonadal Function*

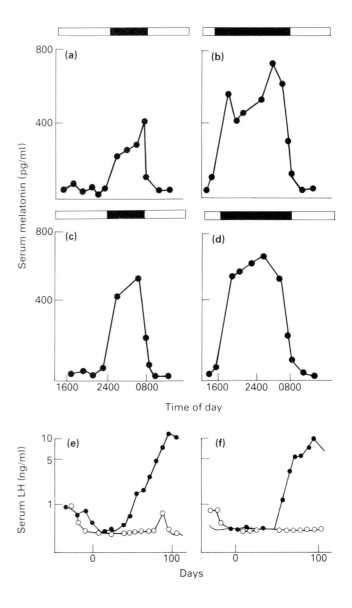

**Fig. 5.30.** The effects of melatonin on LH secretion in ewes. (a), (b) Serum melatonin profiles of ewes with intact pineals exposed to artificial long (a) and short (b) photoperiods. Note the increased duration of the melatonin signal during the longer night of (b). (Darkness is indicated by the solid black bar over each graph). (c), (d) Serum melatonin profiles of pinealectomized ewes receiving programmed infusions of melatonin designed to mimic the patterns of long and short photoperiods (shown in a and b). (e) Reproductive response (LH secretion) of pineal-intact ewes to artificial long (○—○) or short (●—●) photoperiods (remember that the ewe is an autumn, or 'short-day' breeder).(f) Reproductive response of pinealectomized ewes to programmed infusion of melatonin mimicking long (○—○) or short (●—●) photoperiodic profiles in serum. Clearly, this is virtually identical to the pattern in (e).

melatonin which reproduce the signal normally generated by the pineal under either long or short photoperiods (Fig. 5.30c–f). One important question yet to be answered is how does melatonin work and, in particular, how is it that the same signal leads to opposite neuroendocrine responses in different species?

<table>
<tr><td>b Effects of coitus on fertility</td><td>In humans and other primates, sheep, rats and many other mammals, ovulation is said to occur 'spontaneously'. Thus, it depends on an endogenous event timed by the ovary, an oestradiol surge, which results in an ovulatory discharge of LH that may or may not be subject to additional, circadian controls. The neurally mediated variable controlling ovulation to be discussed next in this chapter concerns the 'induced' or 'reflex' ovulators, such as the cat, rabbit and ferret. These animals remain in behavioural oestrus for long periods of time without ovulating until they copulate with a male. Stimulation of the cervix and vagina during coitus evokes the reflex release of an ovulatory surge of LH via afferent, sensory pathways, which presumably gain access to the GnRH release mechanism. Even in these species it seems that the hypothalamo-pituitary axis must be primed with high levels of oestrogen for the neural input to be effective.</td></tr>
</table>

Although the data supporting it are less than convincing, there are several reports of reflex ovulation in women. Ovulation has variously been observed to follow coitus very early in the follicular phase—even during menstruation—while termination of abnormally long follicular phases has sometimes been ascribed to coitus-induced, acute LH release. The subject has not been studied systematically however, and these reports should therefore be viewed sceptically, if with interest.

Similarly, in the rat, pregnancy and pseudopregnancy, i.e. prolongation of luteal life, are also dependent on coitus. In this case, cervical stimulation appears to set up a diurnal (=during the day) peak in prolactin secretion, in addition to the normally occurring nocturnal peak. Thus, two daily peaks in prolactin concentration are measurable in the blood and this seems essential for corpus luteum formation.

<table>
<tr><td>c Effects of social interaction on fertility</td><td>Recent studies of primates living in social groups have revealed that the social context in which individuals interact can change their endocrine status.</td></tr>
</table>

In groups of talapoin monkeys, consisting of four males and four or five females, only the dominant male is sexually active. He mounts females and ejaculates, and directs aggression to subordinate males but does not receive any aggression from them (Fig. 5.31). Measurement of plasma testosterone yields some fascinating data. Before being placed with females the (future) dominant and subordinate males

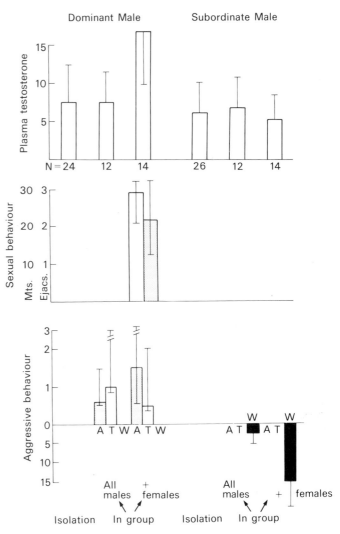

**Fig. 5.31.** Changes in sexual and aggressive behaviour and plasma testosterone levels in dominant and subordinate male talapoin monkeys during a 12-week period in isolation, a 6-week period in an all-male group and a 7-week period in a social group with oestradiol-treated females. A = attacks, T = threats, W = withdrawals—all measures of aggressive behaviour; Mts = mounts, Ejacs = ejaculations—both measures of sexual behaviour. Note only the dominant male's plasma testosterone increases in the social group, and that only *he* is sexually active and *giving* aggression to, but not receiving aggression from, other males.

have very similar testosterone levels. The same is true when these males are placed together in a group. However, when the *females* treated with oestradiol are also present in the group, the *dominant male* displays sexual activity with the females, is aggressive to subordinate males and *his* plasma testosterone levels, but not those of the subordinate male, *rise significantly* (Fig. 5.31).

That the behavioural interaction determines the change in plasma testosterone and not vice versa is apparent if all the males are taken out of the group and each replaced *alone* in turn with the females. In this situation, each male is sexually active and shows a significant rise in plasma testosterone. Addition of the dominant male to a group consisting of the subordinate male alone plus females results in the latter male's plasma testosterone declining. These data have been interpreted to suggest that subordination, or the receipt of aggression, causes the testosterone levels to decrease. Conversely, *giving* aggression and/or displaying sex, as in the dominant male, is associated with increased plasma testosterone. The significance of the elevated testosterone levels in the dominant male, and the mechanism by which it is achieved, are not entirely clear. As we will see (Chapter 7), testosterone levels have but a tenuous relationship to sexual motivation and even the subordinate male has sufficient levels of the hormone to maintain his sexual behaviour. It has been suggested that the hormone may enhance the attractiveness of the male to the female by behavioural (e.g. posture) and/or non-behavioural (e.g. coat quality and odour) means, although experimental evidence for either is generally lacking.

A similar consequence of the dominance hierarchy is seen in the females of the group. The dominant female receives more sexual attention from the male than does the subordinate female, and hardly any aggressive behaviour from other members of the group (Fig. 5.32). An intriguing consequence of this state of affairs is seen if both the dominant and subordinate females are challenged with an oestradiol surge. Only the dominant female displays an LH surge in response to it. If plasma prolactin levels are measured, it is seen that they are much higher in the subordinate females (hyperprolactinaemia). Thus, it seems likely that only the dominant female in this case has the capacity for positive feedback, and, if she had ovaries, she would cycle normally and be potentially fertile. Not so the subordinate female who has lost the capacity to respond to an oestrogen surge with an LH surge and who, if her ovaries were intact, would have anovulatory and irregular cycles and probably display amenorrhoea, as is seen in some women with hyperprolactinaemia. That prolactin is the key to this change in positive feedback capacity was demonstrated by lowering prolactin in the subordinate female using a dopamine agonist, and raising it in the dominant female using a dopamine antagonist. This treatment effectively *reversed* the LH response to an oestradiol surge in the two females.

These two examples emphasize the important and far-reaching consequences of social interaction in determining levels of sexual activity as well as endocrine and reproductive status.

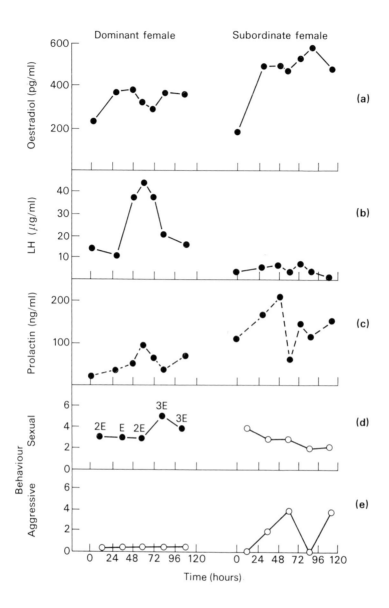

**Fig. 5.32.** Sexual and aggressive behaviour received by the dominant and subordinate female talapoin monkeys in a social group, and changes in plasma LH and prolactin levels when challenged with an oestradiol surge. (a) Increase in plasma oestradiol concentrations following the oestradiol surge. (b) Changes in plasma LH concentrations which follow the oestradiol surge. (c) Plasma prolactin concentrations in the females. (d) Sexual interaction with males ○——○ indicates mounts without ejaculation; ●——● etc. indicates mounts with ejaculations. (e) level of aggressive behaviour received by the females. Note the dominant female receives high levels of sexual behaviour and low levels of aggression, has low plasma levels of prolactin and shows a surge of LH in response to an oestradiol surge. The converse is true of the subordinate female.

**8 Summary**

The critical event underlying ovulation in the menstrual or oestrous cycle is an oestradiol-induced gonadotrophin surge. The oestradiol surge is not preceded by an obvious endocrine or neural trigger, and appears to arise spontaneously within

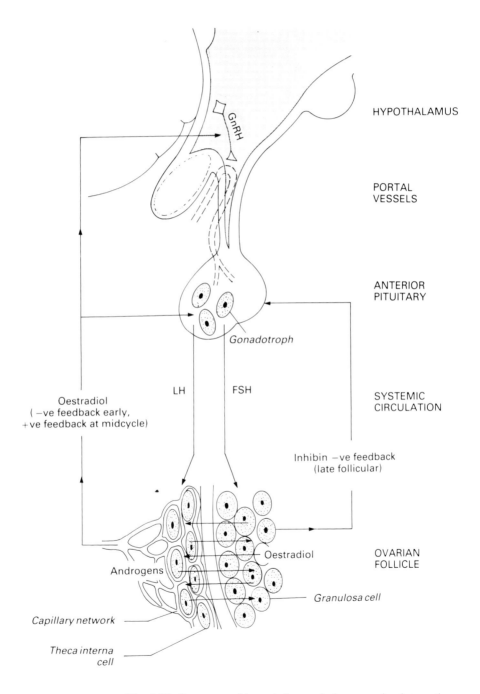

**Fig. 5.33.** Summary of hypothalamo-pituitary–ovarian interactions during the follicular phase of the cycle.

151     *The Regulation of Gonadal Function*

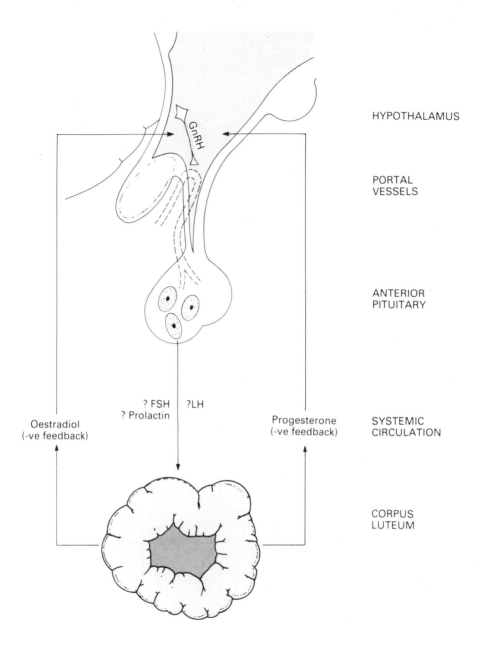

HYPOTHALAMUS

PORTAL
VESSELS

ANTERIOR
PITUITARY

? FSH    ?LH
? Prolactin

Oestradiol
(-ve feedback)

Progesterone
(-ve feedback)

SYSTEMIC
CIRCULATION

CORPUS
LUTEUM

**Fig. 5.34.** Summary of hypothalamo-pituitary–ovarian interactions during the luteal phase of the cycle.

the ovary. This leads to the conclusion that the ovary and not the hypothalamus or pituitary times the cycle. However, the oestrogen surge from the ovary acts on both pituitary and hypothalamus to induce the release of an ovulating surge of LH (Fig. 5.33). This capacity to respond to an oestradiol surge is lacking in male animals other than primates. However, in both males and females similar negative feedback effects of

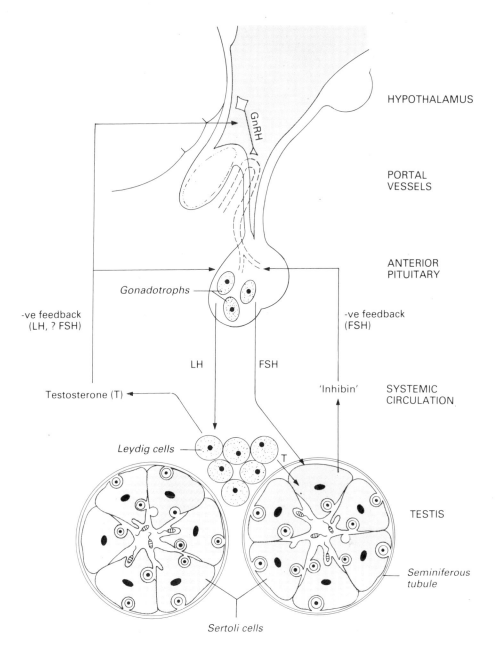

HYPOTHALAMUS

PORTAL
VESSELS

ANTERIOR
PITUITARY

Gonadotrophs

-ve feedback
(LH, ? FSH)

-ve feedback
(FSH)

LH          FSH

Testosterone (T)

'Inhibin'          SYSTEMIC
CIRCULATION

Leydig cells

T

TESTIS

Seminiferous
tubule

Sertoli cells

**Fig. 5.35.** Summary of hypothalamo-pituitary testis interactions in the male.

gonadal hormones on the pituitary and hypothalamus are observed (Figs 5.34 and 5.35).

The feedback effects of gonadal hormones are adequate, in themselves, to explain all the basic features of the reproductive patterns in both males and females. However, the hypothalamo–pituitary–gonadal axis is not a closed system. External influences clearly modulate its activity to render the

153     *The Regulation of Gonadal Function*

basic reproductive pattern susceptible to environmental changes, such as time of day or year, and proximity of a potential reproductive partner or rival. By such mechanisms the efficiency of the reproductive process is increased and the survival of the species promoted.

**Further reading**

Ciba Foundation Symposium 62. *Sex, Hormones and Behaviour.* Excerpta Medica, 1979.

Everitt BJ & Keverve EB. Reproduction In *Neuroendocrinology* (Eds Lightman SL, Everitt BJ). Blackwell Scientific Publications, 1986.

Ganong F, Martini L (Eds). *Frontiers in Neuroendocrinology.* Raven Press, 1978. (Especially Chapter 9 by E Knobil and T Plant.)

Haymaker W, Anderson E, Nauta WJH (Eds). *The Hypothalamus.* Springfield, 1959.

*Journal of Reproduction and Fertility*, 1979. *Inhibin, FSH and Spermatogenesis.* Supplement 26.

Kreiger DT, Hughes JC (Eds). *Neuroendocrinology.* Sinauer Associates Inc. Sunderland, 1980.

Short RV (Ed). Reproduction, *Br Med Bull* 1979; **35** (2). Scientific Section, British Council.

Stumpf WE, Grant LD (Eds). *Anatomical Neuroendocrinology.* Karger, 1975.

Yen SSC, Jaffe RB (Eds). *Reproductive Endocrinology.* Saunders, 1978.

# Chapter 6
# Maturation of the Hypothalamo-Pituitary-Gonadal Axis

In earlier chapters we have seen how the gonads and genitalia develop during fetal life, mature at puberty and function in the adult. The regulation of adult function is, as we saw in the last chapter, complex. Now we return to the subject of puberty and examine how this adult pattern of endocrine activity is achieved. Our account will concentrate mainly on human puberty.

## 1 Hormonal changes at puberty

Summarized in Fig. 6.1 are the physical changes occurring through puberty which were described in Chapter 1. They are largely controlled by gonadal and adrenal steroids, together with adenohypophyseal hormones in some instances. It is important to remember here that although the absolute tempo of these changes may vary individually, their sequence is remarkably consistent—so much so that, should it deviate from the pattern shown, a clinical investigation is often warranted. We will now examine the endocrinology of puberty which, as will be seen, is a gradual process of gonadal activation to achieve the adult pattern of regulation described in Chapter 5.

The low plasma levels of *gonadotrophins*, which are present at or soon after birth, *rise and fall intermittently* for the first year or two of life, but then remain very low during childhood until the initiation of events leading to puberty. From this time FSH and LH levels rise gradually to reach adult levels (Fig. 6.2). *Prolactin* concentrations also increase in late puberty in girls, but not in boys in whom the plasma levels of this hormone are already similar to those seen in men. This difference may be attributed to the actions of oestradiol which enhance prolactin secretion in females (see Chapter 5). Probably the most intriguing and dramatic change in gonadotrophin secretion occurs in early puberty at night during

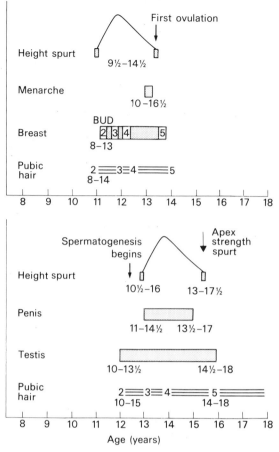

**Fig. 6.1.** Summary of the sequence of events during puberty in girls and boys. The figures below each symbol represent the range of ages within which each event may begin and end. The figures within each symbol refer to the stages illustrated in Figs 1.14, 1.15, 1.16.

sleep. In men and women there is no evidence of a circadian rhythm of FSH and LH secretion (Fig. 6.3d). The same is true of prepubertal children (Fig. 6.3a) in whom levels are uniformly low. From early to mid-puberty, however, a striking increase in the magnitude, and possibly frequency, of LH pulses occurs which reflects a *sleep-augmented LH secretion* (Fig. 6.3b). In late puberty daytime LH pulses also increase (Fig. 6.3c), but less than those still occurring at night, until the adult pattern of higher basal levels with no daily variation is achieved.

*Testosterone* levels in plasma are less than 0.1 ng/ml in boys and girls, *except* during the first 3–5 months after birth in boys when levels similar to those at puberty are found. In early puberty *testosterone* levels rise at night when LH secretion becomes elevated. However, blood samples taken during the day show that increases occur consistently throughout

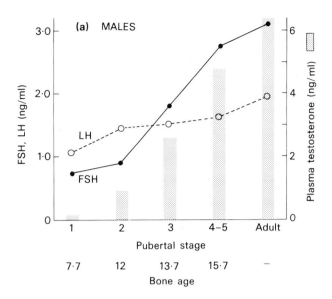

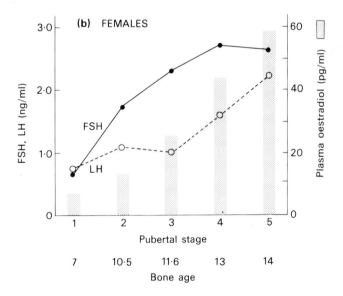

**Fig. 6.2.** Gonadotrophin and steroid hormone plasma concentrations during various pubertal stages in (a) boys and (b) girls. (Bone age is assessed by examining radiographs of hand, knee and elbow, and comparing them with standards of maturation in a normal population. It is an index of physical maturation, and better correlated with the development of secondary sexual characters than chronological age.)

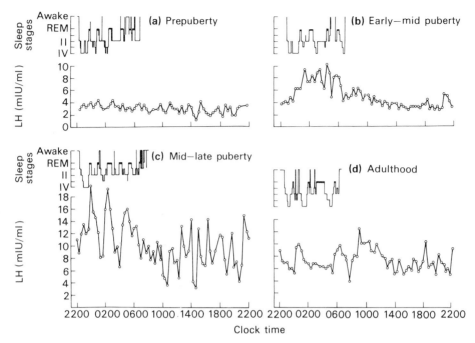

**Fig. 6.3.** Plasma LH concentrations throughout a 24–hour period in: (a) a prepubertal girl (9 years); (b) an early pubertal boy (15 years); (c) a late pubertal boy (16 years); and (d) a young adult male. The sleep pattern for each nocturnal sleep period is depicted in the top left hand corner of each graph. Note the marked daily rhythm in (b) with sleep-augmented LH secretion and the overall higher LH concentrations in (d) compared with (a), but no clear daily rhythm in either. (REM = rapid eye movement or 'paradoxical' sleep.)

puberty, with the greatest changes appearing during pubertal Stage 2, when testosterone concentrations may change from 0.2 to 2.4 ng/ml (Fig. 6.4). There are smaller, but nonetheless consistent, increases in plasma testosterone concentrations in girls between pubertal Stages P1 and P4.

*Oestrogen* plasma concentrations are extremely high in both male and female fetuses at birth (5000 pg/ml) because of the conversion of fetal and maternal C19 steroids by the placenta (see Chapter 10). Indeed, newborn infants may display breast budding and even milk secretion ('witches milk') as a consequence of these high oestrogen levels. *Oestradiol* and *oestrone* levels soon drop to 7 and 20 pg/ml, respectively, and remain low until puberty. In girls, oestradiol levels rise consistently through the stages of puberty to reach concentrations of 50 pg/ml in the follicular phase of the cycle (Fig. 6.2) to 150 pg/ml in the luteal phase seen in mature females. In boys plasma concentrations of oestrone are higher than oestradiol, but both are considerably lower than in girls at comparable stages of puberty. In the male about half the oestradiol is

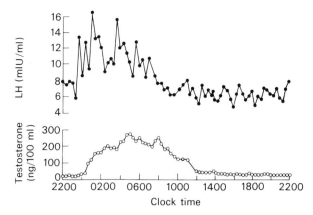

**Fig. 6.4.** A similar figure to Fig. 6.3(b) showing plasma LH and testosterone concentrations in a 12-year-old boy. Note the night-time rise in testosterone concentrations, coincident with the rise in LH.

derived from extraglandular aromatization of testosterone, and a quarter, or less, from testicular secretion.

*Adrenal androgens*, notably dehydroepiandrosterone and its sulphate, begin to show a progressive increase in plasma concentration by 8 years of age (6–8 years skeletal age) which continues until 13–15 years in boys and girls. This change in adrenal androgen output *precedes* changes in gonadotrophins and gonadal steroid secretion by a considerable time period. *Adrenarche*, as it is called, is therefore one of the first pubertal endocrine events although its significance in relation to hypothalamo-pituitary–gonadal activation is obscure. The only clear, somatic function of these adrenal androgens appears to be the promotion of pubic and axillary hair growth. The adolescent growth spurt does *not* depend on them. Neither *hyper*secretion nor *hypo*secretion of adrenal androgens seems to be associated consistently with either early or delayed puberty, and there is little evidence to suggest that these hormones are concerned with timing the onset of puberty in normal children.

So far in this account we have seen evidence of the progressive activation of the gonads (and adrenals) which results in elevated steroid secretion during the pubertal period. These events are in turn dependent on increased trophic stimulation by FSH and LH. In the next three sections, we will examine the control of these processes, the nature and timing of the trigger which induces them and the regions of the CNS through which they may be mediated.

**2 Mechanisms underlying puberty**

There are two distinct, but not mutually exclusive, hypotheses concerning the process of *pituitary–gonadal activation*. The first assumes that the pituitary and gonads are fullly capable of functioning in a mature, adult manner at birth, and all that is required for successful pubertal activation is an increased output of hypothalamic GnRH to drive the system. This hypothesis therefore places emphasis on the CNS and the final common pathway of communication with the pituitary, the GnRH neurons. Few workers would dispute that these neurons, according to any theory, must play a crucial role in puberty. The second hypothesis acknowledges that these hypothalamo-pituitary changes must occur, but, in addition, ascribes special importance to maturation of the feedback effects of oestradiol or testosterone on gonadotrophin secretion and to changes in the pituitary responsiveness to GnRH.

Experimental evidence in favour of the first view comes from the observation that the rise in pulsatile LH release occurs even in the absence of the gonads. More recently studies on immature female rhesus monkeys up to 2 years old has provided overwhelming evidence to favour the first view. Using an externally situated pump to deliver pulses of GnRH to the young monkey's circulation at 1–1.5-hour intervals (as described in Chapter 5), ovulatory menstrual cycles were initiated and maintained in these females (Fig. 6.5a). The data have been interpreted to suggest that a most important event in the initiation of puberty is *activation of the hypothalamic mechanism* which delivers GnRH pulses to the anterior pituitary. Having done this, the pituitary and ovary are able to respond instantly, and maintain their 'conversation' in terms of negative and positive feedback, requiring only the hypothalamus as a source of releasing hormone to control synthesis and release of the gonadotrophins. Switching off the GnRH pump is followed by re-entry into the immature, prepubertal state, suggesting, rather surprisingly, that exposure of the hypothalamus to adult levels of circulating steroids does not contribute to the attainment of a 'mature' pattern of functioning. This experiment represents a convincing demonstration that puberty could arise *solely* as a consequence of a maturational event within the CNS, translated to the pituitary-gonadal system as a stream of GnRH pulses.

Even more convincing data demonstrating that changes in gonadotrophin output at puberty occur independent of gonadal steroids is shown in Fig. 6.5b. Here a number of male rhesus monkeys were gonadectomized at birth and it can be seen that LH levels vary in the adult range for the first 100 days of life. Thereafter, LH concentrations fall and remain at low or undetectable levels for the next 2.5 years or so, following which pulsatile discharges recur heralding the onset of puberty. Clearly, the altered ouput of gonadotrophins is most likely driven by activity in GnRH neurons and so directs our attention firmly towards the CNS when searching for the mechanisms underlying puberty.

There are few comparable clinical studies which have addressed this issue. Gonadotrophin levels in agonadal children also increase around the expected time of puberty and in the absence, therefore, of gonadal steroid influences. Furthermore, precocious puberty may occur in children as young as 2 years of age. Here it is also apparent that pituitary and gonads can function in an adult manner when activated pathologically, often as a result of a CNS tumour (see below). In such cases, however, there is little opportunity of investigating other maturational changes which might occur in the pituitary or gonads, since the clinician is usually presented with a *fait accompli.*

What then of the second hypothesis? The principle to be established here is that the negative feedback regulation of FSH and LH secretion in boys and girls operates at a very low threshold or set point, i.e. it is very sensitive prepubertally, and that the set point *increases*, i.e. becomes less sensitive, at puberty. Therefore, this process will contribute to the rising concentrations of gonadotrophins and sex steroids in the circulation during puberty. Let us consider the evidence that bears on this popular and enduring hypothesis (summarized in Fig. 6.6). The results of determinations of fetal hormones have suggested that the hypothalamo-pituitary–gonadal axis develops functional autonomy during gestation. The first 6 months or so of life are associated with fluctuating, and often high, plasma levels of FSH, LH, testosterone (in boys) and oestradiol (in girls, see above). However, by about 1–2 years, the system has quietened down and gonadotrophin and steroid concentrations are very low—a characteristic of the prepubertal child. Thus, negative feedback regulation of gonadotrophin secretion by sex steroids *exists before birth* and does appear both prenatally and prepubertally to be very sensitive since it operates at low circulating levels of oestrogens or androgens. However, in these studies on the apparent exquisite sensitivity of the pituitary to steroids, it is not clear whether changes in binding-protein levels or affinity and metabolic clearance of steroids have been properly taken into account.

In contrast, it *is* clear that the *pituitary's sensitivity to GnRH does* increase during puberty. Thus, the LH response to injected GnRH, which is small and rather similar in prepubertal boys and girls, increases markedly during puberty, and is even higher in the adult. A dissimilar pattern of response to GnRH is seen if FSH is measured under the same conditions. First, there is little difference in the FSH response to GnRH in prepubertal, pubertal and adult *males*, suggesting changes in pituitary sensitivity do not occur. Second, prepubertal *and* pubertal *females* release more FSH in response to GnRH than do males at any stage of sexual maturation. This implies that girls, and women, have a larger, more readily releasable pool of FSH than males, a finding which may have

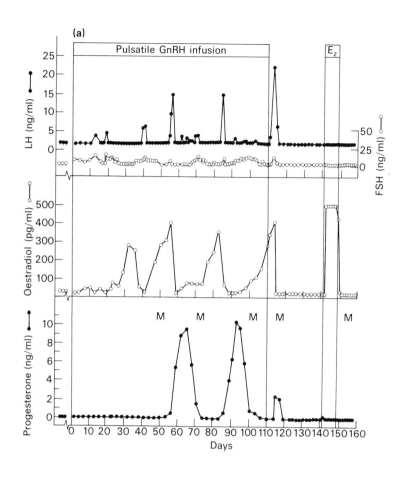

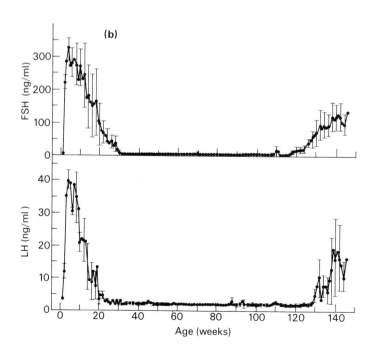

clinical significance in the higher incidence of idiopathic precocious puberty in the former.

Why should the pituitary show a gradual increase in sensitivity to GnRH? There are probably two major factors responsible: first, a form of self-priming of GnRH (referred to in Chapter 5 in the context of a more acute event), whereby exposure of gonadotrophs to GnRH increases the FSH and LH response to subsequent exposure; second, the progressively rising levels of oestradiol (in girls) and testosterone (in boys) probably enhance this effect of GnRH on gonadotrophs (see Chapter 5). Both factors will clearly follow the augmented GnRH output associated with puberty.

*Positive feedback* effects of oestradiol do not appear to be readily elicited in prepubertal or early pubertal girls. Thus, the capacity to evoke a gonadotrophin surge appears to develop late in the pubertal process, and even then, not efficiently, because well over half the cycles occurring in early postpuberty are anovulatory, a proportion which decreases to one-fifth after 5 years or so. The reason for this delay in the

Fig. 6.5. (a) Induction of ovulatory menstrual cycles in an immature rhesus monkey by the infusion of GnRH (1 μg/min for 6 minutes once every hour). The period of GnRH infusion is shown by the horizontal bar (Day 0–110); levels of LH, FSH, oestradiol and progesterone were undetectable in blood samples prior to GnRH infusions. Note that the first oestradiol surge did not elicit a full LH surge (c. Day 25). However, subsequent oestradiol surges induced both LH surges and evidence of corpus luteum formation (progesterone peaks). These LH surges, as well as ensuring menstrual periods, occurred at 28-day intervals. Cessation of GnRH infusions (immediately after the LH surge c. Day 112) was followed by prompt re-entry into a non-cyclic, prepubertal state. Implantation of an oestradiol-containing silastic capsule subcutaneously (between 140 and 150 days as indicated by bar) to produce surge levels of oestradiol in blood *did not* induce an LH surge in the absence of exogenous GnRH. The occurrence of menstruation is denoted by M. (After Wildt L, Marshall G & Knobil E. *Science* 1980; **207**: 1373–1375.)

(b) Circulating FSH and LH levels in the plasma of three male rhesus monkeys which were bilaterally orchidectomized at approximately 1 week of age. Blood samples were collected immediately before castration and thereafter at weekly intervals. Data for the first 142 weeks of postnatal development are shown (0 = day of birth). It can be seen that LH and FSH pulses occur during the first 20 weeks or so of life when plasma levels are in the adult range. Thereafter LH and FSH secretion ceases and levels remain low or undetectable until about week 120. The first visible change is an increase in FSH secretion followed soon after by increasing, pulsatile secretion of LH. It is important to emphasize that these changes occur *in the absence of testosterone* in the circulation. Thus, altered gonadotrophin secretion is a function of altered hypothalamic GnRH output independent of any alteration in steroid feedback regulation of this hypothalamo-pituitary system. (From Plant T. *Endocrinology* 1980; **106**: 1451–1454.)

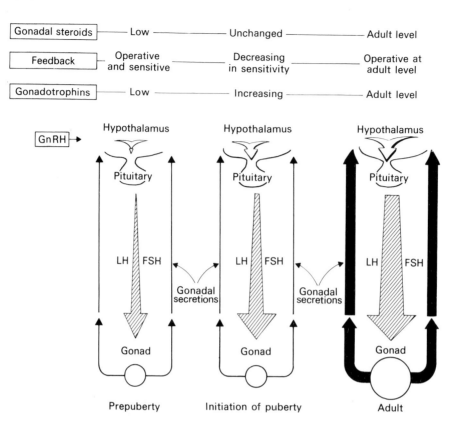

| Gonadal steroids | Low | Unchanged | Adult level |
| Feedback | Operative and sensitive | Decreasing in sensitivity | Operative at adult level |
| Gonadotrophins | Low | Increasing | Adult level |

GnRH →

Hypothalamus          Hypothalamus          Hypothalamus

Pituitary             Pituitary             Pituitary

LH   FSH              LH   FSH              LH        FSH

              Gonadal                Gonadal
              secretions             secretions

Gonad                 Gonad                 Gonad

Prepuberty        Initiation of puberty        Adult

**Fig. 6.6.** The so-called 'Gonadostat' theory of puberty. This summary diagram illustrates the proposed decreasing sensitivity of the negative feedback mechanism to sex steroids during puberty which underlies increasing FSH and LH output. The latter then drives gonadal steroid production to reach adult levels. The evidence supporting this hypothesis is not particularly conclusive.

appearance of positive feedback capacity is not fully understood. Clearly, the anterior pituitary critically depends on continued oestrogen exposure in order to synthesize and store sufficient gonadotrophin to respond to an oestrogen surge (see Chapter 5). It is likely, therefore, that since adequate follicular plasma concentrations of oestradiol are not achieved until pubertal Stage 5, positive feedback capability will not be achieved until then either.

It seems clear that the single most important event underlying pubertal gonadal activation is the augmented secretion of GnRH by hypothalamic neurons. Evidence has increasingly moved towards the view that a *neural process* underlies this event and away from the view that maturational changes in feedback mechanisms are primarily responsible. Moreover, the latter will follow the former quite naturally. We will now consider the difficult question of the nature of the neural process that initiates the pubertal changes.

**3 The timing of puberty**

*a Secular trend towards earlier puberty*

The intrinsic and extrinsic factors responsible for triggering and timing the onset of puberty have proved elusive to define. One phenomenon illustrated in Fig. 6.7 may, however, be important in directing our attention towards particular factors. There has apparently been a *secular trend towards an earlier menarche* in girls, and puberty in boys, in Western Europe and the USA. While the outlying 17-year age at menarche in Norway in 1840 (Fig. 6.7) probably represents atypical data from an isolated sample, it is generally agreed that in the past 100 years or so, the age at which girls first menstruate has decreased from between 14 and 15 years to between 12 and 13 years. The normal range is considerable, for example a clinician would normally investigate diagnostically girls who had never menstruated by 16 years, but not much earlier. What factors have changed in the past century which might have contributed to the earlier attainment of sexual maturity and do they give any insight into the mechanisms controlling the initiation of puberty?

*b Environmental factors*

Clearly, there are many potential answers to this question. Health care and personal health have improved during this time, along with living conditions and socio-economic standards. Undoubtedly these have contributed to the attainment

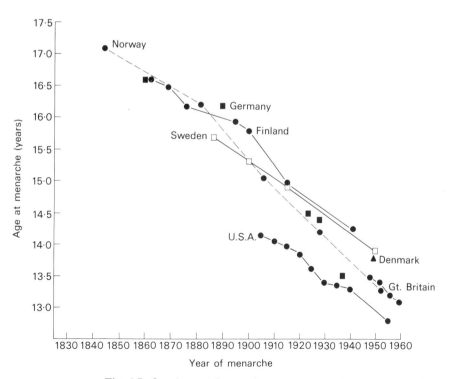

**Fig. 6.7.** Secular trend towards an earlier age at menarche in girls from Western Europe and the USA.

of earlier puberty, and our longer life expectancy: indeed, the latter ensures that the majority of women now survive to experience the menopause (Chapter 14). But from clinical and experimental studies, two factors have been implicated more consistently in this secular trend to earlier puberty, *and* in the mechanisms which normally underlie the onset of puberty. These are *photoperiod* and *nutrition.*

### i Photoperiod

In Western society we are undoubtedly exposed to longer photoperiods now than in the nineteenth century, due to the advent of electricity. Indeed, our daily dark periods may be consistently as short as 7 hours all the year round, as if we were in constant 'long days' or summer photoperiod. As we have seen in Chapter 5 differing daily and annual lighting regimes clearly affect the hypothalamo-pituitary–gonadal axis in the adults of many species, but is there evidence that they can affect the maturation of this system in prepubertal subjects? Experiments on the rat suggest that they can. Thus, exposure to *constant light* accelerates puberty, measured as early opening of the vagina, a key hormone-dependent event in this and many other rodents. There is little direct evidence that photoperiod affects reproductive maturation in the human.

Plasma concentrations of melatonin have been shown to change at the onset of puberty in girls and, since the synthesis of this hormone depends on the length of the night (see Chapter 5), this has been taken by some to indicate a photoperiodic influence, but this awaits confirmation. Interestingly, blind girls experience an *earlier menarche* than sighted girls in the general American population, arguing against any maturational effect of light on the reproductive system.

### ii Nutrition

It seems reasonable to suggest that nutritional factors might have important effects on sexual maturation. Indeed, if cultures such as the nomadic Lapps, that have not experienced such major improvements in living standards and nutrition are examined, it is seen that between 1870 and 1930 there was little or no trend towards an earlier menarche. Experimental studies which emphasize the importance of nutrition on reproduction in the adult are abundant. For example, maintaining female rats on a low-protein diet, such that their body weight is held consistenly below normal (about 80%), results in the cessation of oestrous cycles. Subsequent, sudden exposure of these animals to high-protein food restores the ability to discharge LH after oestradiol challenge. The practice of 'flushing' sheep, exposing them to rich pasture, induces an earlier onset of oestrus and lambing, a phenomenon exploited by farmers. Adolescent girls with the complicated syndrome of anorexia nervosa and, therefore, exhibiting a very low food (particularly carbohydrate) intake

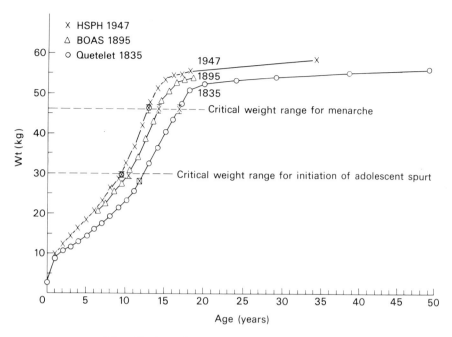

**Fig. 6.8.** Age plotted against body weight in three populations of girls in 1835, 1895 and 1947. Note the constant weight (30 kg) at initiation of the growth spurt and (47 kg) at menarche. (Populations: Belgian girls in 1835 from data of Quetelet, 1869; American girls in 1895 from data of BOAS, and in 1947 from data of Reed and Stuart, 1959.) Copyright American Academy of Pediatrics 1972.

and body weight, have irregular menstrual cycles and often amenorrhoea. These examples suggest that food intake, or a reflection of it—*body weight*—is highly correlated with reproductive efficacy in the adult. Is it similarly associated with the onset of puberty?

*c Body weight*

In Fig. 6.8 it can be seen that although age at menarche has changed considerably during the past 100 years, the *body weight at menarche has remained surprisingly constant* at about 47 kg. Similar constancy is seen in the weight at onset of the adolescent growth spurt (Fig. 6.8). These data have led to the suggestion that, at least in girls, a critical weight must be attained before the growth spurt and activation of the hypothalamo-pituitary–gonadal axis can occur. According to this view, *body weight*, or more correctly a *critical metabolic mass* related to body weight, may *trigger the onset of puberty*, that is, *time its onset*. A similar suggestion for a critical weight of 55 kg underlying sexual maturation in boys has also been made.

The earlier occurrence of puberty today than a century ago may, therefore, be explained by earlier attainment of a critical weight due to improvements in nutrition, health care and social living conditions. Evidence in support of this view is, at first sight, abundant. Moderately obese girls experience an earlier menarche than lean girls. Malnutrition is associated with delayed menarche. Primary amenorrhoea is extremely common in ballet dancers at professional schools who are in the very low range of weight for height and of relative fatness for their age. The amenorrhoea frequently seen in girls with anorexia nervosa has its onset when body weight drops somewhat below 47 kg, while in some anorexic girls who begin to re-feed, the re-occurrence of menstruation is associated with attainment of a 47 kg body weight. This is an impressive array of data supportive of what, teleologically, may be regarded as a very logical signal to the onset of puberty, namely, attainment of a body size sufficient to cope with the demands placed on it by adult reproductive activity, for example, pregnancy.

However, much controversy surrounds this hypothesis, particularly that part of it which assumes a *causal relationship* between a critical metabolic mass, derived from assessment of lean body weight, body fat and total body water, and the time of onset of puberty. First, the studies quoted are retrospective rather than prospective, and depend on derived, rather than direct, measurements. It has subsequently proved quite difficult to *predict* age of menarche from a knowledge of body weight and weight gain. Second, many anorexic girls who re-attain their critical body weight do not begin having menstrual cycles—although this may simply mean that anorexia nervosa is a far more complicated syndrome than just a disorder of food intake and consequent weight loss. Third, menarche is a rather late event in puberty, and may, therefore, be much removed from the *critical factors* which determine the *onset* of those endocrine changes described above, even if menarche itself is often related to body weight.

*d Summary*

The concept of a metabolic signal, conveying information about the state of development of the body, which acts as a trigger to the onset of puberty is an attractive one. It must be emphasized that conclusive data on this subject are not readily available, but there is considerable correlative data supportive of such a view. One thing seems clear however, the CNS is the most likely structure to receive the trigger and initiate, orchestrate or in some way regulate, the subsequent processes which define the attainment of sexual maturity. What do we know of the areas of the CNS concerned with puberty?

**4 The central nervous system and puberty**

Since data on the mechanisms of puberty are somewhat imprecise, it is perhaps not suprising to discover that the neural sites underlying them are also poorly defined. It is, however, generally and probably correctly assumed that areas of the CNS concerned with the control of onset of puberty,

**Table 6.1.** Neurological lesions associated with advanced or delayed puberty in man.

| Site of lesion | Puberty | | Type of lesion |
| --- | --- | --- | --- |
| | Precocious | Delayed | |
| a Hypothalamus | | | |
|   i Anterior | | ✓ | Hamartoma |
|   ii Middle and posterior including | | | Germinoma |
|     mammillary bodies | ✓ | | Teratoma |
| | | | 3rd Ventricle cyst |
| b Pineal gland | | | |
|   i Parenchymatous | | ✓ | i Glandular tissue tumour (rare) |
|   ii Non-parenchymatous | ✓ | | ii Non-glandular tissue tumour |
| c Pituitary gland | | | |
| | | ✓ | Craniopharyngioma |
| | | | Chromaphobe adenoma |
| | ✓ | | Gonadotrophin-producing |

Explanation of terms:
  Hamartoma—hyperplastic growth formed of nerve cells, fibres and glia.
  Germinoma—tumour of germ cell origin.
  Teratoma—tumour of partially developed embryonic tissues.
  Craniopharyngioma—tumour of Rathke's pouch originating from the pituitary stalk.
  Chromaphobe adenoma—prolactin secreting tumour of anterior pituitary.

and indeed the control of reproduction in the adult, exert their effects ultimately through the hypothalamic GnRH neurons. These neurons are, therefore, the most likely final common pathway of communication between the rest of the brain and the anterior pituitary gonadotrophs.

An intriguing clinical condition known as Kallman's syndrome points to the crucial importance of GnRH in the attainment of sexual maturity. This syndrome is characterized by delayed puberty and anosmia. The delayed puberty is the result of an isolated gonadotrophin deficiency, which is itself the result of a GnRH deficiency. Puberty can be initiated in these subjects by giving exogenous GnRH, which causes the release of FSH and LH from the pituitary. This demonstrates both the responsiveness of the pituitary to GnRH in this syndrome, and the dependence of FSH and LH synthesis and release on this peptide.

Additional information on areas of the CNS concerned with puberty comes almost exclusively from clinical cases of advanced or delayed puberty associated with an underlying neuropathology (Table 6.1). However, understanding the mechanisms by which such lesions in the CNS affect the onset of puberty is far from simple.

Generally the *site* rather than the *type* of tumour determines its effects on puberty, unless the tumour is gonadotrophin-producing. In the *hypothalamus*, lesions of the *anterior area*

169    *Hypothalamo-Pituitary-Gonadal Axis*

are associated with *delayed puberty*. Lesions of more *posterior areas*, from the median eminence to the mammillary bodies, are consistently associated with *precocious puberty*. Tumours involving the *pineal gland* were once thought to affect the onset of puberty indirectly, by mechanically compressing the underlying hypothalamus (Fig. 5.1). However, this now seems unlikely since *such tumours may be associated with advanced or delayed puberty* depending upon the involvement of *non-glandular* or *glandular* tissue, respectively, within the pineal. To what extent these effects are related to the production of gonadotrophins (particularly by teratomas) or melatonin (by glandular tumours of the pineal) is unclear. It is not surprising that tumours of the *pituitary* are associated with early or late puberty according to the resultant increase or decrease in gonadotrophin production, respectively. Interestingly, chromaphobe adenomas (hyperplastic lactotrophs) which produce high levels of prolactin, are associated with a delay of puberty which may be induced subsequently with bromocriptine—a dopamine agonist (see Chapter 5).

Lesions of the hypothalamus, pineal and pituitary clearly have major effects on the onset of puberty. The fact that the lesions are often large and of complicated and widespread origin means that accurate interpretation of their effects is difficult. Unless the tumours involved are gonadotrophin or prolactin producing, little is known of the *ways* in which they might alter GnRH secretion, and so exert their puberty advancing or delaying action. It follows, therefore, that identification of those areas of the CNS critically involved in integrating the endocrine changes associated with the initiation of puberty has not been achieved and is an important area for future research.

**Further reading**

Bullough VL. Age at menarche: a misunderstanding. *Science* 1981; 213: 365–366.

Everitt BJ, Keverne EB. In *Neuroedoncrinology* (Eds Lightman SL, Everitt BJ) Blackwell Scientific Publications, 1986.

Grumbach MM, Grave GD, Mayer FE. *The Control of the Onset of Puberty*. John Wiley, 1974.

Tanner JM. *Growth at Adolescence*. Blackwell Scientific Publications, 1966.

Tanner JM. *Foetus into Man; physical growth from conception to maturity*. Open Books, 1978.

Yen SC, Jaffe RB (Eds). *Reproductive Endocrinology*. W. B. Saunders & Co., 1978. (Particularly Chapter 10.)

# Chapter 7
# Actions of Steroids in the Adult

In the foregoing chapters we have established the pivotal role of the gonads and their steroid secretions in reproductive events. We have been concerned mainly with steroid action in two areas: first, in the generation and maintenance of sexual differentiation during fetal, neonatal and pubertal life; second, in the regulation of gonadotrophin secretion by the hypothalamic–pituitary axis. In this chapter we will examine in more detail the remaining actions of the steroids in the adult male and non-pregnant female, namely the regulatory effects of steroids on the physical features of the adult that ensure the attainment of full reproductive capacity and the effects of steroids on the sexual behaviour of the adult male and female during social interactions that lead to copulation.

The effects of sex steroids may conveniently be thought of as falling into two broad categories. Some steroid actions are *determinative,* others are *regulatory.* Determinative actions involve essentially qualitative changes, which are irreversible or only partially reversible. Examples of this type of action are provided by the effect of androgens on the development of the Wolffian ducts and the generation of male external genitalia, the mild enhancement of these sexually distinct features by the low prepubertal androgen levels in males, and the changes in hair pattern, baldness, voice tone, penile and scrotal size and bone growth that occur at puberty. These actions constitute part of a progressive androgenization which establishes a clear and distinctive male phenotype. It represents the completion of a process initiated with the expression of the Y chromosome. In the female an active determinative role for steroids first occurs during the prepubertal and pubertal period, when oestrogens stimulate body

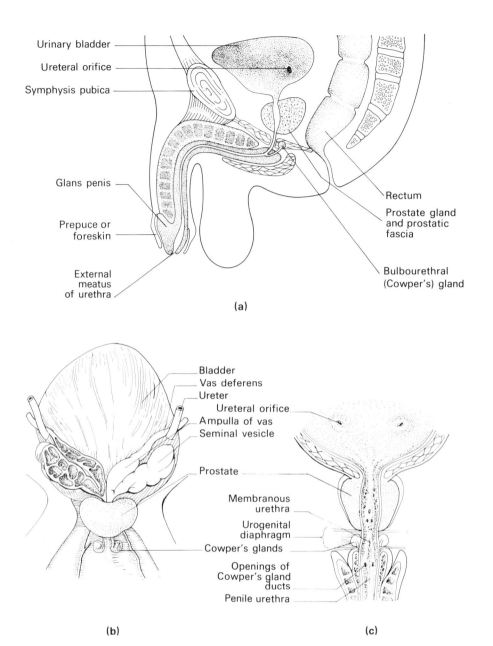

**Fig. 7.1.** View of human male accessory sex glands: (a) midsagittal section through pelvis; (b) posterior view of dissected pelvic contents; (c) coronal section through (b) viewed anteriorly.

growth, breast development and distribution of body fat, which result in the development of characteristic female body contours. Growth of female internal and external genitalia are also dependent on oestrogens, whilst androgens stimulate growth of secondary sex hair.

In contrast, the regulatory actions of steroids are reversible, and involve quantitative, rather that qualitative, changes to established accessory sex organs and tissues. These actions are not concerned with *establishing* the individual as a male or female, but with the *efficacy* with which their reproductive tracts and genitalia subserve the requirements of reproduction. In the male the regulatory actions of testicular steroids influence the activity of the accessory sex glands, body size and metabolism, erectile capacity and in some species the more exotic secondary sexual characters, such as antlers in deer. These actions may be continuous or show seasonal variations. In the female it is the regulatory action of oestrogens and progestagens that results in the external manifestations of the menstrual and oestrous cycles. But these external changes are accompanied by cyclic changes in the vagina, cervix, uterus, oviducts and in some species, for example the ferret, unusual swelling of the vulva or, in various female primates the sexual skin.

## 1 Androgens and the male reproductive system

In Chapter 3 we saw that testosterone was essential for the maintenance of spermatogenesis. Neutralization of testosterone by an antibody, or by synthetic anti-androgens such as cyproterone, blocks or reduces the effectiveness of sperm production. Deprivation of testosterone also has profound and immediate effects on the accessory sex glands of the male's genital tract (Fig. 7.1). After castration the prostate, seminal vesicles and epididymides, or their equivalents in various species (see Table 7.1), involute, their epithelia shrink and secretory activity ceases (Fig. 7.2). Direct measurement of their metabolic and synthetic activity shows a dramatic fall, and seminal plasma is no longer produced. If the castrate animal is provided with exogenous testosterone, the involuted organs are fully restored, both in size and secretory activity. Not surprisingly these target organs for androgen activity are found to possess androgen receptors. This reversible regression of accessory sex gland activity occurs naturally in seasonally active males, such as sheep and deer. The seasonal appearance of secondary sexual characteristics, such as antlers, and the behavioural interactions during which they

**Table 7.1.** Relative size of principal male accessory sex glands.

| Species | Prostate | Seminal vesicle | Ampulla | Cowper's or Bulbo-urethral glands |
|---------|----------|-----------------|---------|-----------------------------------|
| Man   | +++ | ++  | +   | ±   |
| Cow   | +   | +++ | ++  | ±   |
| Dog   | +++ | −   | −   | −   |
| Pig   | ++  | +++ | −   | ++  |
| Horse | +++ | +   | +++ | ±   |
| Sheep | ++  | +   | ++  | ++  |

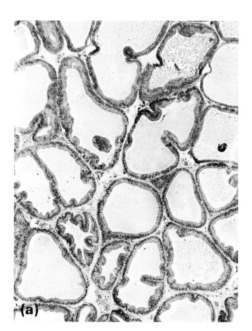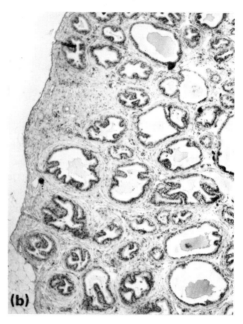

**Fig. 7.2.** Sections through prostate from: (a) intact rat; and (b) rat 18 days postcastration. Note reduction in luminal content, reduced height of luminal secretory epithelium and relative expansion of proportion of connective tissue.

are used, are similarly dependent on the actions of androgens.

The androgen which normally mediates this activity is testosterone. However, since the enzyme $5\,\alpha$-reductase is present and active in these target tissues, it is almost certain that testosterone is normally converted locally to $5\,\alpha$ dihydro-testosterone, which itself exerts the principal androgenic effects. In Chapter 5 we mentioned that the local activity of androgens is enhanced by prolactin. Indeed, androgen-dependent prolactin receptors are present in both prostate and seminal vesicles, and prolactin itself is detectable in seminal plasma at higher levels than in blood. However, the role of prolactin in the male accessory glands is not entirely clear.

In addition to their effects on secondary sex organs androgens also have anabolic or myotrophic effects, a reflection of which is the characteristically more muscular appearance of males that progressively establishes itself after puberty. Androgens also increase kidney and liver weight, and depress thymus weight.

**2 Oestrogens, progestagens and the female reproductive system**

In the female the shifting balance of hormones during the ovarian cycle affects oviducal, uterine, cervical and vaginal activity, as well as exerting more generalized physiological effects. The effects of steroids on these target organs may be studied either by correlating changes in a normal cycle with changing steroid levels, or after the injection of exogenous steroids into intact or castrate females.

<table>
<tr><td>

*a The oviduct
(Fallopian tube)*

</td><td>

The oviduct is a thin muscular tube covered externally with serosal tissue and peritoneum. A ciliated, secretory, high columnar epithelium overlies the stromal tissue internally. The oviduct is the site of fertilization and therefore the egg passes along it from the fimbriated ostium towards the spermatozoa passing in the opposite direction, from the isthmic junction with the uterus (Fig. 4.1). A few days later the fertilized eggs reverse the path taken by the spermatozoa to enter the uterus (Chapters 8 & 9). Clearly then, the oviduct has a role in gamete transport; it must also provide a suitable environment for fertilization and the early growth of the conceptus.

</td></tr>
</table>

After ovariectomy oviducal cilia are lost, secretion ceases and muscular activity declines, indicating an important steroidal influence on the oviduct. Subsequent injections of oestrogen restore both the ciliated, high columnar epithelium and secretory activity, and increase spontaneous muscle contractions. When progesterone is imposed on this oestrogen background the numbers of cilia decline, and the quantity of the oviducal secretion also declines, the small volumes of fluid that are produced having a lower sugar and protein content. Raising the progesterone to oestrogen ratio may also exert a mildly depressant effect on oviducal musculature, particularly relaxing the sphincter-like muscle at the utero-tubal junction (although data on this are somewhat conflicting).

There is considerable evidence to suggest that sudden changes in the oestrogen : progesterone ratio as achieved, for example, by administration of exogenous pharmacological doses of steroids during a cycle, can cause disturbances of gamete and conceptus transport. Thus, if high doses of the synthetic oestrogen, stilboestrol, are taken within 72 hours of fertilization, the pregnancy rate is reduced, probably by causing premature expulsion of the conceptus from the oviduct via the uterus into the vagina. However, the same dose given earlier, just before, or immediately after ovulation can result in *tube locking* of the conceptus, in which prolonged spasm of the oviducal musculature prevents passage of the conceptus to the uterus. Pregnancy may then ensue, but frequently it is not uterine, but occurs in the oviduct (*ectopic*).

<table>
<tr><td>

*b The uterus*

</td><td>

The uterus shows even more prominent steroid-dependent cyclic changes in structure and function than the oviduct (Fig. 7.3). During each cycle, the uterus first prepares to receive and transport the spermatozoa from the cervix to the oviduct, and subsequently it prepares to receive the conceptus from the oviduct and nourish it. The uterus consists of an

</td></tr>
</table>

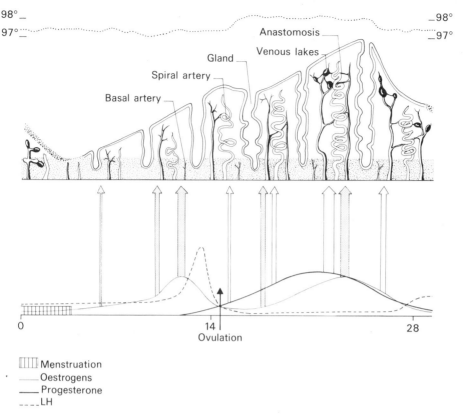

98° _

97° _

_ 98°

_ 97°

Anastomosis

Venous lakes

Gland

Spiral artery

Basal artery

0

14
Ovulation

28

|||||| Menstruation
······ Oestrogens
——— Progesterone
- - - - LH

**Fig. 7.3.** Changes in human endometrium during the menstrual cycle. Underlying steroid changes indicated below and basal body temperature indicated above. Thickness of arrows (oestrogens stipple; progestagens white) indicate strength of action. (After Netter.)

outer peritoneal and serosal investiture, over a thick *myometrium* of smooth muscle arranged in distinctly oriented layers. Internally the *endometrium* consists of a stromal matrix over which lies a simple low columnar epithelium with glandular extensions penetrating into the stroma (Fig. 7.4).

After ovariectomy, the uterus hypotrophies, and its blood supply is reduced. Administration of oestrogen reverses this effect, with massive increases in RNA and protein synthesis, cellular division and growth. In the normal cycle the period during which oestrogen rises rapidly coincides with the latter half of the follicular phase of the menstrual cycle or, in non-primates, with the whole of the abbreviated follicular phase. During this phase a similar *uterotrophic* effect of oestrogen is observed. The myometrium increases both its contractility and excitability, showing increased spontaneous activity. Meanwhile in the endometrium stromal size increases, partly due to stromal cell proliferation (leading to the expression *proliferative phase* of the uterine cycle) and partly due to

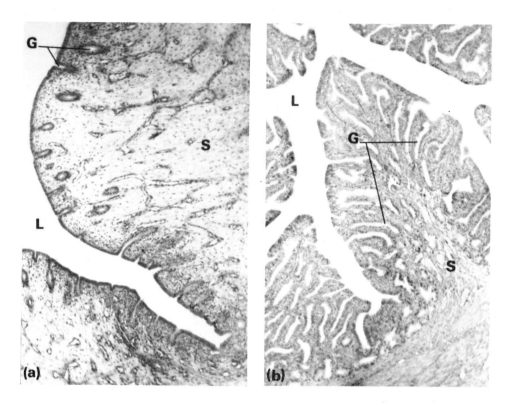

**Fig. 7.4.** Sections through rabbit endometrium at: (a) oestrus; and (b) progestational phase. Note dramatic increase in invaginations of the glandular epithelium (G) from luminal surface (L) into stromal tissue (S).

oedema. The height and synthetic activity of the surface epithelium leads to a large increase in surface area. In primates and large farm animals this also involves an increase in the numbers, and size, of the glandular invaginations of the stroma, but in rabbits and rats, the principal proliferation is luminal. The oestrogen-primed epithelial cells secrete a fluid of a characteristically watery constitution, but it does contain some proteins, notably proteolytic enzymes. These changes reach a maximum at the time of the oestrogen surge.

The oestrogens act by binding to the oestrogen receptors which are present in abundance in the uterine tissue. One of the most crucial actions of oestrogens over this period is to induce the synthesis of intracellular receptors for progesterone. At the beginning of the oestrogenic phase of the cycle progesterone-binding receptors are at a low level, and progesterone therefore has little effect on the uterus. However, after the oestrogen surge and ovulation have occurred the uterus is primed to bind progesterone, and so the progestagenic or *secretory phase* of the uterine cycle begins.

In the rabbit and mouse, there is a rapid extension of the

177     *Actions of Steroids in the Adult*

epithelial proliferation into glandular regions at this time (Fig. 7.4). Progesterone stimulates synthesis of secretory material by the glands of most species so that they become distended with a thick secretion rich in glycoprotein, sugars and amino acids. In some, but probably not all, species the release of this glandular secretion into the lumen requires a secondary peak of luteal-phase oestrogen imposed upon the progesterone background (see also Chapter 9). Stromal proliferation also increases under the influence of progesterone, the stromal cells becoming larger and plumper. This is particularly marked in rodents and primates. Within the stromal tissues of primates, the characteristic spiral arteries become fully developed (Fig. 7.3). Progesterone also acts on the myometrium causing further enlargement of cells, and, in contrast to oestrogens, progesterone depresses the excitability of the uterine musculature. It is important to re-emphasize that these actions of progesterone will only occur in an oestrogen-primed uterus, an example of *receptor regulation* (see Chapter 2).

With the withdrawal of steroid support at the end of the luteal phase of the cycle, the elaborate secretory epithelium collapses. In most mammals, the endometrium is resorbed, and a thin stromal layer overlain with epithelium replaces it, ready for entry into a new cycle as oestrogens rise. In man, apes and Old World monkeys the endometrial tissue is shed as the *menses* via the cervix and vagina, together with blood from the ruptured arteries. The spiral arteries contract to reduce bleeding.

*c The cervix*

The cervix is traversed by the spermatozoa at coitus and the neonate at parturition (see Chapters 8 and 12). In many species, including humans, the spermatozoa must actively swim through the cervix (Chapter 8). The properties of the cervix show marked steroid-dependent changes during the cycle and these can be crucial to normal fertility.

During exposure to oestrogen in the follicular phase the muscles of the cervix relax, and the epithelium becomes secretory. However, during the luteal phase, when progesterone levels are elevated, secretion is reduced and the cervix is firmer. Cervical mucus may be collected for examination during the human cycle and tested in a variety of ways for its steroid-dependent properties. The test of greatest functional significance is that of *sperm penetration,* in which the capacity of spermatozoa to swim into and through a smear of mucus on a slide is assessed. Characteristically, sperm penetration is low in the luteal phase of the cycle, and reaches a maximum around the time of ovulation (Fig. 7.5). These effects can be mimicked by administration of exogenous steroids, oestrogens enhancing sperm penetration whilst progesterone, even in the presence of oestrogens, depresses penetration. Thus,

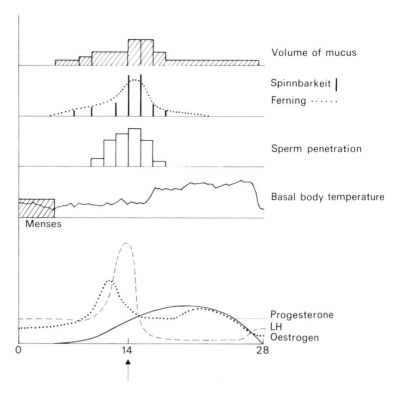

**Fig. 7.5.** Changes in properties of cervical mucus at various days of human cycle (blood hormone levels plotted). Parameters changing under influence of high oestrogen and low progesterone are: volume of mucus; stretchability (spinnbarkeit) of mucus; crystallization pattern of dried mucus (ferning) and *in vitro* tests of the ability of spermatozoa to penetrate mucus. Note luteal oestrogen does not induce changes due to elevated progesterone. Progestagenic contraceptives prevent normal periovulatory changes.

continuous administration of progestagens throughout the cycle, or the local release of progestagens from capsules placed in the uterus, suppresses sperm penetration even at the time of ovulation and the oestrogen surge. Use of this property is made in the low-dose progestagenic contraceptives. The steroids act via an effect on the amount and nature of the glycoproteins secreted by the cervical epithelium. Under progestagen dominance, small volumes of thick mucus are secreted and strands of mucus may be stretched to only short lengths before the threads snap–a low *spinnbarkeit* (Fig. 7.5). If mucus from oestrogenic cervices is allowed to dry on a slide, it forms, due to its distinctive molecular composition, a characteristic pattern known as *ferning* (Fig. 7.5). These tests of cervical mucus are important, since a hostile, impenetrable mucus will reduce sperm progress towards the oviduct and thus fertility.

179      *Actions of Steroids in the Adult*

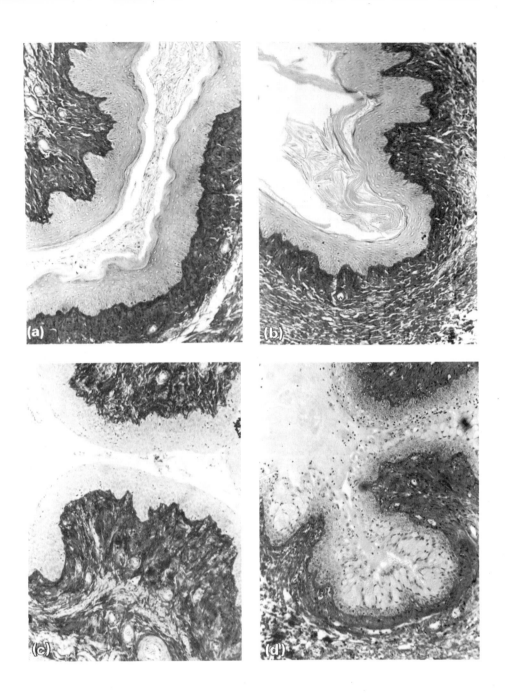

**Fig. 7.6.** Sections through rat vagina. (a) Pro-oestrus: rising oestrogen causes nucleated cells to be shed into lumen and a keratin layer to develop on the surface epithelium; (b) this process is completed at oestrus with a heavily keratinized layer and some desquamation of keratin; (c) during dioestrus, under a low oestrogen:progesterone ratio, the epithelial cells are nucleated and non-keratinized, and leucocytes are visible in the lumen; (d) in pregnancy the surface epithelium is glandular and leucocytes abound.

## d The vagina

The vagina shows marked structural changes during the cycle in some species. In the guinea-pig, for example, a membrane completely closes the vaginal os throughout most of the cycle, and only breaks down under the influence of rising plasma oestrogen concentrations at oestrus. In many mammals, including humans, oestrogens induce an increased mitotic activity in the vaginal surface columnar epithelium with a tendency to keratinize. This change is particularly marked in rodents, in which the oestrous cycle can be staged quite accurately by examination of the different cell types present in smears from the vaginal epithelium (Fig. 7.6). The fluids within the vagina also change during the cycle, and one effect of this is to vary the metabolic substrates available to the bacterial flora there. Cyclic changes in the vagina, induced by oestrogen and progesterone, result in the generation by bacteria of differing proportions of volatile aliphatic acids. These give distinctive odours to vaginal secretions and may have marked behavioural consequences, as will be seen later in this chapter.

## e Other tissues

Several other features of the female's anatomy and physiology may change with the ovarian cycle. Oestrogens have general effects on the cardiovascular system and metabolism that may be revealed cyclically or in women taking oral contraceptives. They depress appetite, and are mildly anabolic. Oestrogens also appear to be closely related to the reduced capillary fragility, blood cholesterol levels and incidence of thrombosis seen in premenopausal women when compared to men. Oestrogens may also increase the ability of the cardiovascular system to withstand high blood pressures. The mechanisms by which oestrogens act in this way are unclear. However, maintained *elevated* levels of oestrogens, such as occurs in pregnancy or during treatment with some contraceptive pills, may actually cause hypertension and, via effects on lipid metabolism, an increased blood clotting rate (Chapter 14).

Progesterone, in contrast to oestrogens, is mildly catabolic in humans, and has been suggested to increase appetite. It also has two general actions that may become manifest during the menstrual cycle. Progesterone elevates basal body temperature (Fig. 7.3); this is not a consequence of increased catabolism or stimulation of the thyroid but appears to involve a direct action on the areas of the hypothalamus concerned with thermoregulation. The rise in temperature occurs only *after* ovulation and thus is only useful as a basis for establishing the regularity of a woman's cycle if the *rhythm method* of contraception is contemplated (Chapter 14).

Progesterone also shows some affinity for aldosterone receptors in the kidney, presumably because of similarities of stereochemical structure. However, after binding by progesterone, the receptor is not activated and in consequence natriuresis ensues. As a result a compensatory rise in aldoste-rone output occurs to restore sodium retention. In women, sodium retention may be enhanced in the luteal phase by a direct stimulatory effect of luteal oestrogen on angiotensino-gen production. The consequence of these events may be a net retention of sodium and water towards the end of the luteal phase, which contribute to some symptoms characteristic of the premenstrual period, for example, heavy, tender breasts.

These examples of the widespread cyclic changes in struc-ture and function of many of the somatic tissues of the female, and not just her reproductive organs, show how the ovary and its secretions play such a dominant part in normal day-to-day physiology. Ovarian steroids also have profound and cyclic effects on the behaviour and mood of the females of some species, and this, together with comparable effects of testicu-lar hormones in males, is the subject of the next section.

## 3 Hormones and the regulation of sexual behaviour

We have already discussed in Chapter 1 how hormones may influence the developing brain and so determine the patterns of sexually dimorphic behaviour observed subsequently dur-ing childhood and maturity. The most obvious of sexually dimorphic behaviours in adulthood are those associated with sexual interaction, the result of which is coitus and the consequent juxtaposition of male and female gametes in the female's reproductive tract. We have seen in this chapter how steroidal hormones condition the adult reproductive system for its role in gamete transport. We now turn our attention to the role of the same hormones in regulating the expression in the adult of the patterns of sexual behaviour laid down neonatally and rehearsed postnatally in the prepubertal period.

Stimuli which elicit sexual behaviour, for example, members of the opposite and sometimes the same sex, surround most of us most of the time, but sexual interaction occurs only sporadically. What determines when these sexual stimuli induce sexual activity? The answers to this question, particularly in primates and man, are complex, but if we examine non-primate species we see rather clearly what *hormones* do in this context. A *castrated* male or female rat, cat, dog, etc., does not display sexual activity when placed with a member of the opposite sex. Treatment with the appropriate hormones results in the activation of sexual behaviour, and in fact renders the sexual stimuli effective in some way. Thus, in these species, *hormones increase the probability that an appro-priate set of sexual stimuli will induce sexual activity.* In primates, including man, this profound controlling influence of sex hormones has been modified considerably such that social and volitional factors become increasingly important. Undoubtedly, this is one reflection of the increasing size and

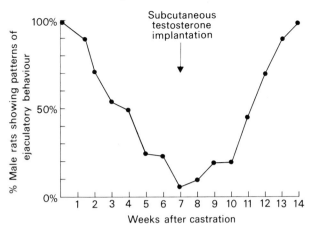

**Fig. 7.7.** Effects of castration and testosterone replacement on patterns of *ejaculatory* behaviour in the male rat. (Ejaculatory behaviour consists of a prolonged intromission and characteristic manner of terminating the intromitted mount.) Note that appreciable levels of the behaviour persist for some weeks after castration, and that, in this sample of rats, a comparable time was required for restoration of the behaviour by subcutaneously implanted testosterone.

complexity of the brain—particularly the neocortex—as the phylogenetic scale is ascended. We will first discuss the *ways* in which hormones affect sexual behaviour in various species, and then go on to consider *how* and *where* they exert their effects. Although much remains to be discovered about *how*, it will become apparent that hormones alter the expression of sexual behaviour by acting both on *the genitalia* and within *the brain*.

*a Sexual behaviour of the male*

In male non-primate mammals, sexual behaviour is clearly hormone-dependent, as it declines after castration and this decline is reversed by treatment with testosterone (Fig. 7.7). However, one feature of this relationship between testicular androgens and sexual behaviour remains difficult to explain. The decline in sexual behaviour after castration is often slow, taking several weeks to reach its nadir (Fig. 7.7) although plasma testosterone is virtually undetectable within hours. Similarly, treatment with testosterone after long-term castration, although it will effectively reverse these changes in sexual behaviour, does so with a long latency (Fig. 7.7). Furthermore, not all behaviours are equally affected by removal of testosterone. Thus, mounting persists long after intromission and ejaculation patterns have ceased, perhaps indicating degrees of 'dependence' of these behaviours on the hormone. The time period between castration and testosterone-replacement therapy also affects the re-instatement of sexual behaviour. Much less hormone is required to restore some elements of sexual behaviour if it is given soon after

183　　*Actions of Steroids in the Adult*

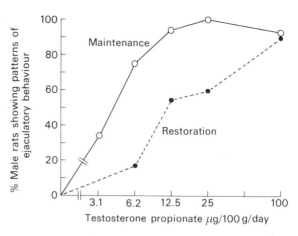

**Fig. 7.8.** Differential testosterone requirements for 'maintaining' ejaculatory behaviour in newly castrated rats, and 'restoring' the same behaviour in long-term castrates. Note that much more testosterone propionate was required in the latter group; treatment, as in Fig. 7.7, was given for several weeks to cause this change.

castration rather than after several weeks (Fig. 7.8). This suggests that an insensitivity to the hormone develops progressively in the target tissue(s) that mediates its behavioural effects. Could peripheral effects of testosterone on the sensitivity of the genitalia to tactile stimuli, and perhaps on the capacity for penile erection, also contribute indirectly to changes in sexual behaviour, and thereby play a key role in determining the disappearance and reinstatement of intromission and ejaculatory patterns of behaviour? Such an action could be in addition to any effects of the hormone on the brain which may underlie the male's 'motivational state'.

Some insight has been gained into this complicated problem, as a result of the fortuitous finding in the rat, that testosterone exerts its *behavioural* effects on the brain after aromatization to oestradiol (Chapters 1 and 2). After castration, therefore, small amounts of oestrogen administered to male rats will reverse the central or 'motivational' deficits that follow testosterone deprivation, but regression of seminal vesicles, prostate and penis will continue with the attendant deficits in related elements of sexual behaviour such as erection and intromission. Conversely, dihydrotestosterone given to castrate male rats has potent stimulatory effects on sex accessory glands and penis, but little or no effect on central 'motivational' components of sexual behaviour because unlike testosterone it *cannot* be aromatized to oestrogens in the brain. *Combining* extremely small amounts of oestradiol with dihydrotestosterone results in identical effects on behaviour *and* the periphery to those which follow testosterone treatment. These findings suggest that testosterone serves as a prehormone. Its androgenic effects on somatic

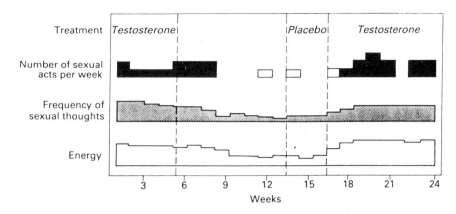

**Fig. 7.9.** The effects of testosterone replacement in a hypogonadal man aged 40, castrated one year earlier for testicular neoplasm. Sexual activity, ejaculation, sexual thoughts and energy all decline about 3 weeks after stopping testosterone treatment. There is no response to placebo, but a rapid response within one or two weeks of restarting testosterone treatment. (After Bancroft 1983.)

structures depend on reduction to dihydrotestosterone, the peripherally active androgen, and its effects on the brain depend on aromatization to oestradiol. Unfortunately, this precisely worked out chain of events is not universally applicable, for example in primates aromatization does not seem to represent an obligatory event for the central actions of androgens.

In male monkeys and in men, castration may eventually be followed by a reduction in sexual activity, but the time course ranges from months (in monkeys) to years (in men), and often the reduction does not proceed to total loss. In those subjects in whom total loss does occur, testosterone treatment effectively reverses the changes following castration, restoring sexual activity to the previous usual level. Indeed, testosterone therapy has been used successfully to treat low levels of sexual arousal and activity in hypogonadal men, and withdrawal of the hormone is reliably followed by a reduction in these measures (Fig. 7.9).

Conversely, *anti-androgens,* such as cyproterone acetate or medroxyprogesterone acetate, have been reported to *decrease sexual arousal* (measured as frequency of penile erections and/or sexual fantasies) in men watching 'blue movies'. These compounds, in addition to any putative action in the brain, markedly decrease the capacity for erection. This again raises the problem of determining the relative importance of peripheral and central mechanisms in controlling sexual arousal. Do erections occur only as a result of sexual excitement or do they also contribute to its occurrence? A decrease in erectile ability may contribute to a much higher threshold of sexual arousal and hence decreased sexual

185 *Actions of Steroids in the Adult*

activity in addition to any anti-androgen-induced change in 'central state'. Although the sites of action of anti-androgens remain to be determined (see below), they are nevertheless used clinically to treat individuals with unusual patterns of sexual behaviour such as exhibitionism or paedophilia, which society has deemed undesirable.

*b Sexual behaviour of the female*

In non-primate female mammals sexual behaviour is more clearly hormone-dependent than in males. In Chapters 4 and 5 we described how sexual receptivity or 'heat' occurs cyclically, and is closely coordinated with the period of ovulation which ensures that copulation occurs at a time when fertilization is most likely. (Reflex ovulators, for example the cat, rabbit and ferret have a somewhat different mechanism for achieving this end: the female comes into heat and stays in that state until she copulates, and this event itself *triggers* ovulation, see Chapters 4 and 5.) Ovariectomy in these species is followed by a prompt and usually complete abolition of *receptivity* (they will no longer accept mounts by the male) and *proceptivity* (they no longer 'solicit' his sexual attention). Restoration of these elements of the female's sexual behaviour is achieved equally rapidly by treatment with oestradiol and, in some species, progesterone (which must be given by injection in appropriate sequence with oestradiol in order to mimic the hormonal events seen during the oestrous cycle; Chapter 5, Fig. 5.28). These behavioural events may be affected additionally by a circadian rhythm (for example in the rat, heat occurs in the dark phase of the day/night cycle) and/or a seasonal rhythm (for example in the ewe, oestrous cycles begin in the autumn), events discussed in more detail in Chapter 5.

Similarly dramatic changes in behaviour are seen in large domestic animals like the sheep and pig. Thus, the active nudging, blocking and nibbling behaviours displayed by ewes, or the immovably solid posture displayed by sows in heat, disappear after ovariectomy. Reinstatement of oestrous levels of these behaviours follows *appropriate* treatment with ovarian hormones, for example, progesterone *followed by* oestradiol in the ewe.

It is now well recognized that the strict relationship between ovarian hormones and sexual behaviour in non-primate mammals is largely lost, or rather altered, in primates. Superficially, this does not appear to be the case since, if sexual *interaction* between a male and female monkey is measured, it is often seen to follow a cyclic pattern (Fig. 7.10a). Ovariectomy is followed by a major reduction in sexual interaction, which is restored by treating the female with oestradiol—as it would be in rodents. However, if the behaviours displayed by the male and female are measured carefully, we see that it is the *male's* behaviour which declines

(a)

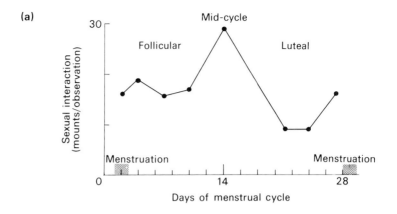

(b)

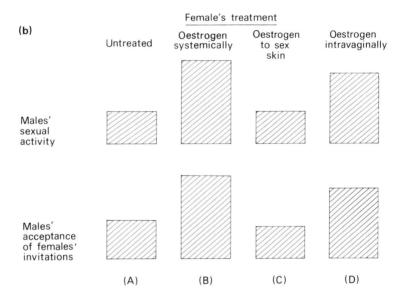

**Fig. 7.10.** Sexual interaction, measured as mounts by the male, between pairs of male and female rhesus monkeys during the menstrual cycle. Note the high follicular-phase levels of interaction which peak at mid-cycle and fall in the luteal phase. The premenstrual rise is characteristic. This pattern of interaction *seems* to be more due to fluctuating interest in the female by the male, as is revealed more clearly in (b), a block diagram summarizing the results of experiments which demonstrate (A) that ovariectomy of the female depresses the male's sexual activity and reduces his *acceptance* of her sexual invitations. In (B) it can be seen that treating the *female* with oestradiol, by subcutaneous injections, increases both these parameters of the male's behaviour. This effect is lost if the oestradiol is smeared, as a cream, on the female's perineum (C), but if it is placed in the vagina (D) an effect identical to that seen in (B) occurs. These data indicate that *oestradiol increases sexual attractiveness by an action on the vagina.*

187    *Actions of Steroids in the Adult*

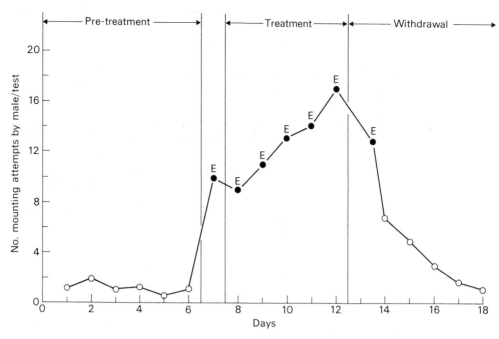

**Fig. 7.11.** Summary of effects on three male rhesus monkeys of a synthetic mixture of aliphatic acids applied to the perineal region (sexual skin) of an ovariectomized female rhesus monkey. Mounting behaviour is markedly stimulated during treatment, and ejaculations (E) occur consistently. Withdrawal of the mixture is followed by a reversal of these changes in the male's sexual behaviour.

markedly after ovariectomizing the female and which increases after treating *her* with oestradiol. The female's behaviour *may* change during the cycle, but it does so less consistently than the male's. One obvious interpretation of these data is that oestradiol is somehow changing the 'sexual stimulus value' of the female to the male, in other words changing her sexual *attractiveness*.

Numerous experiments have pointed quite consistently to the *vagina* as the site where oestradiol, and also progesterone, exerts these effects (Fig. 7.10b), and that the male interprets them as changes in the female's *odour*. Male rhesus monkeys rendered reversibly anosmic, by placing bismuth iodoform paste in their noses, fail to discriminate between oestrogen treated females and females from whom the oestrogen is withdrawn. They copulate with the latter until their anosmia is reversed, at which point they usually cease promptly until oestrogen treatment (systemic or *intravaginal*) is resumed.

Analysis of vaginal secretions using gas chromatography and mass spectrometry techniques has shown them to consist of a mixture of simple aliphatic acids (acetic, propionic, isobutyric, butyric and isovaleric). The concentrations of these acids vary during the menstrual cycle, decrease to very low levels after ovariectomy and increase markedly after oestrogen treatment. Vaginal lavages taken from ostrogen-treated females and placed on the perineum of untreated females stimulate the sexual activity of males with whom they are paired. Similar effects follow the placement of *a synthetic mixture of aliphatic acids* (in correct proportions) on to an ovariectomized female's perineum (Fig. 7.11). The effectiveness of this mixture can be enhanced by the addition of phenolic compounds (phenylpropanoic and para-hydroxy-phenyl-propanoic acids) also present in vaginal secretions, but ineffective in stimulating sexual activity in the male when applied alone to the female. Thus, the female monkey's vagina clearly produces odour cues which are dependent on oestradiol. The acids do not appear to be a glandular secretion or exudate, but are instead the *product of microbial action on vaginal secretions*. Penicillin added to incubations of vaginal lavages prevents production of the fatty acids, and it seems that the most likely mechanism of action of oestrogens is to increase the availability of substrate (cornified cells and mucus) from which the vaginal flora produces fatty acids.

Progesterone, the luteal-phase hormone, *decreases the sexual attractiveness* of female monkeys, and this probably largely explains the decrease in sexual interaction during the luteal phase of the menstrual cycle. (Clearly a very different effect of progesterone to that seen in non-primate females.) Progesterone causes the vaginal secretions of oestrogen-treated female rhesus monkeys to lose their sex-attractant properties, as measured by the decreased sexual activity of males paired with females receiving them. Whether this effect of progesterone is due to the loss of attractants in the vaginal secretion or the presence of some aversive odour has not been determined, but clearly vaginal secretions can both 'turn on' and 'turn off' the male's sexual interest.

It must be emphasized that these effects of odour cues on the sexual activity of males are *not all or none*. Some males are oblivious of changes in the female's odour, others seem to have their sexual activity completely regulated by them. Therefore, the behavioural response of a given male to changes in vaginal odour is variable, and differs enormously from the stereotyped behavioural responses of other mammals, and particularly insects, to 'pheromones'. These highly species-specific substances 'release' patterns of aggressive or sexual (and other) behaviours in an invariant way, very different from the effects of aliphatic acids in primates. Furthermore, vaginal secretions from women and other species of monkey will stimulate the sexual activity of male rhesus monkeys. The same aliphatic acids are present in human vaginal secretions and they vary in concentration during the menstrual cycle.

Do ovarian hormones affect sexual behaviour in women?

189     *Actions of Steroids in the Adult*

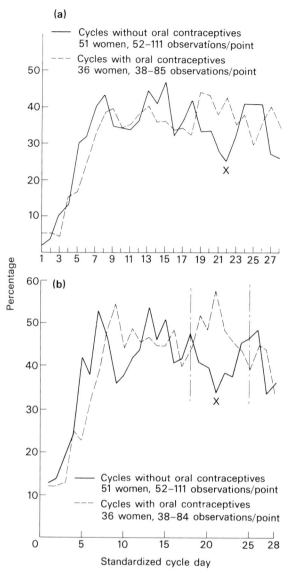

**Fig. 7.12.** (a) Changes in coital frequency in a group of women during menstrual cycles without (——) and with (---) oral contraceptives. The luteal trough at X occurring in normal cycles does not occur in cycles during oral contraceptive treatment. (b) When the data is re-analysed in terms of *coitus initiated by the man*, a similar picture is obtained. This suggests that men initiate sexual activity *less* in the luteal phase of the menstrual cycle, and that oral contraceptives prevent this decrease in initiations by the male partner.

There is little consensus that coital activity varies predictably during the menstrual cycle, although some recent reports seem more consistent in showing mid-cycle (periovulatory) peaks and luteal troughs of coitus and occasional increases prior to menstruation (Fig. 7.12). It is generally accepted, however, that ovariectomy in women does not normally result in changes in libido, i.e. the desire for sexual activity, although oestrogen-replacement therapy may be necessary to ensure vaginal lubrication and to prevent atrophic changes in the reproductive tract. As with the male, then, direct effects of sex steroids on the genitalia in maintaining levels of sexual activity must always be taken into account before making assumptions about changes resulting from an alteration in 'motivational state'.

It has been suggested, however, that ovarian hormones may exert effects on sexual interaction between men and women in a manner comparable to that described above in rhesus monkeys. In Fig. 7.12, data collected using a 'self-report, diary technique' show that coital activity varied 'classically' during the menstrual cycle in this population of subjects. However, when *initiation* of the sexual interaction was analysed, the cyclic variation was accounted for by changes in the man's behaviour and not the woman's! This was interpreted as evidence of cyclic fluctuations in sexual attractiveness, and suggested to be evidence of an olfactory communication between the couples similar to that seen in pairs of monkeys. Even more interesting was the finding that oral contraceptives were associated with the absence of a luteal trough in initiation by the man (Fig. 7.12). This phenomenon was explained on the basis of a block of endogenous progesterone production by the ovary in women taking oral contraceptives so that it does not exert its anti-oestrogenic (attractiveness-reducing) actions on the vagina.

There is, as yet, no direct evidence for such a view, or, indeed, that sexual activity in men is affected by cyclic variation in vaginal odours. But the data summarized in Fig. 7.12 are fascinating evidence of how little we understand the influence of ovarian hormones on sexual interaction in man. The fact that synthetic forms of these hormones are taken by large numbers of women to control their fertility suggests the importance of investigating their behavioural effects. In a study by the Royal College of General Practitioners in 1974 comparing 20 000 women taking oral contraceptives with a similar number of controls, loss of libido was a major symptom in the pill user's group, being *four times* more common than in controls. The mechanism underlying this change in sexual behaviour, apparently at variance with the data shown in Fig. 7.12, is also obscure, although it will be further discussed below.

A logical conclusion from the majority of data is that ovarian hormones have little effect on the sexual activity of female monkeys and women, even though they have pronounced effects on sexual interaction, largely by affecting the sexual interest of the male. Clearly, this situation is very different from the strict dependence of oestrous behaviour on

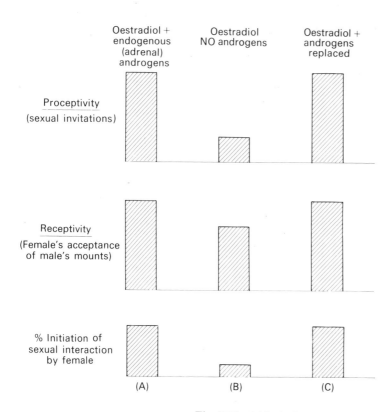

**Fig. 7.13.** A block diagram summarizing the results of experiments on adrenal androgens and proceptive and receptive behaviour in female rhesus monkeys. All females were ovariectomized and received oestradiol benzoate throughout the experiment, ensuring their attractiveness to the males (A). Removal of endogenously secreted androgens from the adrenal by suppression with dexamethasone or adrenalectomy with glucocorticoid replacement, caused large decreases in proceptive behaviour and initiation of sexual interaction by the female, and smaller decreases in receptive behaviour (B). These decreases in the females' sexual activity were reversed by treatment with testosterone proprionate or androstenedione (C).

ovarian hormones in female non-primates. But does this necessarily mean that proceptive and receptive behaviours of females are independent of hormonal regulation in primates? The answer to this question seems to be no, and, surprisingly, it is *androgens* of adrenal as well as ovarian origin that are the hormones which affect sexual activity in female primates most consistently. Thus, as we have seen, the oestrogen-treated, ovariectomized female monkey is both attractive to the male and willing to solicit and accept his sexual interest. However, suppressing her adrenal cortex either chemically, using dexamethasone, or surgically, by adrenalectomy but providing appropriate gluco- and mineralo-corticoid replacement therapy, results in a marked reduction of proceptive and receptive behaviour (Fig. 7.13) despite the continuing presence of oestradiol. These changes in the female's behaviour are reversed by treatment with testosterone (Fig. 7.13), suggesting an important role for this hormone. This has been confirmed recently by the much more specific and less invasive procedure of passively immunizing female rhesus monkeys against their own testosterone. These females showed much lower levels of proceptivity and receptivity which could be increased by subsequent treatment with an androgen not bound by the antiserum. Although the decreases in sexual behaviour in females deprived of androgens are considerable, it should be emphasized that the *strict dependence of sexual behaviour on steroid hormones seen in female non-primates has no obvious parallel in primates.* Androgen-deprived female monkeys may still accept mounting attempts by males (Fig. 7.13) just as androgen-deprived males may continue to display sexual interest in females. It is the incidence of such acceptances and interest which falls.

Clinical studies have suggested that androgens affect sexual activity in women. It has long been known that testosterone, most commonly the orally active methyl testosterone, can increase 'libido' in some women. Conversely, bilateral drenalectomy has been reported to cause a dramatic decrease in sexual interest in women undergoing surgery to control (steroid-dependent) breast cancer. The discovery that women taking oral contraceptives have low urinary levels of androgen metabolites has also been taken to explain the frequently reported loss of libido referred to above. But a double-blind trial on women suffering loss of libido for the first time after taking oral contraceptives, in which they received pills *with and without* an androgen (androstenedione) supplement, revealed no evidence of an ameliorative effect of the hormone.

Many criticisms of some of these studies have been made on the grounds that proper control groups were not investigated and that the behavioural analysis was poor. They remain the only studies to implicate androgens in the control of sexual activity in women, and should be viewed with caution.

*c Summary*

This account of steroid hormones and sexual behaviour should reveal two important points. First, there is convincing evidence that ovarian (or adrenal) and testicular hormones exert a less dramatic, controlling effect on the expression of sexual behaviour in primates than in 'lower' mammals. This conclusion is particularly true of human sexual behaviour where, although steroids undoubtedly influence it, other factors such as mood, time, place, partner preference and novelty are also important determinants. Second, this is a complicated and controversial field of study which, so far as human sexual behaviour is concerned, has only become socially, scientifically and ethically acceptable during the very recent past. Unqualified assertions as to the importance, or otherwise, of hormones on human sexual behaviour should be viewed, therefore, with caution. Some of the data reviewed, particularly those on female primates, demonstrate how important *peripheral* actions of hormones can be in determining sexual interaction. This subject has certainly not been explored to an equivalent degree in females of other species, or in males, and so our picture of this aspect of hormonal regulation is incomplete. However, we now turn our attention to the evidence of hormone action in the brain to control the expression of sexual behaviour.

*d Central actions of hormones in the control of sexual behaviour*

The techniques available to the neuroendocrinologist to study the neural mechanisms underlying sexual behaviour include the stereotaxic placement of lesions, hormones and also drugs at discrete neural loci; electrical stimulation in these areas and autoradiographic methods designed to localize the target neurons bearing receptors for one or other hormone. Although these techniques are sophisticated our understanding of the neural areas and mechanisms underlying sexual and other behaviours, as well as control of the pituitary, is limited by their relative 'crudeness', compared to the extremely complex nature of the structure that they are used to investigate. For example, electrolytic and other lesions destroy not only the neurons in a part of the brain, but also axons traversing that area which may have little or nothing to do with it functionally. Implanted or injected hormones may diffuse away from the original site (although this can be controlled). While advances in techniques and their application have increased our understanding of some neuroendocrine mechanisms controlling sexual behaviour, it should be borne in mind that sex hormones are taken up in many areas of the brain but it is not clear if they are all involved in reproductive functions, and, if not, what determines their selective actions at each site?

It has been a consistent finding that lesions, including those specific to neuronal cell bodies, placed in the medial preoptic area and adjacent anterior hypothalamus (Figs 5.2 and 5.3) of *males* of many non-primate and some primate species abolish sexual behaviour. This area of the hypothalamus is sexually dimorphic, being larger in the male than in the female (Chapter 1). The testes do not atrophy as a result of these lesions, therefore the pituitary–gonadal axis is unaffected, and the behavioural change is not simply secondary to a loss of testosterone. Indeed, treatment with testosterone does not restore sexual behaviour in lesioned animals. Conversely, implantation of testosterone into the preoptic–anterior hypothalamic areas, which are rich in androgen receptors, restores sexual behaviour in castrate male rats (Fig. 7.14). These dramatic effects of testosterone in the anterior hypothalamus (whether or not it is subsequently aromatized to oestradiol in the rodent) argue strongly that this area is of primary importance in the hormonal regulation of sexual behaviour. This does not in any way minimize the important peripheral effects of androgens, but instead points to the fact that neuroendocrine integration in this behavioural system requires the actions of hormones at different levels. The maintenance of sexual behaviour requires the action of testosterone in the hypothalamus, but the hormone also acts in the periphery (and in the spinal cord), presumably to facilitate the transduction and transmission of sensory stimuli to behaviourally relevant areas of the CNS or of the motor responses to these stimuli.

It has become clear that, while both preoptic area lesions and removal of testosterone by castration decrease the display of sexual behaviour, they do so in different ways. The castrated male is sexually inactive. By contrast, the preoptic area-lesioned male shows much interest in the female and attempts to mount her. It is at this point that his deficit is revealed since he cannot intromit. This sort of observation has led to the suggestion that the preoptic area is important for the *performance* of copulatory responses and that the *motivational* effects of testosterone are not expressed solely at this site. A similar conclusion follows the fascinating observation that preoptic area-lesioned male rhesus monkeys make little attempt to copulate with females, but masturbate to ejaculation at other times. Clearly, the capacity for sexual arousal and the performance of sexual reflexes are separable neurally.

In female non-primate mammals, notably the rat, a similar hypothalamic action of oestradiol exists. However, the ventromedial nucleus rather than the more anterior preoptic areas seems to be the principal site of action of the hormone. Thus, oestradiol implanted in the ventromedial nucleus (Figs 5.2 and 5.3) in amounts sufficient to saturate a proportion of the oestrogen receptors there increases markedly the receptive behaviour of ovariectomized females. Such treatment, then, is quite adequate as a background 'priming' for the subsequent actions of progesterone in those species, such as the rat, which require it together with oestradiol to induce oestrous levels of proceptive and receptive behaviour. The site of action of progesterone has proved more elusive to define, but a number of studies now point to the ventromedial hypothalamic area as most likely. It is rich in progesterone receptors which are actually induced by prior treatment with oestradiol. This probably explains why the sequence of oestradiol followed by progesterone is so critical in this species for oestrous behaviour to occur.

As we have seen androgens, notably testosterone, rather than oestrogens seem to underlie proceptivity in female

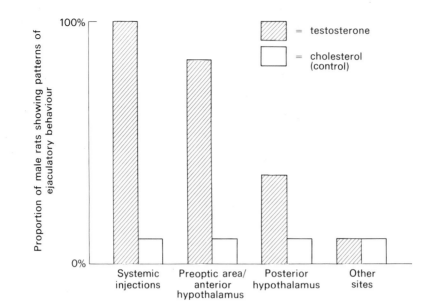

**(a)**

Proportion of male rats showing patterns of ejaculatory behaviour

= testosterone

= cholesterol (control)

Systemic injections | Preoptic area/ anterior hypothalamus | Posterior hypothalamus | Other sites

**(b)**     Site of testosterone implant

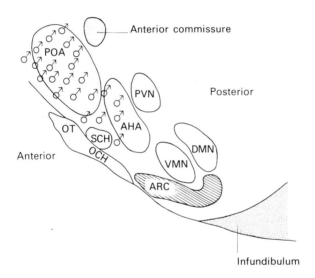

Anterior commissure

POA

PVN

Posterior

OT

AHA

SCH

OCH

DMN

Anterior

VMN

ARC

Infundibulum

**Fig. 7.14.** The effects of testosterone implanted in the CNS on the sexual behaviour of castrated male rats. (a) Testosterone placed in the preoptic–anterior hypothalamic continuum induced high levels of ejaculatory patterns, comparable to those following systemic treatment with the hormone. Testosterone placed in the posterior hypothalamus or elsewhere in the brain was without significant effect. Important controls were provided by the fact that cholesterol had no effect on sexual activity in any site, so arguing for some degree of hormonal specificity. More recently it has been shown that, at least in rats, oestradiol has similar effects. (b) Sites (♂) in the hypothalamus where testosterone induced increases in sexual behaviour in male rats. POA = preoptic area; AHA = anterior hypothalamic area; SCH = suprachiasmatic nucleus; PVN = para-

monkeys. Their behavioural effects are also apparently mediated by the anterior hypothalamus as is the case for oestradiol in female rats. Thus, implanting testosterone in an area extending from the ventromedial nucleus to the preoptic area reverses the behavioural effects of androgen deprivation which follow adrenalectomy (Fig. 7.15). No comparable effects of intrahypothalamic oestradiol have been reported in the monkey.

Relatively little is known about the ways in which hormones alter neural activity to bring about changes in sexual behaviour. Experiments on rats have revealed that oestradiol, for example, exerts its effects on receptivity after a delay of 26–48 hours. The finding that inhibitors of protein synthesis completely prevent these actions of oestradiol has led to the widely held view that the hormone, which is clearly accumulated in neuronal nuclei, modulates neural activity via an action involving gene expression. However, the nature of the products of this action has not been determined although they may include progesterone receptors. Nor is it clear whether the passage of time between exposure to oestradiol and its behavioural effects is necessary for materials synthesized in the neuronal cell body (e.g. enzymes, receptor molecules or peptidergic transmitters themselves) to be transported to other regions of the neuron (for example, terminals, dendrites).

There is considerable interest in the nature of the neurochemical mechanisms in the hypothalamus with which sex steroids interact when exerting their behavioural effects. There is little definitive data, but GnRH-containing neurons have been suggested to be one indirect target of steroid action (see Chapter 5). Thus, GnRH infused into the dorsal midbrain, a site to which preoptic GnRH neurons project, enhances the display of receptive lordosis postures in the female rat. Conversely, GnRH antibodies infused into the same site block the actions of GnRH. Although the peptide has been shown to have some effect in male monkeys, the precise importance of hypothalamic GnRH neurons in the regulation of sexual behaviour in male or female mammals remains uncertain.

Neurons in the arcuate nucleus containing pro-opiomelanocortin-derived peptides, particularly β-endorphin, richly innervate the medial preoptic area (see Chapter 5, Fig. 5.2) and infusion of β-endorphin into the latter site profoundly inhibits sexual responses in male rats. Levels of β-endorphin in the preoptic area vary markedly with the steroidal environ-

---

ventricular nucleus; ARC = arcuate nucleus; VMN = ventromedial nucleus; DMN = dorsomedial nucleus; OT and OCH = optic tract and chiasm.

197    *Actions of Steroids in the Adult*

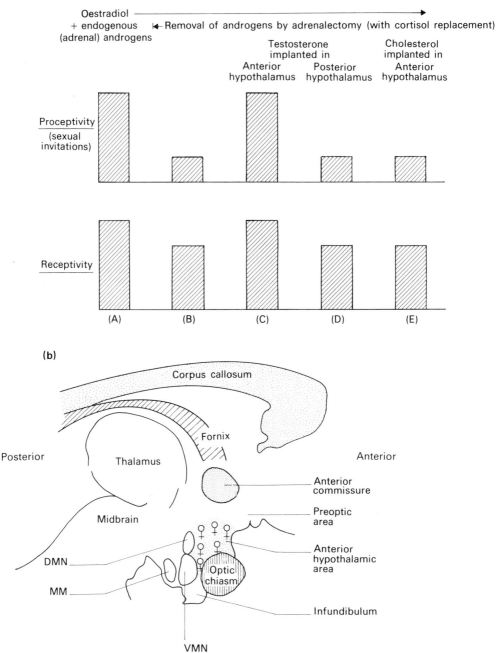

**Fig. 7.15.** (a) A block diagram summarizing the results of experiments in which testosterone propionate was implanted in the CNS of androgen-deprived female rhesus monkeys. All females were ovariectomized and received injections of oestradiol benzoate throughout the experiment, so they remained attractive to the males. Adrenalectomy was followed, as shown in Fig. 7.13, by decreased levels of proceptive and receptive behaviour (compare A with B). These changes in sexual activity were reversed by placing testoster-

ment, suggesting that this neural system may be important in mediating some of the behavioural effects of testosterone withdrawal. The fact that the same peptide reduces GnRH secretion in both males and females has been taken to indicate that it may have an important, pivotal role in mediating the inhibition of reproduction in a number of different situations.

*e Summary*

A great deal is known about the strict, controlling role of hormones in the sexual behaviour of non-primate species and about the hypothalamic sites where their actions are exerted. But even in these species, many aspects of the neural regulation of sexual behaviour are not understood: for example, what are the functions of structures in the limbic forebrain, such as the amygdala and septum, which bind large amounts of steroid hormones in males and females and which are well known to be concerned with the expression of motivated and emotional behaviour? In the primates, although the effects of hormones and their sites of action in males and females have been described to some extent, it is clear that much remains to be understood about the ways in which social interactions modify, and are modified by, these basic neuroendocrine mechanisms. Thus, not all members of a social group of monkeys engage in sexual interaction with equal ease. Indeed, often only the dominant male copulates with females with any great frequency, and even then, preferentially with only the more dominant females. Subordinate males and females risk aggressive attacks when soliciting sexual interaction and, although they do copulate, this behaviour occurs less overtly and with a much reduced frequency compared to higher ranking individuals. Clearly, gonadal hormones are not the only determinants of sexual activity. Furthermore, social interaction, as we have seen in Chapter 5, may profoundly influence the activity of the hypothalamo-pituitary–gonadal axis and hence fertility as well.

If such factors are complicated and poorly understood in non-human primates, it is hardly surprising that so little is known about the interactions between hormones (which do influence human sexual activity), social environment and individual variables (such as personality, mood and early history) which determine patterns of sexual behaviour in man.

---

one in the anterior hypothalamus (C) but not in the posterior hypothalamus (D). Cholesterol in the anterior hypothalamus was without effect (E). (b) Diagram of a sagittal section of the rhesus monkey's hypothalamus to show sites (♀) in the anterior hypothalamic area where testosterone implants reversed the behavioural effects of adrenalectomy. Abbreviations as in Fig. 7.14; MM = mammillary nucleus.

199    *Actions of Steroids in the Adult*

**Further reading**

Bancroft J. *Human Sexuality and its Problems*. Churchill Livingstone, 1983.

Bermant G, Davidson J. *Biological Bases of Sexual Behaviour*. Harper & Row, 1974.

Blandau RJ, Moghissi K (Eds). *The Biology of the Cervix*. University of Chicago Press, 1973.

Ciba Foundation Symposium 62. *Sex, Hormones and Behaviour*. Excerpta Medica, 1979.

Everitt BJ, Keverne EB. In *Neuroendocrinology*, Chapter 18 (Eds Lightman SL, Everitt BJ). Blackwell Scientific Publications, 1986.

Hafez ES, Evans TN (Eds). *Human Reproduction*. Harper & Row, 1973.

Hutchison JB. *Biological Determinants of Sexual Behaviour*. John Wiley & Sons, 1978.

Kreiger DT, Hughes JC (Eds). *Neuroendocrinology*. Sinauer Associates Inc., Sunderland, 1980.

Money J, Museph H. *Handbook of Sexology*. Elsevier, 1977.

Pfaff DW. *Oestrogens and Brain Function*. Springer Verlag, 1980.

Zuckerman Lord, Weir BJ (Eds). *The Ovary*. 2nd Edn. Academic Press, 1977.

# Chapter 8
# Coitus and
# Fertilization

In the foregoing chapters we have considered reproductive function in the male and the non-pregnant female. The central concept underlying all of our discussion is the coordinating role played by the gonadal steroids in production of mature gametes and conditioning of sexual behaviour and function. In this way the chance of fertilization is maximized. If successful fertilization occurs then cyclic sexual patterns must be replaced by pregnancy and nidatory patterns. In this chapter we will consider how the oocytes and spermatozoa travel from their site of production to the site of fertilization in the oviduct. Then we will consider the events of fertilization itself.

## 1 Transport of spermatozoa

In Chapter 3 we left the spermatozoa in the lumina of the seminiferous tubules. Human spermatozoa, a few microns in length, must travel through some 30–40 cm of male and female reproductive tract, or more than 100 000 times their own length, to reach the oviduct. During this long and hazardous journey, several major obstacles must be overcome, including transport *between* individuals at *coitus*. Much less than one in a million of the spermatozoa produced ever complete the journey. It is not just that the journey itself is difficult, but also the spermatozoa must successfully undergo a series of changes in both the male and female genital tracts before they gain full fertilizing capacity. These changes are termed *maturation* in the male tract, and *capacitation* and *activation* in the female tract.

### a Maturation

Spermatozoa are released from their close association with the Sertoli cells into the highly characteristic fluid of the

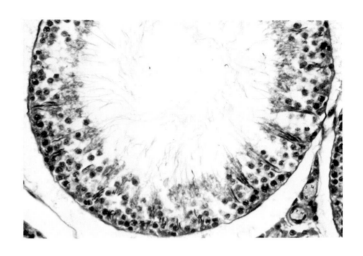

**Fig. 8.1.** Section through a rat testis which has 'blown up' 2 days after ligation of vasa efferentia. Note distended lumen of seminiferous tubule containing spermatozoa, regardless of stage of cycle compared to those shown in Fig. 3.10. The associations between the spermatogenic cells and the Sertoli cells are particularly well observed. The adluminal portions of the Sertoli cells are forced apart from each other by fluid.

seminiferous tubules. This fluid is secreted by the Sertoli cells such that a continuous flow rich in spermatozoa washes towards the *rete testis* (Fig. 1.12). If the *vasa efferentia* draining the rete testis are ligated, then fluid outlet is blocked and the testis literally 'blows-up' with accumulating fluid (Fig. 8.1). Normally as the fluid passes through the rete testis, the composition of its ions and small molecules changes. The change is probably mediated mainly by diffusional equilibration through the tubule walls since the absence of inter-Sertoli cell junctions renders the blood–testis barrier much less complete in this region. On entering the vasa efferentia and the *epididymis* (Fig. 8.2), the composition and volume of the fluid shows further major changes. The spermatozoa are concentrated by fluid absorption, from $50 \times 10^6$ per ml on entry to some $50 \times 10^8$ per ml on leaving the epididymis. In addition, the epididymis adds secretory products including carnitine and glycerylphosphorylcholine, and also modifies the glycoprotein profile on the surface of the sperm. The spermatozoa experience major environmental changes in their passage through the vasa efferentia and epididymis (which takes around 6–12 days in most species) and these changes have profound effects on their behaviour.

Spermatozoa entering the vasa efferentia are quite incapable of movement (beyond an infrequent twitch) and, when inseminated into females, cannot attach to and fertilize an

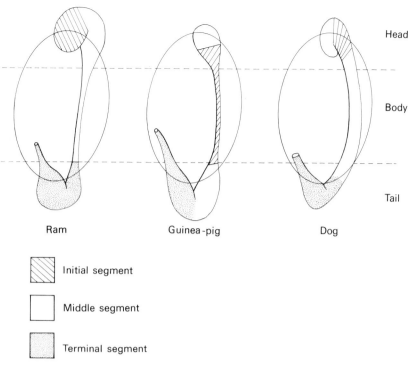

Head

Body

Tail

Ram                    Guinea-pig                Dog

Initial segment

Middle segment

Terminal segment

**Fig. 8.2.** Vasa efferentia (varying in length and from 10 to 20 in number with the species) connect the rete testis to a long, highly convoluted tube—the epididymis. Anatomically, the coils of the tube form a head (caput), body (corpus) and tail (cauda). Three histological segments, which usually do not correspond with the anatomical divisions, can be seen. An initial segment of high, ciliated epithelium and smallish lumen, a middle segment with wider lumen and shorter cilia and a wide terminal segment of low cuboidal, relatively poorly ciliated epithelium with more smooth muscle investiture in the underlying stroma. In the initial and middle segments, micropinocytosis occurs—absorption occurs here and possibly also some secretion in the more terminal parts of the middle segment. The terminal segment is not clearly absorptive but stores spermatozoa at a high density in the lumen. See Fig. 1.12 for illustration of the human excurrent ducts.

egg. However, with their passage from the rete testis through the vasa efferentia and epididymis to the vas deferens, they acquire both the ability to swim progressively and also fertilizing capacity. These changes in functional capability are accompanied by changes in the biochemistry and morphology of the spermatozoa. For example, as spermatozoa mature on passage through the epididymis, they use up endogenous reserves of metabolic substrates and become dependent on exogenous available substrates such as fructose, which they metabolize by glycolytic activity rather than by the oxidative metabolism characteristic of testicular spermatozoa.

Testicular, but not mature, spermatozoa can also use glucose as a substrate for conversion into amino acids and lipids. Spermatozoa utilize selected phospholipids and cholesterol as substrates during maturation and the residual phospholipids contain an increased proportion of unsaturated fatty acids. Structurally, the spermatozoa show a completion of the process of nuclear condensation, loss of the cytoplasmic drop (the small tag of cytoplasm remaining where the residual body was shed) which migrates down the tail and is discarded, an increasing negativity of surface charge, the acquisition of a coat of glycoprotein derived from epididymidal secretion, and also small changes in acrosomal morphology. It is not yet clear which, if any, of these biochemical and morphological changes are critical for the functional maturation of the spermatozoa. It is clear, however, that the whole process of maturation is crucially dependent upon adequate stimulation of the epididymis by androgens.

If the androgens are removed by castration, secretion ceases and epididymidal regression occurs. Injection of testosterone restores secretory activity. It seems probable that the major portion of the natural androgens stimulating epididymidal function in the initial segment of the epididymis (Fig. 8.2) is derived not from the circulation but from the rete testis fluid itself, since ligation of the vasa efferentia results in considerable functional and structural regression of the epididymis. Testosterone bound to androgen-binding protein is carried into the head of the epididymis in the fluid at concentrations approaching those of testicular venous blood (between 30 and 60 ng/ml; dihydrotestosterone is also present at about half this level). Within the epididymis, intracellular receptors take up the androgens, and $5\alpha$-reductase converts testosterone to dihydrotestosterone to yield very high tissue levels of this more active androgen. There is some evidence to suggest that the epididymis may even engage in a little androgen synthesis itself.

The spermatozoa pass from the tail of the epididymis into the vas deferens as a very densely packed mass, and their transport is no longer a result of fluid movement but due to the muscular activity of the epididymis and vas deferens. The absence of a massive fluid flux is important, since patients or animals with a ligated vas deferens (*vasectomy*) do not accumulate masses of fluid behind the ligature as occurred with ligation of the vasa efferentia (see Fig. 8.1). Spermatozoa, however, do build up behind the ligature and these must be removed either by phagocytosis within the epididymis or by leakage through the epididymidal wall. The normal non-ligated vas deferens serves as a storage reservoir of spermatozoa. In the absence of ejaculation, spermatozoa dribble through the terminal *ampulla* of the vas deferens into the urethra and are washed away in the urine.

*b Semen (= spermatozoa + seminal fluid)*

The seminal fluid in which the spermatozoa are carried to the female tract is largely derived from the major accessory sex glands (Fig. 7.1), with only a small contribution from the epididymis. Different species have bewilderingly different patterns of accessory sex gland structure (see Table 7.1) and function (Table 8.1), but certain general features emerge. Seminal fluid cannot be essential for effective sperm function, since spermatozoa taken directly from the vas deferens can fertilize eggs. However, spermatozoa require a 'fluid vehicle' for their normal transport and the seminal plasma supplies this, either exuberantly in half-litre volumes in the boar, or with a more conservative 3 ml or so in the human. In addition to providing a transport medium, the seminal plasma may also provide nutritional and protective factors for the spermatozoa (Table 8.1). Whilst these factors are not strictly essential, they may become crucial in cases of marginal fertility.

*c Coitus, genital reflexes and sexual responses*

In mammals, fertilization is internal and the male gametes must be introduced into the female tract at coitus. Coitus itself is of variable duration (minutes in man, hours in the camel) but is accompanied in most species by extensive physiological changes not just in the genitalia but also in the body as a whole.

It is only relatively recently that research into human sexual physiology has become an accepted and important part of the study of reproduction. Even today, medico-social opinion on the ethics of studying human sexual activity in the laboratory varies considerably. Not until the mid 1960s was the first complete and carefully conducted study of humans, observed during heterosexual interaction or while masturbating, published. Masters and Johnson, as a result of their observations, proposed a model, or cycle of sexual responses in men and women which has become a widely accepted framework for subsequent research. Their so-called 'EPOR' model is as follows: (i) An initial *excitement phase* (E) during which stimuli from any source, psychogenic or somatogenic, raise sexual arousal or tension. This sexual arousal becomes intensified in the (ii) *Plateau phase* (P) and subsequently easily reaches a level at which orgasm occurs. If the level of stimulation is inadequate, however, orgasm does not occur and sexual arousal gradually resolves or subsides. (iii) The *orgasmic phase* (O) itself is described as the few seconds of involuntary climax in which sexual tension arising earlier is relieved, usually in an explosive wave of intense pleasure and often accompanied by myotonia. (iv) Lastly, the *resolution phase* (R) when sexual arousal is dissipated and pelvic haemodynamics resolve to the unstimulated state. This occurs in minutes after orgasm but may take up to hours

**Table 8.1.** Composition of semen of man and domestic animals.

| Constituent | Species | Volume (ml) | Approximate concentration | Principal source | Function |
|---|---|---|---|---|---|
| Spermatozoa (no./nl) | Pig | 150–500 | 20–300 | Testis | — |
| | Cow | 2–10 | 300–2000 | | |
| | Dog | 2–15 | 60–300 | | |
| | Man | 2–5 | 50–150 | | |
| | Sheep | 1–2 | 2000–5000 | | |
| | Horse | 30–300 | 30–800 | | |
| Fructose (mg/ml) | Man, sheep, cow | | 1.5+ | Seminal vesicle and ampulla | Anaerobic fructolysis |
| | Pig | | 0.3 | | |
| | Dog, horse | | 0.01 | | |
| Inositol (mg/ml) | Man, cow, horse | | 0.4 | Testis and epididymidal fluid (seminal vesicle—bulls and boar) | Preserves seminal osmolarity? |
| | Pig | | 4 | | |
| | Sheep | | 0.1 | | |
| Citric acid (mg/ml) | Man, pig, sheep | | 0.1–0.3 | Seminal vesicle (stallion, ram, boar, bull) prostate (man and dog) | Ca$^{2+}$ chelator (limits rate of Ca$^{2+}$-dependent coagulation to prevent seminal 'stones'?) |
| | Cow | | 0.6 | | |
| Acid phosphatase (activity/ml) | Man | | 2470 | Prostate | Cleaves choline from glycerylphosphorylcholine for use in phospholipid metabolism |
| | Cow | | 6 | | |
| | Pig | | 2 | | |
| Glycerylphosphorylcholine (mg/ml) | Man | | 0.5 | Epididymis | See above |
| | Sheep | | 18 | | |
| | Cow | | 4 | | |
| | Dog, pig | | 2 | | |
| | Horse | | 38 | | |

pH = 7.2 to 7.8; also contains significant amounts of prostaglandins (seminal vesicle) and zinc + magnesium (prostate).

otherwise. The specific physiological changes occurring during these phases will be discussed below.

In the male an *absolute refractory period* occurs after orgasm during which time sexual re-arousal and orgasm is impossible. Its duration depends on age somewhat, being very abbreviated in boys, and also on a variety of situational factors such as novelty of partner or context. The female is generally said not to experience an absolute refractory period and may, therefore, be repeatedly aroused and orgasmic, although good data on this point are lacking.

i The male

Penile *erection* can be elicited by tactile stimulation of the penis and adjacent perineum. This simple reflex has, as its afferent limb, the internal pudendal nerves but to date its efferent limb cannot be described with confidence. Data from neurological cases, for example cord transection patients, reveal the parasympathetic outflow from S2, 3 and 4 to be essential for erection, but the sympathetic outflow from T12–L3 may also be important. Erection can also be mediated by psychogenic stimuli, such as visual cues and erotic imagery, so emphasizing the important role of events integrated in the brain which influence, by way of projections descending in the spinal cord, somatic efferent and autonomic projections to the genitalia. The recent discovery of vasoactive intestinal polypeptide (VIP) as a putative transmitter in the local ganglia innervating penile vascular tissue suggests that afferents to the penis exert their control over erection via these neurons. Indeed, infusion of VIP intravenously results in erection.

The likely sequence of haemodynamic events underlying the change from flaccidity through tumescence to erection of the penis is as follows. First a decrease in the resistance to arterial pressure and flow is achieved by relaxing the arterial pads on the dorsal artery and the arteries supplying the corpus cavernosum and bulbo-urethral glands (Fig. 8.3). Consequently, the arteries dilate so enabling the arterial pressure head to reach the spaces of the corpus cavernosum. Closure of the arterio-venous shunts immediately prevents draining and then the smooth muscle of the cavernous tissue relaxes, so decreasing resistance to the major increase in blood volume there. Finally, the venous 'bleed' valves close—indeed the veins themselves are compressed by the rapidly developing turgor.

This sequence of events, then, changes the intracavernous space from a low-volume, low-pressure system into a large-volume, high-pressure one. In man, all this is achieved by increasing the *inflow* of arterial blood but not its *through flow*. The corpus spongiosum (Fig. 8.3) does not increase in turgor as much as the corpora cavernosa, thereby avoiding compression of the urethra. Coincident with erection, the testes are

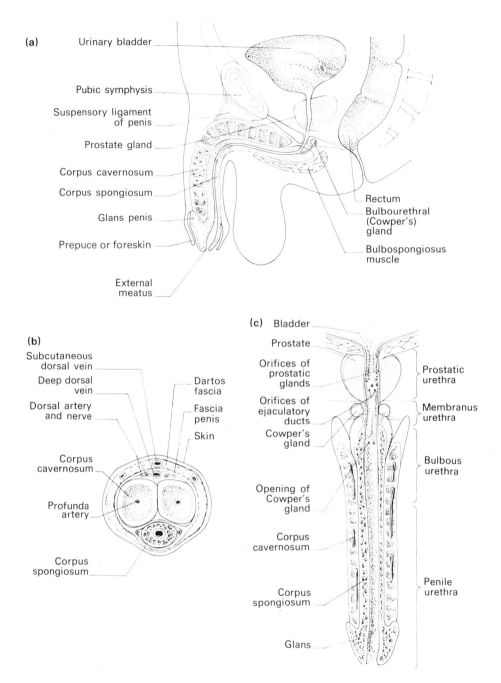

**Fig. 8.3.** Structure of the human penis: (a) mid-sagittal section through pelvis; (b) transverse section through shaft of penis—note fibrous sheaths enclosing corpora which allow generation of hydrostatic pressure required for erection; tears in the fibrous capsule lead to failure of erection; (c) coronal sections through urethra (anterior view) showing entry ducts of prostate, ejaculatory ducts (vasa deferentia + seminal vesicles) and Cowper's glands. Penile secretory glands of Littre open into the roof of the penile urethra. (After Netter.)

drawn reflexly towards the perineum, and the scrotal skin thickens and contracts due to activity in the dartos muscle.

With maintained tactile stimulation, the penile circumference at the coronal ridge increases and drops of fluid, bearing spermatozoa, ooze out of the tip of the penis. The human testes may increase their volume by as much as 50% due to vasocongestion. With further stimulation, a sequence of contractions of the muscles of the prostate, vas deferens and seminal vesicle is induced and the components of the seminal plasma and spermatozoa are expelled into the urethra. This *emission* is mediated largely by sympathetic fibres from T11–L2 via the hypogastric plexus, and administration of drugs that interfere with the *a*-adrenergic system (as in treatment for hypertension) lead to '*dry orgasms*'; erection is not impaired, but emission is. *Ejaculation*, whereby semen is expelled from the posterior urethra, is achieved by contraction of the smooth muscles of the urethra and striated muscles of bulbocavernosus and ischiocavernosus. It is usually associated with contractions of the pelvic floor musculature innervated by the pudendal nerves (S2–4). The passage of semen back into the bladder is normally prevented by contraction of the vesicular urethral sphincter, and failure of this can lead to *retrograde ejaculation* into the bladder.

The composition of the early and late fractions of the human ejaculate reflects the sequential nature of the contractions and the relative lack of mixing of the various seminal components within the urethra. The early fraction is rich in acid phosphatase (prostatic), the mid-fraction in spermatozoa (vas deferens) whilst the late fraction is rich in fructose (seminal vesicle).

Concomitant with penile erection in man, erection of the nipples and increases in heart rate and blood pressure occur as sexual excitement rises. Immediately prior to ejaculation, skin rashes may develop over the epigastrium, chest, face and neck together with involuntary muscle spasms. At ejaculation, the cardiovascular changes, skin rash and muscle spasms intensify and are often accompanied by hyperventilation, contractions of the rectal sphincter and vocalizations. Associated with ejaculation, invariably so far as is known, is *orgasm* whereby sexual tension and arousal are released and an intense sensation of pleasure occurs.

ii The female

At coitus, frictional stimulation of the glans penis is provided by movements against the external genitalia and vaginal walls of the female (Fig. 8.4). The human female undergoes a remarkably similar sequence of reflex responses to that observed in the male and its elicitation is also dependent on tactile and psychogenic stimulation. Tactile stimulation in the perineal region, and in some women particularly on the glans clitoris, provides the primary afferent input reinforced

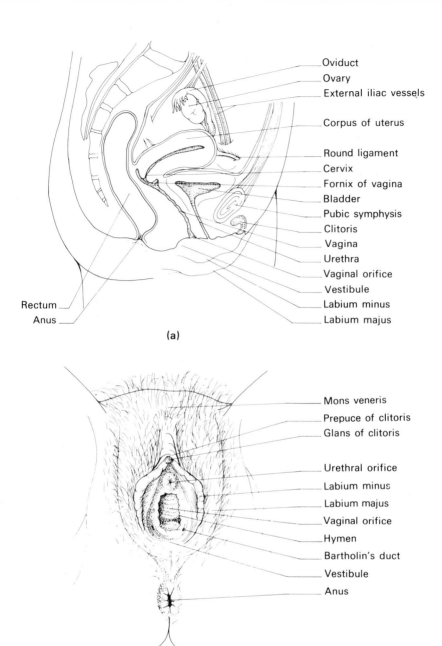

Oviduct

Ovary

External iliac vessels

Corpus of uterus

Round ligament

Cervix

Fornix of vagina

Bladder

Pubic symphysis

Clitoris

Vagina

Urethra

Vaginal orifice

Vestibule

Labium minus

Labium majus

Rectum

Anus

**(a)**

Mons veneris

Prepuce of clitoris

Glans of clitoris

Urethral orifice

Labium minus

Labium majus

Vaginal orifice

Hymen

Bartholin's duct

Vestibule

Anus

**(b)**

**Fig. 8.4.** Structure of human female genitalia; (a) mid-saggittal section through pelvis—note ventroflexure of uterus which tends to lift upwards, moving the cervical os anteriorly in the tenting effect during sexual stimulation. The smooth muscular wall of the vagina is overlain by extensive highly vascularized stroma covered in a stratified squamous epithelium. It is *not* a secretory epithelium— vaginal fluids are derived either by transudation, or from secretions from cervix and Bartholin's glands plus minor vestibular glands at the vaginal vestibule; (b) view of genitalia from exterior. (After Netter.)

by vaginal stimulation after penile penetration. A vascular response may cause engorgement of the corpora of the clitoris as in the male, with consequent *clitoral erection,* although the extent of clitoral involvement varies considerably, being less in women with a small or 'hidden' clitoris. Vaginal lubrication occurs by *transudation* of fluid through the vaginal wall which vasocongests and becomes purplish red, the vagina expands and the labia majora become engorged with blood. With increased stimulation, the width and length of the vagina increase further and the uterus elevates upwards into the false pelvis lifting the cervical os to produce the so-called *tenting* effect in the mid-vaginal plane. At orgasm, frequent vaginal contractions occur and uterine contractions beginning in the fundus spread towards the lower uterine segment. Systemically, the female also may experience increased heart rate and blood pressure and manifest skin flushes, vocalizations and muscle spasms (notably rhythmic contractions of the pelvic striated musculature), and intense sensations of pleasure. Post-orgasmically, clitoral erection is lost, the labia and the vaginal os detumesce and the uterine and vaginal walls relax to their original positions. It has not proved possible to differentiate between the orgasms which follow clitoral or vaginal stimulation in terms of their physiological manifestations, although some reports suggest that differences occur at a behavioural level.

It is perhaps interesting that descriptions of orgasms written by men and women with respect to their own genitalia cannot be identified with any particular sex, suggesting there are no gender-specific reactions at a peripheral level. The time course of sexual responses in the female is generally longer than in the male. Indeed, ejaculation and orgasm are the end result of coitus within minutes in men (an average of 4 min in the western world according to one study). Conversely, relatively few women appear to achieve orgasm from coitus, various surveys giving figures of 30–50%. Larger numbers achieve orgasm by clitoral stimulation, but even then a significant number, 10–20%, do not achieve orgasm despite being highly sexually aroused. This has led to convincing suggestions that orgasm in women is not a reflexive event, as it appears to be in men, but is instead more a learned activity and its attainment a task that must be worked for and concentrated upon. Cross-cultural studies appear to support such a view and suggest that in societies where women are expected to enjoy sex and have orgasms, then they are more likely to do so than in societies where they are not (e.g. in the Arapesh and on Inis Beag, an offshore Irish island).

Factors, social and physical, affecting the capacity to respond sexually to a partner by experiencing orgasm have more than passing interest. It has become clear that the ability to communicate sexually and to give and receive sexual pleasure within a relationship is very often a cornerstone of its success. Efforts to define at least some of those factors are likely to have considerable value when such communication and the relationship itself are becoming poor.

## 2 Transport in the female tract

The semen is ejaculated within the vagina in man, rabbit, sheep, cow and cat, and although much of it is deposited there, spermatozoa are also deposited on the cervical os and perhaps into the cervical canal. In the pig, dog and horse a direct deposition in the cervix and uterus occurs. In many species studied, the *semen coagulates* rapidly during or immediately after deposition in the female tract. The coagulation may become gelatinous (man, pig and horse), fibrous (guinea-pig) or calcareous (mouse) and results from interaction of a substrate with an activated enzyme. In man, coagulating enzymes derived from the prostate interact with a fibrinogen-like substrate derived from the seminal vesicle. It is possible that the coagulum retains spermatozoa in the vagina preventing their physical loss or perhaps buffering them against the hostile acidity of the vaginal fluids (pH 5.7). In some cases of ejaculatory disorder, coagulation takes place within the urethra or (in retrograde ejaculation) within the bladder, and obstruction to urinary flow can result. In normal circumstances in the human female tract, the coagulum is dissolved within 20–60 minutes by progressive activation of a proenzyme derived from the prostatic secretion of the ejaculate.

Within a minute or so of mating in all species studied, spermatozoa can be detected in the cervix or the uterus. It is not at present clear how, or indeed whether, any of the spermatozoa deposited in the vagina actually enter the cervix. Perhaps the ciliated surface of the cervical os wafts them towards the cervical canal. However, in the human over 99% of spermatozoa do not enter the cervix and are lost by leakage from the vagina. The few successful spermatozoa may survive for many hours deep in the cervical crypts of the mucous membrane and are nourished by the mucus (Table 8.2). Further progress to the uterus depends on the consistency of the mucus. Only in the absence of progesterone domination does the mucus permit sperm penetration (see Chapter 7), and even then abnormal spermatozoa are prevented from passing further up the female tract.

Few unequivocal studies have been undertaken on the transport of spermatozoa through the uterus to the oviduct— and indeed, such studies are difficult to undertake. It does seem that neither genital phenomena associated with orgasm in the female nor any major activity of the musculature of the female genital tract is required for effective transport, since

**Table 8.2.** Estimates of the survival of viable fertile gametes in the female genital tract.

| Species | Spermatozoa (hours) | Egg (hours) |
|---|---|---|
| Man | 30–45 | 6–24 |
| Cow | 30–48 | 12–24 |
| Sheep | 30–48 | 15–25 |
| Horse | 140–150 | 15–25 |
| Pig | 25–50 | 10–20 |
| Mouse | 6–12 | 6–15 |
| Rabbit | 30–36 | 6–8 |

although uterine activity is often present in the preorgasmic and orgasmic phases, absence of movement is not incompatible with gamete transport and fertility. Neither is it likely that the prostaglandins present in the semen are required as stimulants to the female tract, since these are absent from semen of many species and also in artificial insemination. Probably the spermatozoa move through the uterus and into the ampulla of the oviduct by their own propulsion and in currents of fluid set up by the action of uterine cilia. Although some sperm are found in the oviducts within minutes of coitus, they are few and often dead. The earliest that living spermatozoa are recovered from the oviduct is 4–7 hours (data for hamster and rabbit). Only 15–25 oviducal living spermatozoa are detected during the early stages of fertilization, and even subsequently the total number present at any one time rarely exceeds 100. It is not at all clear how the flow of sperm to the oviduct is regulated. The cervical crypts may act as a reservoir slowly releasing sperm into the uterus. Additionally the utero-tubal junction may regulate entry to the oviduct by its action as an intermittent sphincter. Having attained the isthmus of the oviduct, spermatozoa tend to linger for a few hours before passing along to the ampulla (Fig. 4.1) and the site of fertilization.

*a Capacitation*

If mature spermatozoa recovered at ejaculation are placed with eggs *in vitro*, fertilization does not occur or at least does so only after a delay of several hours. In contrast, spermatozoa recovered from the uterus or oviduct a few hours after coitus are capable of immediate fertilization of the egg. This observation has lead to the concept of spermatozoa requiring a process of *capacitation* (=attaining a fertilizing capacity) during their journey through the female tract. Capacitation seems to involve the stripping off from the spermatozoal surface of the coat of glycoprotein molecules acquired in the epididymis and seminal plasma. *In vivo*, an oestrogen-primed uterus is optimal for capacitation, but it is possible to mimic

213    *Coitus and Fertilization*

**Fig. 8.5.** (a) Schematized spermatozoon prior to activation; (b) at activation multiple fusions occur between the plasma membrane and the outer acrosomal membrane, first at the tip of the acrosome and at the equatorial region, from which they spread. Concomitant with these changes, the pattern of tail movement changes. The plasma membrane remaining in the equatorial region changes, acquiring the potential to fuse with the plasma membrane of the oocyte.

the conditions *in vitro* by preparing suitable culture media. It seems probable that a specific 'capacitating factor' is not involved, the most critical feature of the capacitating conditions being the proteolytic enzymes and high ionic strength fluid provided by the secretions of the oestrogen-dominated uterus. Under such conditions, the glycoprotein molecules which were adsorbed to the spermatozoal surface during maturation in the epididymis are eluted. Once capacitated, spermatozoa are still not fully competent to fertilize eggs but are capable of undergoing a final change—the process of *activation*.

*b Activation*

Activation is a $Ca^{2+}$-dependent event which involves changes throughout the spermatozoon. Three sorts of change occur at activation. First, the acrosome swells and its membrane fuses at a number of points with the overlying plasma membrane (Fig. 8.5). This process—called the *acrosome reaction*—creates a vesiculated appearance and exposes the contents of the acrosomal vesicle and the inner acrosomal membrane to the exterior. Second, there is a change in the tail movements of the spermatozoa. The regular undulating wave-like flagellar beats are replaced by more episodic, wide amplitude or *'whiplashing' beats* of the tail that push the spermatozoa forwards in vigorous lurches. Third, there is a change in the surface membrane overlying the middle and the posterior half of the spermatozoal head (the equatorial and postnuclear-cap regions). Hitherto, this membrane was unable to fuse with the surface membrane of the egg, but after activation the *membrane becomes 'fusible'*.

The molecular basis underlying the activation changes is not yet established, but a change in the stability of the surface membrane may be critical. It has been proposed that the glycoprotein molecules removed during capacitation had been stabilizing the molecules within the membrane, and that after capacitation molecular movements and interactions occur leading to leakiness and fusibility of the membrane.

Although activation of capacitated spermatozoa *in vitro* does not *require* a specific *activation factor*, there is some evidence to suggest that such a facilitatory molecule may be present *in vivo* since, over the period of fertilization acrosome-reacted spermatozoa tend only to be found in close association with the egg mass. Some workers suggest that a taurine-like molecule in follicular fluid is responsible for activation, whilst others have implicated a glycoprotein of the zona pellucida that not only activates but also binds spermatozoa. It is certainly *important* that the spermatozoa be activated in the close vicinity of the oocyte because once activated their viability is much reduced. Herein probably lies the explanation for the rather complex series of changes in spermatozoa undergone in the female tract. Hours (or even days, in the human) may intervene between deposition of the ejaculate into the female tract and the ovulation of an oocyte. High fertility may be ensured by establishing a reservoir of spermatozoa in the cervix, from which the spermatozoa trickle gradually through the capacitating uterus into the oviduct where they become activated in the presence of an oocyte. If all spermatozoa were fully competent to activate immediately after their release into the female tract, very strict time limits on fertility would operate.

**3 Fertilization**

The ovulated oocyte(s) plus attached cumulus cells is picked up from the peritoneal cavity by the fimbriated ostium of the oviduct (Fig. 4.1), and swept by oviducal cilia along the ampulla towards the junction with the isthmus (Fig. 4.1). Oocyte transport is affected adversely if cumulus cells are lacking and/or if oviducal cilia are malfunctional. However, as in the uterus, there is no evidence that muscular activity in the oviduct is essential for transport *along* the ampulla, although smooth muscle in the meso-salpinx and tubo-ovarian ligaments is involved in the ostial pick-up of oocytes. Furthermore, as we saw in Chapter 7, pharmacological doses of steroids may adversely affect oocyte transport in the oviduct.

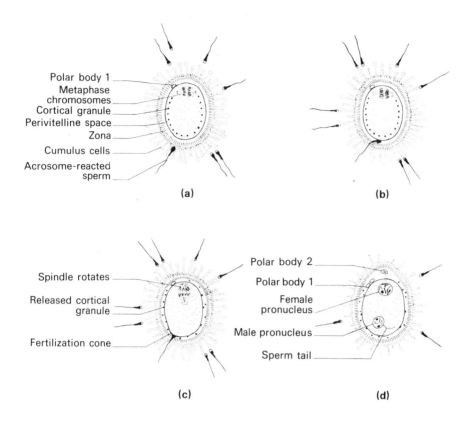

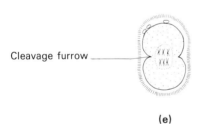

**Fig. 8.6.** (a) Spermatozoa approach ovulated oocyte with its first polar body, chromosomes at second metaphase, cortical granules, and perivitelline space between the plasma (vitelline) membrane of oocyte and the zona pellucida. The cumulus mass is penetrated only by acrosome-reacted spermatozoa. (b) Spermatozoon penetrates the zona pellucida to lie in the perivitelline space, and (c) fuses with the plasma membrane, thus activating release of cortical granules and initiating rotation of spindle. At the point of fusion, the swelling of the fertilization cone occurs. (d) Meiotic division is completed yielding second polar body and female pronucleus. Meanwhile the sperm nucleus decondenses and the male pronucleus forms, sperm tail adjacent. (e) Chromosomes duplicate their DNA, pronuclei migrate together, their membranes break down as the spindle of first mitotic division forms, chromatids separate and move apart, and a cleavage furrow develops. By this stage the cumulus cells have dispersed.

The oocytes and the spermatozoa come together in the ampulla and it is here that the significance of the three events of spermatozoal activation becomes clear as the first phase of fertilization commences.

The acrosome reaction releases the acrosomal contents which include an enzyme *hyaluronidase*. The intercellular matrix of hyaluronic acid which holds the granulosa cells together is digested by the enzyme, thus allowing the spermatozoon to pass towards the zona pellucida (Fig. 8.6a). If inhibitors of hyaluronidase are added to eggs and capacitated spermatozoa, penetration through the cumulus cells does not occur. The spermatozoon attaches now to a specific glycoprotein receptor (of molecular weight *c*. 83 000) on the outer surface of the zona pellucida. A proteolytic proenzyme, *proacrosin*, present on the exposed acrosomal membrane, is activated to yield *acrosin* which digests a path through the zona, along which the spermatozoon passes aided by the whiplash forward propulsion of the activated tail. The spermatozoon comes to lie between the zona and the oocyte membrane in the *perivitelline space* (Fig. 8.6b). The microvilli on the surface of the egg envelop the sperm head, and the sperm membrane in the equatorial region of the head fuses with the surface or *vitelline* membrane of the oocyte. Immediately there is a dramatic cessation of movement by the spermatozoon. This first phase of fertilization, from entry into the cumulus mass to fusion, lasts just 10–20 minutes.

The ensuing events of fertilization last some 20 hours or so and are concerned with two distinctive types of activity. First, the diploid genetic construction of the embryo must be ensured. Second, the developmental programme of the embryo must be initiated.

*a Establishment of diploidy*

The newly fertilized oocyte, now called a *zygote*, confronts two immediate problems. It must prevent further spermatozoa from fertilizing it (the *block to polyspermy*) as this would cause *triploidy* (for one extra sperm) or *polyploidy* (for several). Polyploidy of this sort is described as *androgenetic* since the extra sets of haploid chromosomes are derived from the male. The zygote must also complete its second meiotic division (frozen in second metaphase at ovulation), since failure to jettison one haploid set of female chromosomes would lead to *gynogenetic* triploidy.

Immediately after fusion of the oocyte with the spermatozoon, a series of hyperpolarizing electrical changes are initiated across the vitelline membrane of the zygote and last for several hours. These are due to increases in potassium conductance. A comparable electrical change in fish eggs has been shown to suppress further binding and fusion of spermatozoa, and the same may apply in mammals. Associated with this electrical change is a displacement of bound

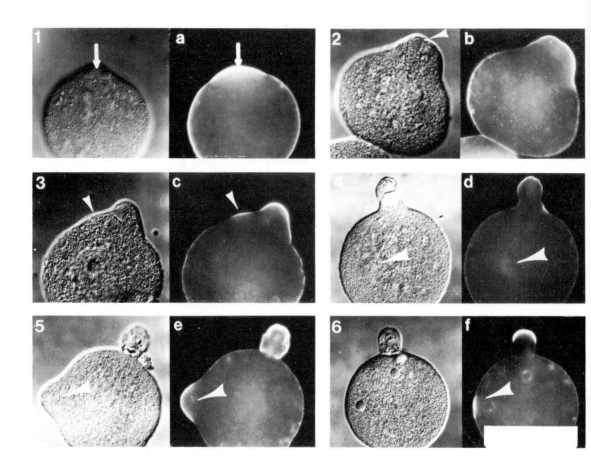

**Fig. 8.7.** Whole mouse eggs at different stages after fertilization examined by differential interference microscopy (1–6) or after visualizing the distribution of actin by use of a fluorescent antibody to it (a–f). (1 & a) The unfertilized egg, note the second metaphase spindle and the heavy staining for submembranous actin that overlies it (arrowed). (2 & b) Within minutes of fertilization the membrane over the spindle indents at the equator of the spindle (arrow) yielding a 'double-bump' structure over the spindle. (3 & c) One of the actin-lined 'bumps' appears to contract down (arrowed) and as a result the spindle rotates. (4 & d) The second 'bump' now starts to restrict at its base leading to the formation of the second polar body. Notice also the fertilizing spermatozoon (arrowed) now being pulled into the egg at the fertilization cone at which actin is evident (arrow). (5 & e) A slightly later stage—the second polar body is now pinched off but the fertilization cone is large and actin-lined (arrow). (6 & f) By about 4 hours male and female pronuclei can be seen clearly. The female pronucleus is smaller and adjacent to the actin-positive second polar body. The male pronucleus is still close to its actin-positive entry point (arrow), but the marked fertilization cone is now gone. (Scale bar in panel f = 60 $\mu$m.) (Courtesy B. Maro, M.H. Johnson, S. Pickering & G. Flach.)

Ca$^{2+}$ into the cytoplasm of the ovum just underlying the surface membrane. This displacement results in fusion of the cortical granules in the cortex of the zygote with the surface vitelline membrane, thus releasing their contents into the perivitelline space (the *cortical reaction*) (Fig. 8.6c). Amongst the contents of these granules are (in many mammalian species) enzymes which act on the zona pellucida to prevent, or impair, further penetration by spermatozoa (the *zona reaction*). In addition, fusion of the granules in some species also effects a change in the composition and properties of the vitelline membrane itself. Spermatozoa will no longer bind to and fuse with its surface. These events occurring rapidly within minutes of fusion, reduce the chances of androgenetic polyploidy.

The avoidance of gynogenetic triploidy takes longer to achieve. Within 2–3 hours after fertilization, meiotic metaphase is reactivated and the second polar body is extruded (Figs 8.6c, d and 8.8). The second metaphase spindle of the unfertilized mouse egg lies just under and parallel to the surface (Figs 8.6a and 8.7a). Sperm penetration activates *actin-containing microfilaments* that lie between the spindle and the surface (Fig. 8.7b–d) to *rotate the spindle* ensuring that on completion of meiosis, cleavage generates a large zygote and a small second polar body. However, in the human and the sheep, rotation is not required, the egg being ovulated with the spindle perpendicular to the surface. Actin-containing microfilaments also seem to be important in the incorporation of the sperm nucleus itself at the site of the *fertilization cone* (Figs 8.6c and 8.7d–f). It is of interest to note that spermatozoa do not bind to the oocyte vitelline membrane immediately overlying the second metaphase spindle, probably due to the absence of microvilli in this region (Fig. 8.8a). This makes sense biologically since the coincidence of a spermatozoon entering the oocyte and a second polar body being ejected could lead to mutual interference and *aneuploidy* (deviation from the normal *euploid* chromosomal number).

The successful attainment of one maternal and one paternal set of haploid chromosomes within the zygote is critical for the success of the conceptus. Aneuploidy causes embryonic abnormality. Fortunately most aneuploid conceptuses die relatively early, although a few survive to term and beyond. Others may develop into tumours such as *hydatidiform mole*, a benign trophoblastic tumour, or *choriocarcinoma*, its malignant derivative. Thus aneuploidy, apart from being of potential danger to the mother, is reproductively wasteful and distressing. Failure of the normal mechanisms for establishing euploidy is increased under certain conditions, for example, exposure of females to alcohol, certain drugs or anaesthetics around the time of ovulation seems to interfere

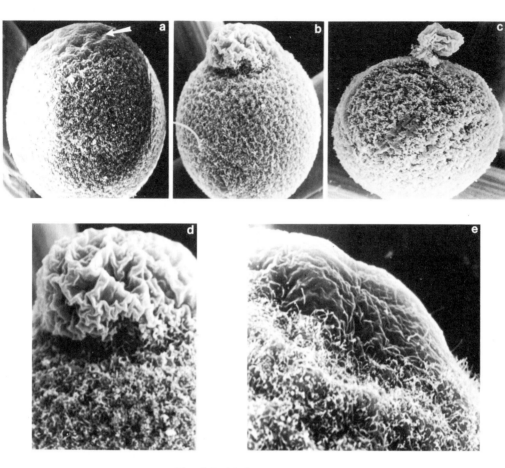

**Fig. 8.8.** (a) Scanning electron micrograph of a newly ovulated oocyte from which zona pellucida has been removed. Note absence of microvilli at one end (arrowed) marking site of metaphase spindle, as shown in (b) where polar body extrusion is underway after fertilization (sperm tail evident); note small size and absence of microvilli on polar body (c). (d, e) show higher power pictures of b & a in the region demarcating the microvillous and non-microvillous zones.

with extrusion of the second polar body. The fertilization of prematurely ovulated oocytes (see Chapter 4) is associated with defective cortical granule release, poor sperm head incorporation and a failure to properly constitute the male pronucleus. The fertilization of 'aged' oocytes that have been ovulated and passed into the oviduct hours before sperm arrival can also be problematic (Table 8.2). In most mammalian species in which ovulation and behavioural oestrus are closely synchronized the problem of ageing oocytes will not be acute. In humans, however, where coitus and ovulation are not necessarily, or even usually, associated, the fertilization of old oocytes may be much commoner. Could the high incidence of genetic abnormality (50% or more of all conceptions) reported for human embryo-fetuses reflect a genetic price to be paid for the increased social and sexual freedoms derived by dissociating mating from fertility or from the use of stimulants such as alcohol? (See Chapter 14.)

*b Initiation of development— ovum activation*

Within 2–3 hours of oocyte–sperm fusion, the second polar body has been expelled and the remaining haploid set of female chromosomes lies in the ooplasm. Over the same period the cytoplasmic contents of the sperm cell membrane (now fused with the vitelline membrane) have passed into the ovum cytoplasm. The sperm nuclear membrane breaks down and over the next 2–3 hours the highly condensed chromatin starts to swell releasing filamentous strands of chromatin into the cytoplasm. The protamines that created such compressed chromatin are released and replaced by normal histones. This decondensation is actively induced by factors in the ovum cytoplasm that develop in the terminal phases of intrafollicular oocyte maturation (Chapter 4), and is probably one part of a sequence of changes to chromatin structure that makes subsequent expression of the male-derived genes possible. This 'reprogramming' of the male chromatin appears to make it particularly susceptible to the insertion of extra genetic material, either by viral infection or by the new experimental technique of *gene injection*. Copies of isolated and purified genes can be incorporated into the DNA, where they replicate together with the host DNA, thereby becoming distributed in all cells of the body. Such genes can be expressed later in the life of the fetus or adult. This technique raises a number of possibilities, including *gene therapy* in cases of genetic disorders.

Between 4 to 7 hours after fusion, the two sets of haploid chromosomes each become surrounded by distinct membranes and are now known as *pronuclei* (Figs 8.6d, e, and 8.7f). The male is usually the larger of the two. Both pronuclei contain several nucleoli. During the next few hours, each pronucleus gradually moves from its subcortical position to a more central and adjacent cytoplasmic position. During this period, the haploid chromosomes synthesize DNA in preparation for the first mitotic division which occurs about 18–21 hours after gamete fusion. The pronuclear membranes around the reduplicated sets of parental chromosomes break down (Fig. 8.6e), the metaphase spindle forms and the chromosomes assume their positions at its equator. The final phase of fertilization has been achieved, *syngamy* (or coming together of the gametic chromosomes) has occurred. Immediately anaphase and telophase are completed, the cleavage furrow forms and the one-cell zygote becomes a two-cell conceptus.

During this whole 18–22-hour period, there is no evidence to suggest that the chromosomes within the one-cell zygote are synthesizing significant amounts of mRNA. The whole process appears to be regulated by information laid down in the cytoplasm during oocyte maturation—the so-called *maternal cytoplasmic inheritance* of the conceptus. The utilization of this inheritance involves the selective translation of different populations of maternal mRNA as well as the post-translational modification of maternally-derived proteins by, for example, phosphorylation and glycosylation. How this complex sequence of molecular events is regulated is at present not clear. However, from the early two-cell stage onwards, the conceptus starts to express its own genes, and thereafter the importance of maternally-derived, non-genetic information rapidly diminishes. Increasingly, the conceptus shapes and controls its own future. It faces a hazardous few days. Somehow this tiny structure, only 70–150 $\mu$m in diameter, must alert the mother to its presence. Failure to do so would result in a continuing maternal cyclicity with loss of the conceptus (with the menses in the human). The way in which the conceptus achieves this task and avoids an early abortion is quite remarkable, and the subject of the next chapters.

*c Parthenogenetic activation*

Although the spermatozoon induces many remarkable changes in the oocyte, it does not appear to be essential for many of them! An oocyte may be *activated parthenogenetically* by a variety of bizarre stimuli such as electric shock, exposure to various enzymes or to alcohol. It seems that these stimuli in some way mimic some component of the spermatozoon, and as a result the oocyte's development programme is activated and it may undergo cleavage, implantation, and development to a stage where a beating heart, somites and forelimbs are present. However, most parthenogenetic conceptuses die fairly early on in this sequence, and the main reason for their death seems to be a deficiency in their placentae rather than in the fetus itself.

**4 Summary**

The interval between the departure of the gametes from the gonads and the successful formation of a two-cell conceptus encompasses an extraordinarily complex sequence of events involving biochemical, behavioural, endocrine, physiological and genetic components. Not surprisingly, aspects of this process often go wrong and infertility results (see Chapter 14). Fortunately, recent biomedical advances mean that most of these events can now be carried out *in vitro*. Mature oocytes can be aspirated directly from the follicles, ejaculated spermatozoa can be capacitated and activated in defined media, fertilization and the formation of a conceptus can be achieved outside the body. These techniques will provide therapy for many otherwise infertile couples. Considerable controversy has surrounded the ethics of use of these procedures; in particular the decisive role that fusion of oocyte and spermatozoon might play in the establishment of a human life has been stressed. It is important to be aware that scientific evidence does not support the view that one single event is decisive for the creation of an individual life. The process is a continuum which starts with the growth of the oocyte and the synthesis of the maternal cytoplasmic inheritance, and continues with the formation and development of the zygote, the transfer of developmental control from maternally-inherited to embryonically-derived genetic information, the signalling by the conceptus to the mother of its presence, and the subsequent acquisition of a distinctive human form capable ultimately of independent existence. The natural losses in mammals are massive over the early events in this sequence, particularly from oocyte atresia and preimplantation death. Human preimplantation losses are increased further by our changed social and sexual habits and the use of various forms of contraceptive. Biology cannot provide the answer to the question 'when does a life begin?' because the question is based on false premises. Ethical decisions cannot therefore be taken on the basis of biological observation alone.

**Further reading**

American Physiological Society, *Handbook of Physiology*. Section 7, Endocrinology Vol. V. Male Reproductive System. American Physiological Society, 1975.

Austin CR. *Ultrastructure of Fertilization*. Holt, Rinehart and Winston Inc., 1968.

Brancroft J. *Human Sexuality and Its Problems*. Churchill Livingstone, 1983.

Edwards RG. *Conception in the Human Female*. Academic Press, 1982.

Gwatkin RBL. *Fertilization Mechanisms in Man and Mammals*. Plenum Press, 1974.

Harper MJK *et al*. *Ovum Transport and Fertility Regulation*. Copenhagen, Scriptor, 1976.

*Human Embryo Research: Yes or No?* Tavistock Publications, 1986.

Kaufman MH. *Parthenogenesis*. Progress in Anatomy I, I–57, 1981.

Masters WH, Johnson VE. *Human Sexual Response* (1966) and *Human Sexual Inadequacy* (1970). Churchill, London.

Money J, Musaph H (Eds). *Handbook of Sexology*. Elsevier, 1977.

Setchell BP. *The Mammalian Testis*. Paul Elek, 1978.

Surani MAH, Reik W, Norris ML & Barton SC. Influence of germ line modifications of homologous chromosomes on mouse development. *J Embryol Exp Morph* 1986; **97**: (Suppl) 123–136.

Wagner G, Green R. *Impotence: Physiological, Psychological, Social Diagnosis and Treatment*. Plenum Press, 1981.

# Chapter 9
# Implantation and the Establishment of the Placenta

The conceptus remains at the oviducal site of fertilization for a further few days (Table 9.1, column 2). It is then transferred through the isthmus of the oviduct to the uterus. Transfer is facilitated by the changing endocrine milieu of the early luteal phase with its rising ratio of progesterone to oestrogens, which affects the oviducal and uterine musculature and relaxes the isthmic sphincter (see Chapter 7). It is probable, however, that the cilia rather than the musculature of the genital tract are the primary active transporters of the conceptus. Thus, pharmacological inhibition of oviducal muscle does not prevent transfer of the conceptus; furthermore, if a segment of oviduct is excised, turned round and replaced such that its cilia beat *away* from the uterus, the conceptus moves only up to the point of oviducal reversal and then stops.

On reaching the uterus the conceptus engages in an elaborate interaction with the mother in which several messages are transmitted in both directions. This interaction, which may be likened to a conversation, has two important components. First, the conceptus establishes physical and nutritional contact with the maternal endometrium at implantation; failure to do so properly would deprive the conceptus of essential nutritional substrates and arrest its growth. The conversation operating during implantation involves short-range signals, and is the major subject of this chapter. Second, the conceptus makes the maternal pituitary-ovarian axis aware of its presence; failure to do so would result in the normal mechanisms of luteal regression coming into operation, causing progesterone levels to fall and loss of the conceptus. Somehow the conceptus must convert the whole of the female's reproductive system from a *cyclic* pattern, in

**Table 9.1.** Times (in days) after ovulation at which various developmental and maternal events occur.

| Species | Cleavage to 4 cells | Conceptus enters uterus | Formation of blastocyst | Time of attachment | Expected time of luteal regression if mating infertile | Duration of pregnancy |
|---|---|---|---|---|---|---|
| *Invasive* | | | | | | |
| Mouse | 1.5–2 | 3 | 3 | 4.5 | 10–21 | 21 |
| Rat | 2–3 | 3 | 4.5 | 4.5–5.5 | 10–12 | 22 |
| Rabbit | 1–1.5 | 3.5 | 3.5 | 7–8 | 12 | 28–30 |
| Man | 2 | 3.5 | 4 | 6(?) | 12–14 | 270 |
| *Non-invasive* | | | | | | |
| Sheep | 4 | 2–4 | 6–7 | 16 | 16–18 | 145–150 |
| Pig | 1–3 | 2–4 | 5–6 | 18 | 16–18 | 114 |
| Cow | 2–3 | 3–4 | 7–8 | 30–45 | 18–20 | 282 |
| Horse | 1.5–2 | 4–5 | 6 | 30–40 | 20–21 | 340 |

which oscillating dominance of oestrogens and progestagens occurs, to a *non-cyclic, pregnant* pattern, in which progestagens dominate throughout. The conversation operating during this *maternal recognition of pregnancy* involves long-range signals, and is the major subject of Chapter 10. Thus, *pregnancy is not initiated with fertilization* but *only when the conceptus has signalled its presence successfully to the mother.*

## 1 Growth of the preimplantation conceptus

During its period in the oviduct, the conceptus continues its cellular division at a rate characteristic for each species (Table 9.1). Each cell or *blastomere* undergoes a series of divisions called *cleavage*, during which the total size of the conceptus remains much the same (Fig. 9.1). Thus, with each cleavage division, the size of the individual blastomeres is progressively reduced, restoring to normal the exaggerated cytoplasmic: nuclear ratio of the oocyte. The large volume of oocyte cytoplasm contains essential materials for the progress through cleavage. However, it is certain that from at least as early as the two-cell stage in the mouse, and probably from the four- to eight-cell stage in the human and large farm animals, the genes of the conceptus itself start to contribute to development. At around the four- to 16-cell stage in most species studied, two other events occur. First, the *cleaving* conceptus changes its morphology by undergoing the process of *compaction* to yield a *morula* (Figs 9.1–3). Second, the conceptus shows a marked quantitative increase in its biosynthetic capacity. Net synthesis of RNA and protein increases, transport of amino acids and nucleotides into the cells rises and changes occur in the synthetic patterns of phospholipids and cholesterol. Shortly after these events, at about the 32- to 64-cell stage in most species, a further morphological change

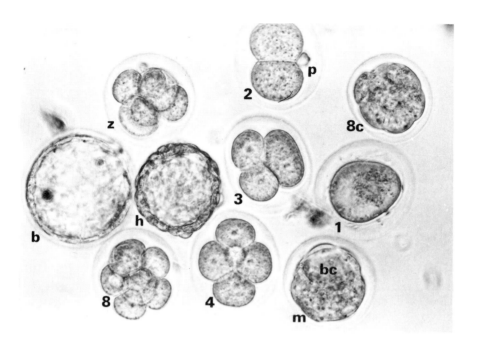

**Fig. 9.1.** Photographs of various pre-implantation mouse stages. (1) Unfertilized egg surrounded by spermatozoa; (2) two-cell with polar body (p); (3) three-cell; (4) four-cell; and (8) eight-cell cleavage stages; (8c) eight-cell compacted morula; (m) 32-cell late morula/ early blastocyst with beginning of blastocoelic cavity (bc); (b) fully expanded blastocyst with inner cell mass (ICM) and trophectoderm; (h) blastocyst which has lost its zona pellucida (z). (Courtesy of Dr P.R. Braude.)

occurs with the transition of the morula to a *blastocyst* (Figs 9.1 and 9.2). The precise form of the blastocyst varies in different species, but in all species the blastocyst contains two distinctive types of cell, an outer rim of *trophectoderm* cells surrounding a *blastocoelic cavity* containing *blastocoelic fluid* and an *inner cell mass* (ICM) which is eccentrically placed within the blastocoelic cavity against the trophectoderm. The trophectoderm cells constitute the first so-called *extra-embryonic* tissue since they do not contribute to the embryo or fetus proper. Instead, they give rise to part of an accessory fetal membrane—the *trophoblast* of the *chorion* which is concerned with the nutrition and support of the fetus. The first 14–18 days of human development are concerned in large part with the further elaboration within the conceptus of various extra-embryonic tissues (Table 9.2 and Fig. 9.7) and only *after* this time can distinct and separate tissues be defined that will give rise *exclusively* either to a single fetus or to extra-embryonic supporting tissues. Thus, only after this time is it appropriate to refer to an embryo (i.e. that which gives rise to a fetus and then a baby), and prior to this stage the total product of

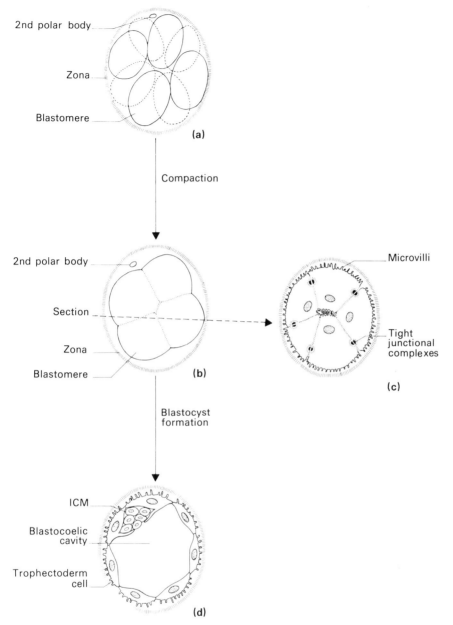

**Fig. 9.2.** (a–c) Compaction: spherical cells become wedge-shaped and, by apposing adjacent surfaces, maximize cell contact; tight junctional complexes develop between outer membranes of adjacent cells—these are punctate at first, but later become zonular forming a barrier to intercellular diffusion between the inside and outside of the conceptus; each cell also becomes polarized, the nucleus occupying a more basal position, microvilli being continuous over the outer surface and at points of contact with other cells basally. (d) Section through a 34- to 64-cell blastocyst; fluid accumulation within the blastocoelic cavity is possible since the tight junctional complexes between adjacent trophectoderm cells prevent its escape (see also Fig. 9.3).

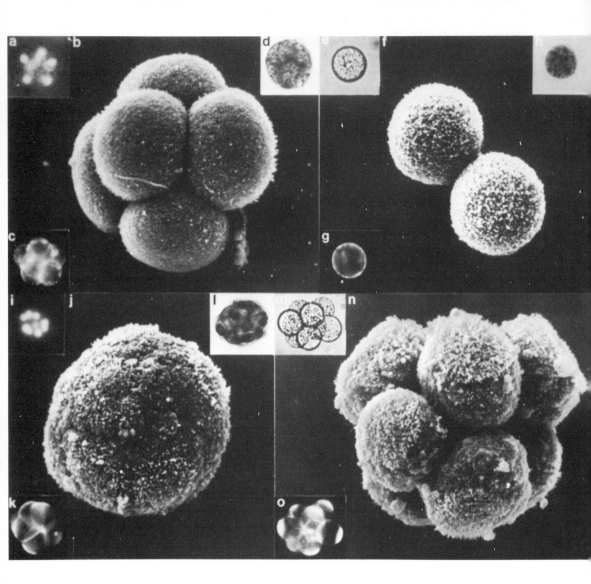

**Fig. 9.3.** The eight-cell mouse conceptus. (a-d) show intact *early* eight-cell mouse stages. Note that individual blastomeres are spherical and distinctly separated from each other whether observed (a) after staining of nuclei, (b) with the scanning electron microscope, (c) after labelling with a fluorescent surface marker, or (d) after they have been allowed to ingest a black dye (speckles in cytoplasm). The spherical form and the symmetry of the cells is shown further when isolated blastomeres are viewed in the same ways (e–h). In contrast, a late eight-cell conceptus has 'compacted' (i–l); this involves flattening of cells on each other so that clear cell outlines cannot be resolved (j). Moreover, the nuclei cluster centrally (i), and the ingested black dye is accumulated not throughout the cytoplasm but is concentrated just above the nucleus and just below the outward facing membrane (l). If this surface membrane is examined more closely it is shown to concentrate binding of the fluorescent surface marker (k). This is observed more readily if the flattening component of compaction is reversed by briefly incubating the conceptus in medium low in $Ca^{2+}$; the cells pull apart (m–o) revealing that they are no longer spherical and radially symmetrical but are highly polarized with a tuft of microvilli facing outwards that stains for fluorescence (n–o). Thus, at compaction, the cells not only flatten on each other but also become epithelial-like as judged by surface morphology, basal or central nuclei, and peripheral concentration of ingested dye.

**Table 9.2.** Derivatives of a typical blastocyst.

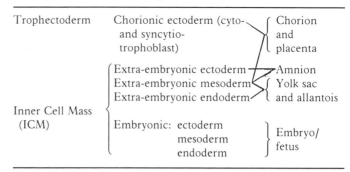

| Trophectoderm | Chorionic ectoderm (cyto- and syncytio- trophoblast) | Chorion and placenta |
| --- | --- | --- |
| | Extra-embryonic ectoderm | Amnion |
| | Extra-embryonic mesoderm | Yolk sac |
| | Extra-embryonic endoderm | and allantois |
| Inner Cell Mass (ICM) | Embryonic: ectoderm<br>mesoderm<br>endoderm | Embryo/ fetus |

fertilization is properly called the conceptus (also pre-embryo or pro-embryo).

Throughout development from fertilization to the blastocyst, the conceptus remains enclosed within the zona pellucida. The zona probably has two functions. First, it prevents the blastomeres of the conceptus falling apart during early cleavage prior to compaction. If the conceptus does become divided into two distinct groups of cells at this stage, *identical* twins result (*monozygotic*—derived from one zygote and therefore genetically identical rather than *dizygotic* twins which are derived from the independent fertilization of two eggs). Second, and conversely, it prevents two genetically distinct conceptuses from sticking together to make a single *chimaeric* conceptus composed of two sets of cells each of distinct genotype (see Chapter 1, p.11). It is during the transition from a morula to a blastocyst that the conceptus enters the uterus (Table 9.1), and it is, therefore, the blastocyst that engages in the conversations vital to its survival and future development. First we will consider how the blastocyst and the uterine endometrium interact locally with one another to effect implantation.

## 2 Implantation

The free-living blastocyst is bathed in the secretions of the uterus and draws from them the oxygen and metabolic substrates that it requires for continued growth and survival. It actively accumulates organic molecules and ions by specific, cellular transport mechanisms, whilst the exchange of oxygen and carbon dioxide is diffusional. There is a limit to the size that a free-living conceptus can attain before such exchanges become inadequate. Before this critical stage is reached, the conceptus develops its own blood vascular system that distributes the essential metabolites being taken up at its extra-embryonic surface throughout the tissues of the growing conceptus. Conceptuses also develop a distinct and highly vascularized region of their extraembryonic surface through which the interchange of materials with maternal tissues is particularly facilitated. This region draws on two major maternal sources of nutrition. During the initial implantation stages of pregnancy, it continues to utilize almost exclusively secreted maternal 'tissue juices' of various sorts (see later for details). However, these become progressively inadequate in most species, and so with the successful conclusion of implantation there develops in the maternal endometrial tissue a corresponding specialized and vascularized region. This zone of adjacent and highly vascularized contact between mother and conceptus is called the *placenta.* In the placenta the two discrete circulations lie sufficiently close that rapid and efficient transfer of materials can occur between them. The placenta, then, is the ultimate outcome of the primary interactions between mother and conceptus which occur at implantation and *the form that the mature placenta takes is determined largely by the pattern and timing of the events of implantation.*

On entering the uterus, the conceptus is positioned for implanting. The location of the implantation sites within the uterus at which maternal tissue and conceptus make intimate physical contact tends to be characteristic for each species. There is evidence that the activity of uterine musculature is important for the movement and final location of the conceptus, since its inhibition leads to abnormal sites of implantation. Moreover, in ruminants there are distinct areas of projecting aglandular uterine mucosa called *caruncles,* that are the only sites at which close attachment of conceptus tissue occurs. The appropriate location and, in *polytochous* species having several conceptuses, spacing of implantation sites is important in minimizing physical and nutritional competition between conceptuses.

At the site of implantation the blastocyst makes contact with the uterine tissue. Interposed between the outer layers of trophectodermal cells and the surface epithelium of the uterine endometrium is the zona pellucida. According to species, this barrier is removed in one of two different ways. In many species some of the trophectodermal cells have a proteolytic activity that locally digests a pathway through the zona pellucida, thus permitting passage of trophectodermal cell processes. In some species (for example, the mouse and rat), the uterine secretions also contain a proteolytic enzyme that causes complete dissolution of the zona pellucida, exposing directly the whole trophectodermal surface to the uterine epithelium. Thus exposed, the trophectoderm and the luminal epithelium establish physical contact and become firmly adherent. This process, which involves close apposition of cells with an interlocking of microvilli, is called *attachment.* Although the consequences of attachment vary considerably in detail among different species, there is a common feature that seems to link all these variations. The initial close, local contact between trophectoderm and uterine epithelium induces vascularization and differentiation in the underlying endometrial stromal tissue that lead to its establishment as the maternal component of the placenta.

The events subsequent to attachment vary with species. In some, implantation is classified as *invasive* in which the conceptus breaks through the surface epithelium and invades extensively the underlying stroma. In other species implantation is classified as *non-invasive* in which epithelial integrity is retained (or at least is only breached locally, transiently or much later in gestation—see later) and the epithelium becomes incorporated within the placenta. In invasive implantation, tissues of the conceptus may erode not just the maternal epithelium but also the stromal cells, connective tissue and, in man, even the blood-vessel walls. The stromal response to invasive implantation seems to anticipate the event by developing protective devices which limit and control the depth of invasion. In non-invasive implantation, it is the uterine epithelium and not the underlying stroma which limits and controls the invasion of the conceptus.

*a Invasive implantation (man and all primates except lemurs and lorises, dog, cat, mouse, rabbit)*

The blastocysts of these species are nurtured at first by uterine secretions but, compared with species in which implantation is non-invasive, this free-living phase is relatively short-lived (Table 9.1, column 4). In consequence, invasive conceptuses tend to be smaller at attachment (see below). When intimate contact is made with the maternal epithelium, only relatively few trophectodermal cells are involved. Yet within an hour or so it is evident that the stromal tissue underlying the epithelium is affected by the contact and, moreover, this effect is observable over an extensive area of endometrium. Thus, a *primary signal* from the conceptus has been effectively *transferred* through the epithelium to the stroma and then *amplified*. One of the earliest of the visible responses to this event is an increased vascular permeability in the stromal tissue underlying the conceptus, followed by oedema and compositional changes in the intercellular matrix, changes in the morphology of the stromal cells and a progressive sprouting and ingrowth of capillaries (Fig. 9.4). The stromal reaction set in motion by the blastocyst signal is particularly marked in primates and rodents where it is called *decidualization*, and it prepares the major endometrial component of the placenta — the *decidua*. In other species the reaction may be less marked but is functionally equivalent to the decidua. It is clear that the rapid progression of changes in the endometrium underlying the attachment site is organized and stimulated by a message derived from the blastocyst. The nature of that message has not been established. It could be a low molecular weight secretory product of the blastocyst itself

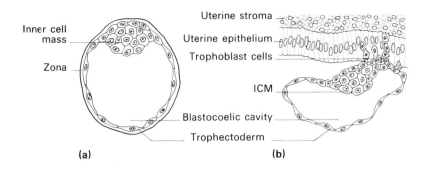

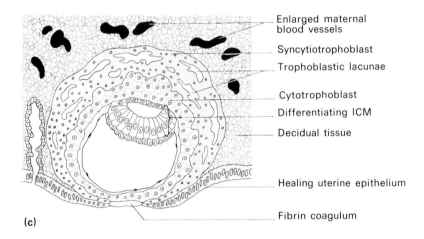

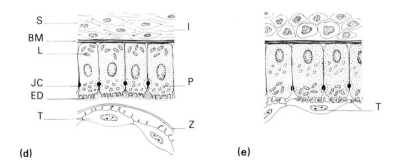

**Fig. 9.4.** Scheme of conceptus uterine relations during invasive implantation (human). (a) 4–5 days: free-living blastocyst within zona pellucida; (b) 6 days: zona is lost and trophectoderm (overlying ICM) transforms to trophoblastic tissue, attaches to epithelial cells, and inducing changes in these and underlying stromal tissue, then starts to penetrate epithelium; (c) 9 days: decidualization in underlying stromal tissue spreads out rapidly from attachment site, trophoblast erodes surface epithelium, invades and destroys adjacent

that diffuses rapidly through the endometrial tissues, but more probably the blastocyst stimulates the adjacent uterine tissue to secrete a 'decidualizing molecule' which then spreads through the stroma. Thus, non-specific stimuli to the endometrial tissue can, at this stage and in the complete absence of a blastocyst, set off a full decidualization response (a so-called *deciduoma*). The nature of the decidualizing molecule(s) has been the subject of considerable experimental work, and evidence is available which suggests that both histamine and prostaglandins are involved. Both molecules are present in the progestational endometrium, and will induce decidualization if injected locally. Moreover, decidualization is reduced considerably in the presence of inhibitors of histamine and prostaglandin. In the rat (and probably also the human) a maximal decidual response is only elicited by the conceptus if the endometrial epithelium and stroma have been primed with progesterone and subsequently exposed to elevated oestrogen levels. Such a hormonal regime occurs normally in the luteal phase and causes exquisite sensitivity and responsiveness of the endometrium to the implantation signals.

Within a few hours of the initiation of implantation, the surface epithelium underlying the conceptus becomes eroded (Fig. 9.4) and trophectodermal processes seem to 'flow'

---

decidual tissue and becomes embedded. Surface epithelium heals over; (d) and (e) details of early phase. (d) Part of blastocyst in zona (Z) with microvillous trophectoderm (T) lies free in uterine lumen adjacent to epithelial cells linked by zonular junctional complexes (JC), with surface microvilli interdigitating with a thick surface secretion which is electron dense (ED). Within epithelial cells are pinocytotic vesicles (P), central nuclei and basal lipid droplets and mitochondria (L). The cells rest on a basement membrane (BM) underlain by condensed fibrous connective tissue. Stromal cells (S) are spindle-shaped, and lie in an extensive extracellular matrix (I) abundant in collagen fibres. (e) At attachment, zona has been lost, microvilli become flat and interdigitate; trophoblast and epithelial cells become very closely apposed. Epithelial nuclei assume basal position and mitochondria are apical, with dispersed fat droplets in between. In stroma, oedema and increased vascular permeability are accompanied by loss of collagen fibres and swelling of stromal cells, which subsequently develop extensive endoplasmic reticulum, polysomes, enlarged nucleoli, lysosomes, glycogen granules and lipid droplets and become 'decidual cells'. In the rat, extensive intercellular gap junctions are also seen. Nuclei frequently become polyploid. Peripherally, sprouting and ingrowth of maternal blood vessels occurs. Note: in the human endometrium, some decidual-like changes may occur in stromal cells in the absence of a conceptus during the late luteal phase. These changes are often called 'predecidualization' ((b) and (c) after Hamilton, Boyd and Mossman).

**Table 9.3.** Classification of implantation and placental forms in several species.

| Species | Depth of invasion | Extent and shape of attachment (see Fig. 9.5) | Maternal tissue in contact with conceptus (see Fig. 9.6) | No. of layers of chorionic trophoblast (see Fig. 9.6) | Histological classification (see Fig. 9.6) |
|---------|-------------------|-----------------------------------------------|----------------------------------------------------------|-------------------------------------------------------|--------------------------------------------|
| *Invasive* | | | | | |
| Man | Interstitial | Discoid | Blood | 1 | Haemomonochorial |
| Rabbit | Eccentric | Discoid | Blood | 2 | Haemodichorial |
| Rat/mouse | Eccentric | Discoid | Blood | 3 | Haemotrichorial |
| Rhesus monkey | Eccentric | Bi-discoid | Blood | 1 | Haemomonochorial |
| Dog | Eccentric | Zonary | Capillary endothelium | 1 | Endotheliochorial |
| Cat | Eccentric | Zonary | Capillary endothelium | 1 | Endotheliochorial |
| *Non-invasive* | | | | | |
| Sheep | Central | Cotyledonary | Epithelium | 1 | Epitheliochorial |
| Pig | Central | Diffuse | Epithelium | 1 | Epitheliochorial |
| Cow | Central | Cotyledonary | Epithelium | 1 | Epitheliochorial |
| Horse | Central | Diffuse | Epithelium | 1 | Epitheliochorial |

between adjacent epithelial cells, isolating and then dissolving and digesting them. Some trophectodermal cells fuse together and form a syncytium (*syncytiotrophoblast*), whilst others retain their cellularity (*cytotrophoblast*) and may serve as a proliferative source for generating more trophoblastic cells. The uterine glandular tissue and the decidual tissue most adjacent to the invading trophoblastic front of the conceptus is destroyed. It releases massive quantities of primary metabolic substrates (lipids, carbohydrates, nucleic acids and proteins) which are taken up by the conceptus and passed to the adjacent developing blood vascular system for distribution to the growing embryo. The decidual tissue thus functions as a large 'yolk reservoir' equivalent to the yolk of a bird's egg. Since the mammalian conceptus does not leave the female tract to develop, it does not carry its yolk with it and the mother only makes the yolk-equivalent (decidualization) if a pregnancy has occurred.

The depth to which the conceptus invades maternal tissues varies with species and also in certain pathological conditions. Those species in which the conceptus invades the stroma so deeply that the surface epithelium becomes restored over it are said to implant *interstitially* (man, chimpanzee, guinea-pig; Table 9.3). In other species the stroma is only partially invaded and the conceptus continues to project to varying degrees into the uterine lumen—so called *eccentric* implantation (rhesus monkey, dog, cat, rat). Indeed, in these species secondary contact and attachment by the conceptus to the

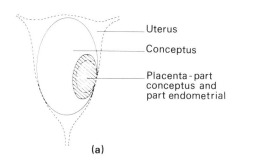

Uterus

Conceptus

Placenta - part conceptus and part endometrial

(a)

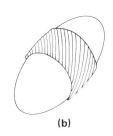

(b)

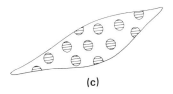

(c)

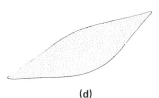

(d)

**Fig. 9.5.** Highly schematic representation of placental morphologies. (a) A discoid human placenta. The egg-shaped sac is the conceptus which has a specialized disc on its surface which forms the placental component. A corresponding area on the endometrium also contributes to the placenta. Discoid placentae are also found in the mouse, rat, rabbit, bat and insectivores. The uterine part is not shown in: (b) zonary (cat, dog, bears, mink, seals, elephant); (c) cotyledonary (cow, sheep, giraffe, deer, goat); and (d) diffuse (horse, pig, camel, lemur, mole, whales, dolphins, kangaroos) placentae.

uterine epithelium on the opposite side of the uterine lumen may result in two sites of placental development—a *bi-discoid* placenta (e.g. rhesus monkey, see Table 9.3)—or a complete 'belt' of placenta round the 'waist' of the conceptus—a *zonary* placenta (e.g. dog, cat; see Table 9.3 and Fig. 9.5b).

The depth of invasion by the conceptus is also influenced by the degree of decidual response. Where the decidual response is inadequate, either pathologically in the uterus, or after implantation at non-uterine *ectopic sites,* invasion is much more aggressive and penetrating. This observation has been taken by some authors to suggest a 'restraining' influence of the decidual tissue over the conceptus, but it could reflect an attempt by the invading trophoblast to overcome the inadequate nutritional support of a defective decidual response. There are also species differences in the degree of proximity ultimately established between the circulations of the invasive conceptus and the mother, as well as in the physical depth of penetration. In some species a relatively high degree of integrity of maternal tissue is retained, but in others, such as man, the maternal blood vessels are eroded to bathe the trophoblast cells in maternal blood (Table 9.3 and

235 *Implantation and the Establishment of the Placenta*

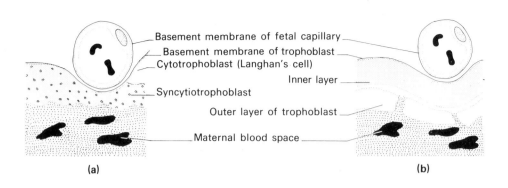

(a)                              (b)

Basement membrane of fetal capillary
Basement membrane of trophoblast
Cytotrophoblast (Langhan's cell)
Inner layer
Syncytiotrophoblast
Outer layer of trophoblast
Maternal blood space

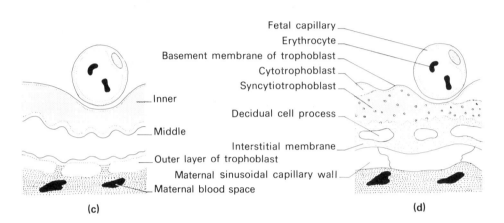

(c)                              (d)

Inner
Middle
Outer layer of trophoblast
Maternal sinusoidal capillary wall
Maternal blood space

Fetal capillary
Erythrocyte
Basement membrane of trophoblast
Cytotrophoblast
Syncytiotrophoblast
Decidual cell process
Interstitial membrane

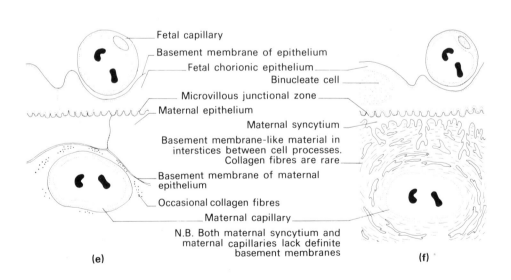

(e)                              (f)

Fetal capillary
Basement membrane of epithelium
Fetal chorionic epithelium
Binucleate cell
Microvillous junctional zone
Maternal epithelium
Maternal syncytium
Basement membrane-like material in interstices between cell processes. Collagen fibres are rare
Basement membrane of maternal epithelium
Occasional collagen fibres
Maternal capillary
N.B. Both maternal syncytium and maternal capillaries lack definite basement membranes

Fig. 9.6a–d). There is little evidence that the number of layers of tissue that ultimately lie between the circulation of mother and conceptus can in any way be related to efficiency of placental transfer. Other properties of the interface are probably of much greater significance (Chapter 11).

With the formation and invasion of the decidua, implantation is completed, a physical hold and a nutritional source of decidual 'yolk' is established and the basis of placental development, leading to adjacent circulations and exchange of nutrients, is initiated.

*b Non-invasive implantation (pig, sheep, cow, horse)*

There are fewer studies of the details of the endometrial response to implantation in these species. Several striking differences are clear, however. Attachment of non-invasive conceptuses is, in general, initiated relatively later than that of invasive conceptuses (Table 9.1). During the prolonged interval prior to attachment, the free-living conceptuses continue to draw on richer and more copious secretions from uterine glands—the non-invasive equivalent of the decidual 'yolk'. This uterine 'milk' is a rich source of ions, sugars, amino acids and lipid precursors, and the conceptus utilizes these resources and grows to a much greater pre-implantation size than the human or rodent invasive conceptus. The growth may be prodigious and rapid. The pig conceptus, for example, elongates from a spherical blastocyst 2 mm in diameter on Day 6 of pregnancy to a membranous, highly convoluted thread some 1000 mm long by Day 12. This growth is almost exclusively confined to the extra-embryonic tissues of the conceptus, and establishes a vast surface area over which exchange of metabolites with the uterine milk can occur.

This large surface area of extra-embryonic chorionic trophoblast also permits an extensive subsequent attachment to uterine epithelial cells. In the horse and pig, attachment occurs at multiple sites over most of the external surface of the conceptus, but in ruminants attachment is limited to uterine caruncles (see earlier), which in sheep and goats may be up to 90 in number. As a result the ultimate form of the placentae in these species differs from that of invasive conceptuses, and is said to be *diffuse* (mare and pig) and *cotyledonary* (cow and sheep) (see Fig. 9.5c and d). Between

**Fig. 9.6.** Schematic view of microstructure at the mature placental interface of various species. (a) Man—haemomonochorial; (b) rabbit—haemodichorial; (c) rat and mouse—haemotrichorial; (d) dog (also bears, cats, mink)—endotheliochorial; (e) mare and pig (also whales, lemurs, dolphins, deer, giraffe)—epitheliochorial; (f) sheep—epitheliochorial—note presence of binucleate cells on fetal side (also seen in cow); recent evidence suggests that they may invade maternal component later in pregnancy, thus making cow and sheep not strict non-invasive implanters (After Steven, 1975.)

the sites of attachment, uterine glands continue to secrete a nutrient 'milk' for the conceptus, especially in the pig and horse. Although penetration and erosion of the epithelium does not occur at implantation in these species (and therefore the conceptus is said to implant *centrally*), junctional complexes are observed between trophoblast and the adjacent epithelial cells (pig and sheep), and invasion of a few conceptus cells into the endometrium may occur either transitorally (horse) or much later in development (sheep). No proper decidual response is induced in the underlying stroma (the 'yolk' is provided by secretion of uterine milk and the epithelium acts to control invasion), but limited stromal changes provide evidence of the recognition of embryonic presence. Thus, vascularity increases and distinct changes in cellular morphology develop. These changes constitute the beginnings of the formation of the placenta and mark the end of the protracted implantation seen in these species.

*c Uterine control of implantation*

The blastocyst provokes a response of considerable magnitude in the uterus. There is, however, evidence to suggest that the uterus also regulates embryonic activity. The uterus can only take part in these interactions if suitably primed by steroid hormones.

Under oestrogen domination, the uterus cannot accommodate an implanting blastocyst. Indeed, it is clearly established for rodents, sheep and rabbits that an oestrogen-dominated uterus is actively *hostile* to the conceptus and will kill it should it leave the oviduct prematurely. In contrast, the cleaving primate conceptus can tolerate, and survive in, an oestrogen-dominated uterus. Indeed, *in vitro* fertilized human eggs are usually transferred directly to the uterus at the four-cell stage. However, although not hostile to cleaving stages, the human uterus, like uteri of other species, requires progestagenic domination if it is to engage effectively in conversations with the implanting blastocyst.

We saw in Chapter 7 that the luteal phase of the cycle in many species was characterized by distinctive patterns of uterine secretions rich in sugars and proteins. Whilst the *elaboration* of these secretory products is dependent upon progesterone action, their *release* in several species requires the presence of oestrogen in the luteal phase. It is possible to deprive female rats or mice of this oestrogen by ovariectomy during the first 2–3 days of pregnancy while progesterone levels are maintained by daily injections. Under such conditions the conceptus enters the uterus as normal, but instead of proceeding to implant, remains free in the uterine lumen. The zona pellucida, which in the rat and mouse is normally dissolved by the action of a uterine proteolytic enzyme, is shed undissolved instead, via the physical activity of the blastocyst itself. The blastocyst may spend many days in the uterus in this quiescent state, its metabolism 'ticking over'. If a single injection of oestrogen is then given, blastocyst metabolism increases, attachment occurs rapidly and decidualization is initiated; moreover, the zonas, which were actively shed to lie free in the lumen, are dissolved.

It appears that oestrogen stimulates the epithelial cells of the uterus in two ways. First, it stimulates the release of a glandular *secretion* of a characteristic composition including specific macromolecules, such as the zona proteolytic enzyme, as well as glucose, amino acids and ions which cause embryonic activation. Second, the oestrogen also acts on the epithelial cells to make them *responsive* or *sensitive* to a blastocyst signal so that they can transmit evidence of the blastocyst's presence at attachment to the underlying stromal cells, and thereby instruct them to commence decidualization. This facility of the rat and mouse to suspend the blastocyst *in utero* for several days is called *delayed implantation*, and it provides us with a clear example of the uterine control over the development of the blastocyst.

Delayed implantation should not be thought of as purely an artificial or experimentally induced phenomenon. It occurs in the rat due to the natural suppression of endogenous oestrogen secretion in females which are suckling young of the previous litter (see Chapter 13 for details). This type of natural delay is often called *facultative* delayed implantation, since it *only* occurs under conditions of suckling. It is shown by mice, gerbils, marsupials and the bank vole, and is clearly useful since the mother delays the growth of her uterine litter for as long as she is suckling the previous litter. If this delay did not occur, the helpless newborn of the second litter would have to compete for milk with older and bigger siblings.

In addition, there are many species in which an *obligatory* delayed implantation is an essential and normal part of their pregnancy (e.g. roe-deer, badger, elephant seal, fruit bat, mink, stoat, wolverine, brown bear). In these species a prolonged period of weeks or months may be spent with blastocysts in delay (or *diapause* as an obligatory delay is also called). Diapause confers distinct biological advantages. The roe-deer, for example, mates in July and August when the adults are well fed and in their prime, ensuring effective competition for mates. The conceptus remains as a blastocyst until January when it reactivates for delivery of young in May and June when the nutritional conditions will have become optimal for the new offspring.

In many species examined, including man, secretion of oestrogen is imposed on the progesterone background of the luteal phase, and it is therefore possible, but not yet proved, that in all species the oestrogen acts to stimulate endometrial release of messages crucial to activation of the conceptus. It has been established that, in at least some of these species, all or part of the oestrogen is *not of ovarian origin,* but instead is synthesized *by the conceptus itself.* In the pig, for example, the luteal phase lacks a significant rise in oestrogen, but if a conceptus is present a clear oestrogen peak is detectable. Pre-implantation stages of the pig, rabbit, sheep, cow and hamster have all been shown to be capable of steroid synthesis and metabolism *in vitro.* Thus, the interactions between the conceptus and the uterus in these species may prove to be even more complex than those in the rat. The conceptus may first produce oestrogens to stimulate the progesterone-primed uterus, the uterine epithelium may then respond by secreting embryotrophic factors and showing increased sensitivity to blastocyst attachment. Finally, the blastocyst may co-respond by attachment and so initiate implantation.

*d Summary*

We have seen that the earliest physical attachment of the conceptus to the uterine endometrium shows considerable interspecies variability in timing, extent and degree of invasiveness. This variation at the earliest stage of pregnancy results subsequently in a correspondingly diverse pattern of placental forms that frequently confuse students of repro-duction. Certain simple rules, however, help to clarify this confusion. The invasive conceptus attaches early when small to yield the prize of uterine decidual nourishment. It's placenta therefore tends to be compact and to have fewer interposed layers between the fetal and maternal circulations. The non-invasive conceptus attaches late when larger, being nourished initially by uterine secretions. Its placenta there-fore tends to be extensive and to be epithelio-chorial. In both cases, the maternal endocrine condition influences the effec-tiveness with which mother and conceptus communicate during implantation.

So far we have considered only the anchoring of the conceptus to the mother. We must now consider precisely how the nourishment of the embryo by uterine decidual and secretion products (so called *histiotrophic* support) is super-seded by the more direct intervascular exchange route of the more mature feto-placental unit (so called *haemotrophic* support).

**3 The change from histiotrophic to haemotrophic support**

The pre-implantation conceptus derives its nutrition from endogenous reserves and via the secretions of the genital tract. As we have seen, during invasive implantation these exocrine secretions are supplemented by decidual material released by the invading trophoblast. For how long does an intermediary fluid phase function as the route for nutritional and excretory exchange and what replaces it? In order to answer this question, we need first to examine briefly the structure of the developing conceptus and the principal routes of metabolic exchange that become available.

The development of the fetal membranes, and the inter-specific variety of their organization, provides one of the most enduring confusions for students of embryology. A simplified scheme to explain the origin of the various fetal membranes is shown in Fig. 9.7, and this figure and its legend should be studied carefully before you read further.

During the histiotrophic phase of growth, the conceptus takes up materials from, and excretes waste products into, the surrounding endometrial fluids. These pass through the shell of trophoblast and distribute by simple diffusion through the cavities and tissues of the conceptus itself. However, as the *mesoderm* of the conceptus becomes more extensive, it develops blood vessels within it, and these link up to form an extensive vascular network. Blood formation occurs in the *yolk sac mesoderm* and a primitive heart forms in the *cardiac mesoderm* (Fig. 9.7f) within the developing embryo itself. Blood can thus be pumped throughout the extensive meso-dermal tissue of the whole conceptus—both its embryonic and extraembryonic parts. This blood passes throughout the chorionic mesoderm where equilibration with maternal se-cretory and decidual fluids occurs. With further development the vascularity in the conceptus becomes particularly marked in the yolk sac mesoderm and where the yolk sac and chorionic mesoderm fuse together (arrowed in Fig. 9.7e), and a corresponding vascularity develops within the endo-metrium adjacent to this site of fusion. Together the two adjacent, highly vascular sites form the *yolk sac* (or *chorio-vitelline*) placenta. The yolk sac, and the yolk sac placenta to which it contributes, is a structure homologous to the yolk-containing sac found in the eggs of reptiles, birds and monotremes (like the platypus). This comparatively primitive origin is reflected in the transitory existence and function of the yolk sac placenta in most mammals, a second exchange site called the *chorio-allantoic placenta* taking over (Fig. 9.7f arrowed). However, in some species, such as most marsupials, the yolk sac placenta functions alone throughout pregnancy and in others such as the rabbit, rat and mouse, it persists and remains functional even after the definitive chorio-allantoic placenta takes over the major role. In higher primates such as man, the yolk sac placenta is never functional at all.

The chorio-allantoic placenta forms as a result of the outgrowth of an endodermal diverticulum from the hindgut region of the developing embryo (Fig. 9.7f). Together with its ensheathing mesoderm containing the pro-umbilical vessels, it contacts the chorionic mesoderm and thereby determines the site (or sites) of chorio-allantoic placentation. As we saw earlier, species with early invading conceptuses have an allantoic outgrowth restricted to discoidal, bi-discoidal or zonary sites, whereas late-attaching conceptuses have a more extensive and diffuse allantoic outgrowth.

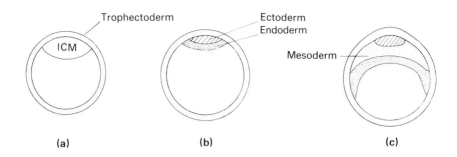

**(a)**            **(b)**            **(c)**

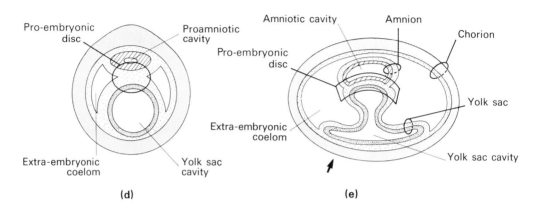

**(d)**            **(e)**

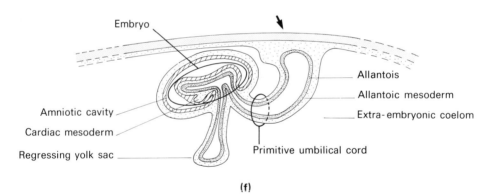

**(f)**

**Fig. 9.7.** (a) Blastocyst with trophectoderm (shaded) and inner cell mass (white). (b) Development has occurred within the inner cell mass, two layers of cells forming, namely ectoderm (striped) and endoderm (heavy stipple). (c) A third mesodermal tissue develops (light stipple). The endoderm spreads out over the trophectoderm which is now transforming to mature trophoblast. (d) The endodermal edges meet to form a hollow 'sphere' containing a cavity—the yolk sac cavity. Mesoderm spreads round between the endoderm and the trophoblast and also separates the ectoderm from the trophoblast. Cavities develop within both ectoderm and mesoderm layers—the pro-amniotic cavity and extra-embryonic coelom (or space) respectively. (e) The extra-embryonic coelom enlarges, but throughout is lined with mesoderm. The amniotic and yolk sac cavities enlarge and change shape. Between them three cell layers persist forming a sandwich, the ectoderm and endoderm forming the

It is important to re-emphasize that the functional placenta incorporates not only a contribution from the conceptus, as shown developing in Fig. 9.7, but also from endometrial tissue. This is shown schematically in Fig. 9.8 for the developing human conceptus which has a discoid, chorio-allantoic placenta. Having considered the general disposition of membranes, fetus and placenta, we will now take a closer look at the organization of the placental interface itself.

*b The placental interface*

The functional chorio-allantoic placenta is characterized by (i) extensive proliferation of the chorionic tissue to give *villi* that penetrate into, and interdigitate with, the endometrial tissue, (ii) by highly developed vascularity of both fetal and maternal components, and (iii) by intimately juxtaposed but physically separate fetal and maternal blood flows. The precise arrangement of vessels in the functional placenta varies considerably, and here we examine the human (discoidal, haemochorial) and sheep (cotyledonary, epitheliochorial) placentae as illustrative representatives.

---

'bread' around a mesodermal 'filling'. It is from this 'trilaminar or pro-embryonic disc' that the definitive embryo will arise (see f below), and all the other tissues are extra-embryonic and form the fetal membranes. Thus, the amnion consists of a layer of extra-embryonic ectoderm fronting the amniotic cavity and a layer of extra-embryonic mesoderm outside it. The yolk sac consists of a layer of extra-embryonic endoderm fronting the yolk sac cavity and a layer of extra-embryonic mesoderm outside it. The chorion consists of a layer of trophoblast fronting the uterine tissue and a layer of extra-embryonic mesoderm within it. *Note:* (1) Blood vessels develop throughout the embryonic and extra-embryonic mesoderm and the embryonic blood starts to flow through them. (2) The point at which the yolk sac mesoderm and chorionic mesoderm fuse is arrowed; this is the site of formation of the yolk sac (or chorio-vitelline) placenta. (f) The trilaminar embryonic disc curls up to give the definitive embryo with its outer ectoderm surrounded by amniotic fluid in the amniotic cavity; the derivatives of this ectoderm include the outside 'skin' of the fetus. Within this curled up disc there is an endoderm-lined cavity (continuous with the yolk sac cavity) that is the primitive gut. The interposed filling of embryonic mesoderm (which will form many of the fetal tissues between the gut and the skin and their derivatives) is highly vascular and includes at the anterior or head end of the embryo the primitive heart (cardiac mesoderm) that pumps blood through the network of embryonic and extra-embryonic vessels. A diverticulum of the endoderm, the allantoic endoderm, develops at the posterior or tail end of the embryo and it grows out surrounded by mesoderm to form the allantois. The allantoic and chorionic mesoderm (both rich in blood vessels) fuse (arrowed) and this is the site of formation of the chorio-allantoic placenta. The connection that the allantois makes between the embryo and the chorio-allantoic placenta will become the umbilical cord (see Fig. 9.8).

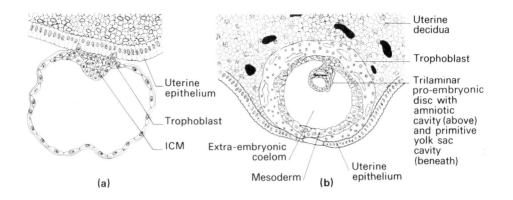

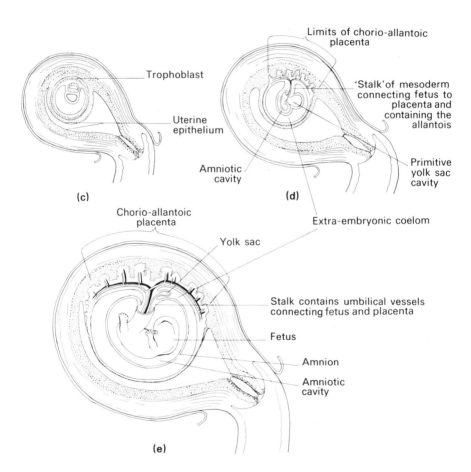

**Fig. 9.8.** Highly schematic views (not to scale) of a human conceptus at: (a) attachment; (b) invasion; and (c)–(e) progressively later stages of pregnancy, to show general relationship of fetus, extra-embryonic membranes, placenta and maternal vessels. Note how the fetus itself ends up floating in amniotic fluid and linked to the placenta by an umbilical stalk. (After Hamilton, Boyd & Mossman.)

i The human placenta

In the mature haemomonochorial placenta of the human, tongues or *villi* of chorionic syncytiotrophoblast containing cores of mesodermal tissue in which fetal blood vessels run, penetrate deeply into the maternal tissue to form a *labyrinthine* network (Fig. 9.9). Each villus, by progressive branching of the main divisions of the umbilical vessels, are separated from the surface by a thin syncytiotrophoblastic layer (Fig. 9.9c). At the tips of the terminal villi, the capillaries are dilated and form tortuous loops (Fig. 9.10). Thus, fetal blood flow through the tips will be slow, allowing for exchange of metabolites with maternal blood.

The branches of the villi are arranged, with a somewhat variable degree of regularity, to form 'fenestrated cups' (rather in the shape of the bowl of a brandy glass). Their terminal villi project inwards into the central space of the cup, between the adjacent villi that form the fenestrated wall of the cup, and outwards into the space peripheral to the cup. Each 'cup' unit is sometimes called a *fetal lobule*. It is suggested that in the human, the maternal spiral artery at the decidual base of the placenta ejects its blood into the space which forms the bowl of the cup, filling the brandy glass as it were. The pressure of the blood causes its circulation through the fenestrated wall of the cup over the fine terminal villi. The blood is then thought to drain back via the basal venous lake into maternal veins (Fig. 9.9b). Up to 200 such lobular units form in the mature human placenta which is a 3-cm-thick 'pancake' of 15–20 cm in diameter. Several lobular units are grouped together to form a cotyledon, the boundaries of which may be seen grossly on the placenta, defined by fibrous septa. This anatomical description is probably somewhat idealized, and the consistency with which such a regular arrangement of maternal vessels and villi occurs in the human placenta is perhaps questionable. The important point to be understood is that maternal blood circulates across the fine terminal villi containing the fetal capillaries. The circulation of maternal and fetal blood within the placenta is established by about 3 weeks of pregnancy.

ii The ovine placenta

The sheep placenta is epithelio-chorial and cotyledonary having some 80 to 90 independent sites of close vascular proximity. Attachment occurs at each uterine epithelial *caruncle;* the fetal chorion at the point of attachment shows specialization as a *cotyledon;* the fetal cotyledon and maternal caruncle together constitute the functional unit of the placenta known as a *placentome* (Fig. 9.11). In the mature cotyledon, an ingrowth of chorionic villi indents and compresses the epithelium of the maternal caruncles. Within the villi a core of vascularized fetal mesoderm develops. Each cotyledon receives one to three branches of the umbilical

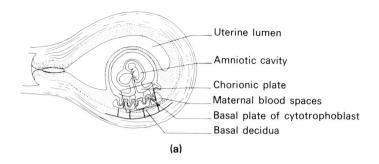

(a)

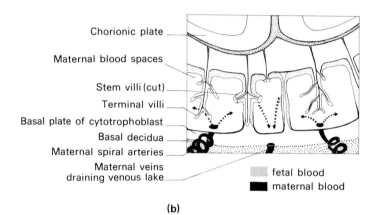

fetal blood
maternal blood

(b)

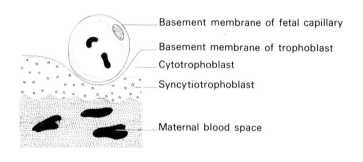

(c)

**Fig. 9.9.** Schematic representation of structures of human placental interface. Stem villi connect chorionic plate to basal plate forming a labyrinthine series of spaces (a). From the stem villi and the chorionic plate smaller villi ramify into the intervillous space forming a network of fine filamentous terminal villi, which are the principal sites of metabolic exchange (b). At these sites only a thin layer of chorionic syncytiotrophoblast separates the fetal blood vessels from maternal blood (c).

vessels. The vessels divide and ramify within the villi where they come to lie under the surface of the trophoblast at the tips of the villi (Fig. 9.11). On the maternal side, the surface epithelium of the caruncle becomes syncytial, and the

**Fig. 9.10.** Low-power view of a cast made from the microvascular capillaries in the terminal villi on the *fetal* side of the placental circulation. Note how the terminal vessels form convoluted knots compared with the larger calibre vessels adjacent. The straight capillaries supplying the knots can be seen (arrow head), as can the terminal dilatations (arrowed) in which blood flow is slower and at which exchange of metabolites between fetal and maternal blood takes place. (Photographs by courtesy of S. Habashi, G.J. Burton & D.H. Steven. *Placenta* 1983; **4**:41–56).

underlying stroma acellular and vascular. In later pregnancy some of the binucleate cells, which form in the fetal cotyledon, penetrate into the maternal stroma (an equivalent late penetration of maternal tissues may occur in the other large farm animals which are nonetheless appropriately classified as non-invasive at implantation). Tortuous, coiling maternal arteries supply each caruncle and split into capillaries between the penetrating terminal villi of the fetal cotyledon. The capillaries then run back along the long axis of the terminal villus towards the tip, where they drain into maternal veins (Fig. 9.11). This organization of fetal and maternal vessels ensures that in the capillary beds, the blood flow is in opposite directions (Fig. 9.11) thereby maximizing opportunities for metabolic exchange.

*c Blood flow in the placenta*

The development of a haemotrophic route of exchange between mother and fetus provides a much more efficient

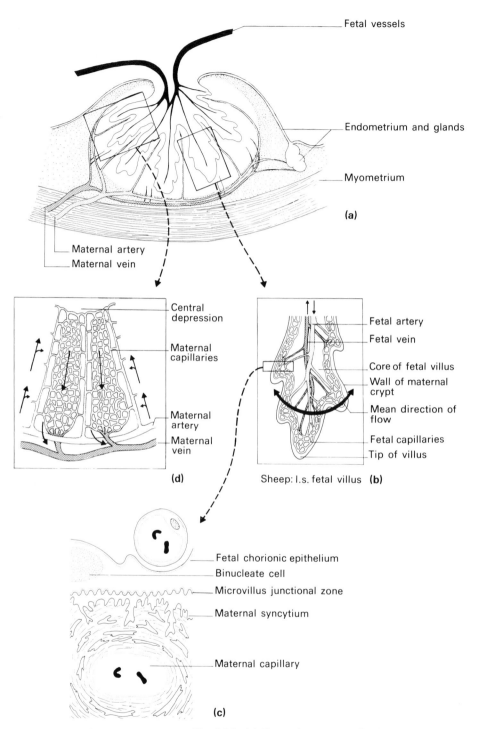

**Fetal vessels**

**Endometrium and glands**

**Myometrium**

**(a)**

**Maternal artery**
**Maternal vein**

Central
depression

Maternal
capillaries

Maternal
artery
Maternal
vein

**(d)**

Fetal artery

Fetal vein

Core of fetal villus
Wall of maternal
crypt
Mean direction of
flow
Fetal capillaries
Tip of villus

Sheep: l.s. fetal villus **(b)**

Fetal chorionic epithelium
Binucleate cell
Microvillus junctional zone
Maternal syncytium

Maternal capillary

**(c)**

**Fig. 9.11.** (a) General structure of a sheep placentome. Fetal villi—
shown in longitudinal section in (b)—interdigitate with maternal
tissue giving close apposition of circulations (c). Maternal blood
flows around the fetal villus as indicated in (d). A similar inter-
digitation of fetal villi with maternal tissue also occurs in the horse
and the cow.

way of establishing metabolic gradients to drive diffusional and carrier-mediated passage of the principal metabolic substrates and excretory products than can be achieved by a histiotrophic route. The effectiveness with which gradients are established, and thus exchange can take place, will depend largely on the effectiveness with which the rates of blood flow through the two halves of the placental circulation are regulated, as well as on the diffusional barriers and special transport mechanisms that might exist between them. The latter are considered in detail in Chapter 11. Here we are concerned with rates of blood flow and the factors affecting them.

Cardiac output is increased by up to 25% in pregnancy in response to the extra peripheral load. Maternal blood reaches the placenta via uterine and ovarian vessels and constitutes about 10% of the total cardiac output by the end of pregnancy. In the human the uterine arteries course along the lateral walls of the uterus giving 9 to 14 branches, each of which penetrates the outer third of the myometrial tissue. At this level anastomosis of these arteries with the ovarian arteries may occur, and from the anastomosis a series of arcuate arteries run within the anterior and posterior myometrial walls of the uterus thereby encircling it. From this enveloping vascular network radial arteries penetrate through the remaining myometrium into the basal endometrial tissue. Here the so-called basal arteries distribute spiral arteries to supply the endometrial decidua. This convoluted or spiral nature of terminal endometrial arteries, together with a tendency to dilate terminally, is a feature common to many species. Both features will tend to diminish the velocity of the blood, which is further diminished to 0.1–10 mm/s in the primate by the extensive intervillous spaces. This sluggish flow may protect the early conceptus from being dislodged by 'spurts' of blood, and also gives ample time for the exchange of metabolites at the placental interface (estimated mean transit time of 15 in the full-term monkey placenta). The perfusion pressure within the intervillous spaces of the primate placenta is also relatively low (4–10 mmHg). A similar slowing of flow occurs on the fetal side of the circulation, where the total cross-sectional area of blood vasculature increases due both to the profusion of vascular branching and the terminal capillary dilatations (Fig. 9.10) at which exchange occurs.

The maternal spiral arteries have a sympathetic innervation, and constrict in response to sympathetic nerve stimulation or sympathomimetic drugs. Reduced placental perfusion can, therefore, result either from local vasoconstriction or from lowered systemic pressure. Transient reductions in perfusion pressure do not seem to have adverse effects on placental exchange or fetal growth, but chronic reductions do, particularly later in pregnancy. Thus, chronic anxiety, heavy smoking and stress during pregnancy result in smaller term babies, probably caused in part via effects on placental perfusion. Similarly, administration of drugs to relieve maternal hypotension, as for example in asthma, will cause an increase in visceral vasoconstriction, which will already be elevated reflexly, and thereby further reduce placental perfusion. During maternal exercise, some reduction in blood flow to the uterus occurs, but the conceptus itself seems to be relatively protected unless exercise is severe and prolonged.

In addition to the adverse effects of impaired flow *towards* the placenta, occlusion or slowed blood flow *at* the placental interface will also reduce the efficiency of metabolite exchange. Such an effect occurs in conditions of increased maternal blood viscosity, for example, sickle cell anaemia, or after the expansion of placental villous structures, with a consequent reduction in the intervillous space available for circulation. This occurs in conditions of increased umbilical vein pressure, such as *erythroblastosis* or vascular occlusion of the fetal liver, both of which cause distension of the villi. Smoking in pregnancy also exerts direct effects on the fetal vasculature in the placenta, there being fewer, narrower and less convoluted capillaries in the terminal villi. Finally, occlusion of the maternal venous drainage, as occurs during uterine contractions at parturition, will reduce flow through the placental interface.

**4 Summary**

The growing conceptus faces a hazardous few days of early life. Not only must it pass from oviduct to uterus but it must also establish adequate nutritional support for its growth and development. The very earliest days are marked by little growth at all, but with the changing female cycle the progestagen-dominated luteal or secretory phase uterus provides a nutritious fluid support that enables the conceptus to undertake one of its first tasks, namely the securing of a physical hold on the mother. The conceptus then establishes its own circulation and stimulates at adjacent endometrial sites the establishment of special maternal circulatory changes. The transition from a histiotrophic to a more efficient haemotrophic route of metabolic exchange has been achieved via this placental structure.

However, a successful pregnancy requires more than this. The uterus will only provide a congenial environment as long as endocrine conditions are correct. Under oestrogenic conditions the uterus may even be hostile to the conceptus, but under progestagenic conditions it will survive (although in some, and maybe all, species some superimposed oestrogen is required for attachment and the initial stages of implantation). However, while the conceptus is establishing a secure physical and nutritional anchorage within the uterus, the luteal phase of the cycle is progressing. In some species, such as the dog, the luteal phase is of the same duration as pregnancy (Table 9.1); in others, such as the rat, rabbit and mouse, coitus is equated with possible pregnancy and so the coital stimulus neurogenically activates a pituitary secretion of luteotrophins that prolongs the normal brief luteal phase of the oestrous cycle into a pseudopregnancy which is about half the length of pregnancy (Table 9.1 and Chapters 4 and 5). In primates and the large farm animals, however, pregnancy greatly exceeds the normal life of the corpus luteum (Table 9.1) and even in the rat, rabbit and mouse, the prolonged luteal phase is not as long as pregnancy itself. Since, as we saw earlier, a uterus dominated by oestrogen and lacking progesterone is hostile to the conceptus in all these species, *the progesterone-dominated state must be extended in some way.* The conceptus must circumvent or neutralize the devices by which corpus luteum regression normally occurs. This achievement is all the more remarkable considering that in species such as the pig and cow, attachment has not even occurred by the expected time of luteal regression (Table 9.1). How does the conceptus achieve this task? How does the maternal organism recognize its presence in the uterus? This is the subject of the next chapter.

**Further reading**

Cole, HH, Cupps PT (Eds). *Reproduction in Domestic Animals,* 3rd Edn. Academic Press, 1977.

Flint, APF, Renfree MB, Weir BJ (Eds). *Embryonic Diapause in Mammals.* Supplement 29. Journals of Reproduction and Fertility Ltd., 1983.

Johnson MH (Ed). *Development in Mammals.* Vol 2. Elsevier, 1977.

Lotgering FK, Gilbert RD, Longo LD. Maternal and fetal responses to exercise during pregnancy. *Physiol Rev* 1985; **65**; 1–36.

Steven DH (Ed). *Comparative Placentation.* Academic Press, 1975.

Wilkinson AW (Ed). *Early Nutrition and Later Development.* Pitman Medical, 1976.

# Chapter 10
# Maternal Recognition and Support of Pregnancy

If the implanting conceptus is to survive, it must signal its presence to the mother and prevent withdrawal of progestagenic support that would normally occur with luteolysis. Moreover, in many species that support has to be prolonged not just briefly but must be sustained over weeks or months. The mechanisms by which the endocrine support of pregnancy is first established and then sustained are the subject of this chapter.

## 1 Maternal recognition of pregnancy

We saw in Chapter 4 that normal luteal survival depended upon *the balance* between a positive *luteotrophic complex* and negative *luteolytic* factors which, in species other than primates, seem to be uterine prostaglandins. If luteal life is to be prolonged, then clearly either normal luteolytic factors must be neutralized or normally dwindling luteotrophic factors must be stimulated. In practice both mechanisms are probably at work.

### a The primate

The primate blastocyst appears to prolong the life of the corpus luteum by production of its own luteotrophic factor which takes over from the inadequate support provided by low levels of pituitary LH. If blood samples from pregnant women 8–12 days postfertilization are compared to those taken from women in the comparable period of a non-pregnant cycle, they are found to contain a glycoprotein called *human chorionic gonadotrophin (hCG)* (Table 10.1). This hormone shows a remarkable structural similarity to LH. It is comprised of two chains, an α-chain, with primary sequence almost identical to that of LH, and a β-chain, which is similar for the first 120 amino acid residues but which has 'a tail' of an additional 30 unique amino acids at the N terminal. In addition, there are

**Table 10.1.** Properties of human chorionic gonadotrophin (hCG) (compare with Table 3.2 LH and FSH).

| | |
|---|---|
| Secreted from | Syncytiotrophoblast (certain tumours also secrete it, particularly those of gastrointestinal tract and teratocarcinomas; care is therefore required in use diagnostically) |
| Acts upon | LH receptors (see Table 3.2) to mimic action of LH (has weak FSH-like activity) |
| Molecular weight | *c.* 40 000 |
| Composition: glycoprotein two peptide chains | 30% Carbohydrate |
| subunit $\alpha$ | 92 amino acids (almost identical to LH with three additional amino acids on N terminal) two carbohydrate chains |
| subunit $\beta$ | 147 amino acids (almost identical to LH with an additional tail of amino acids on N terminal) five carbohydrate chains (including three at amino acids 118, 129, 131) |
| Biological half-life in blood | 300 minutes (cf. LH—30 minutes) |

differences from LH in the attached carbohydrate residues. Synthesis of hCG occurs in the syncytiotrophoblast of the implanting blastocyst and it is released into the maternal circulation. It passes to the ovary where it binds to LH receptors on the luteal cells to exert a luteotrophic action. It is believed that this additional luteotrophic stimulus *supplements* the low endogenous pituitary LH levels during the primate's luteal phase and *overcomes* any luteolytic tendencies. The evidence for this belief comes from two types of experiment.

If non-pregnant women are given daily injections of hCG, starting in the mid-luteal phase, regression of their corpora lutea is prevented or delayed and progestagen levels remain elevated. Conversely, it is possible to prepare highly specific antibodies to the unique terminal $\beta$-chain sequence of hCG (or indeed to CG derived from baboons, rhesus monkeys and marmosets which make bCG, rhCG and mCG, respectively). This antiserum is quite without effect on LH. If the antiserum is administered *passively* through the luteal phase of a pregnant human or monkey cycle, the CG production is neutralized and luteal regression occurs on time (Fig. 10.1). In female monkeys *actively* immunized to their own CG, the antibodies completely neutralize episodic production of CG following a fertile mating while having no effect at all on LH

253    *Maternal Recognition and Support of Pregnancy*

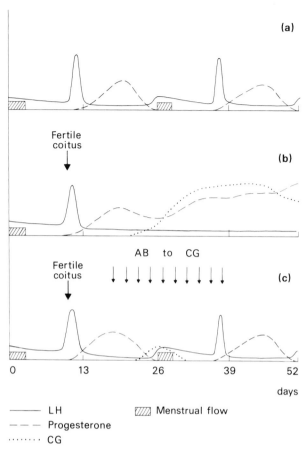

(a)

Fertile
coitus

(b)

AB to CG

Fertile
coitus

(c)

0          13          26          39          52

days

———— L H          ▨ Menstrual flow
— — — Progesterone
········ CG

**Fig. 10.1.** Levels of hormones in the blood during two cycles of a higher primate. (a) Non-pregnant cycles; (b) cycles in which fertile mating occurs; (c) similar to (b) but passive administration of a highly specific anti-CG antibody is given (arrows), depression of CG and loss of pregnancy occurs but there is no effect on LH levels and cyclicity.

and thus on normal cyclicity. Pregnancy, however, is totally suspended.

Thus, since CG levels rise naturally at the time that luteal support is required, since injection of exogenous CG maintains and prolongs luteal function, and since neutralization of endogenous CG prevents prolongation of luteal life, CG is strongly implicated in the prevention of luteal regression during early pregnancy in the primate.

*b Large domestic animals*

In the large farm animals, luteolysis is normally induced by uterine prostaglandins (Chapter 4). We saw that in the absence of these prostaglandins, for example after hysterectomy, luteal life is prolonged or lasts indefinitely. In these species the luteotrophic support by the pituitary is presumably quite adequate and it is the luteolytic agent, $PGF_{2a}$, that must be neutralized. It is not surprising, therefore, that chorionic gonadotrophins have not been detected in the blood of pigs, cows and sheep bearing pre-implantation conceptuses. The pig conceptus does produce oestrogens from about Day 12 onwards, and oestrogens are a major luteotrophic agent in this species (see Table 4.1). It is therefore possible that oestrogens help prolong luteal life in the pig in the way that CGs do in primates. However, in the large farm animals the conceptuses appear primarily to suppress luteolytic activity. In pregnant animals, the levels of prostaglandin $F_{2\alpha}$ and its metabolites in the blood draining the uterus are reduced compared to the non-pregnant animal, and, in the sheep at least, it appears that this reflects reduced synthesis. In the pig oestrogens from the conceptus act on the uterus to suppress $PGF_{2\alpha}$ release as well as exerting a possible direct luteotrophic effect on the corpus luteum. Presumably the conceptus interferes in some way with the synthesis or release of the $PGF_{2\alpha}$, or possibly metabolizes the prostaglandin locally. In this context it is of interest to note that the output of prostaglandins in a normal luteal cycle is increased and occurs earlier in the presence of an intra-uterine device (IUD). It has been suggested that this action of the IUD might be one mechanism for its contraceptive effect (but see Section 3.b.iv, Chapter 14).

## 2 Maternal endocrine support of pregnancy

The successful establishment of pregnancy creates a new and extraordinary parabiotic liaison between mother and conceptus which may last for a period of months in some species. The conceptus is totally dependent upon the mother for protection and nutrition. Her metabolic activities are modified by the conceptus so that there is adequate mobilization of oxygen, salts and organic precursors to supply its needs, and it induces the formation of the vascularized placenta at which the bulk of these substances is exchanged for its own metabolic waste. During pregnancy the mother is also prepared for the *future* requirements of the fetus. Thus, hypertrophy of the uterine musculature, which participates in fetal expulsion at *parturition,* and the development and maintenance of mammary glands for postpartum *lactational nutrition* are both stimulated during pregnancy. This take-over of the mother's metabolism is controlled by the *pregnancy hormones.* In the remainder of this chapter we will consider the nature and origin of these hormones. In Chapters 11–13 we look at their functions.

As we have already seen, the extended secretion of progesterone is critical to the initiation of pregnancy. This absolute requirement for progesterone persists throughout pregnancy and indeed in many species progesterone levels in maternal blood rise continuously as pregnancy proceeds (Fig. 10.2). Moreover, as pregnancy advances the level of oestrogens in

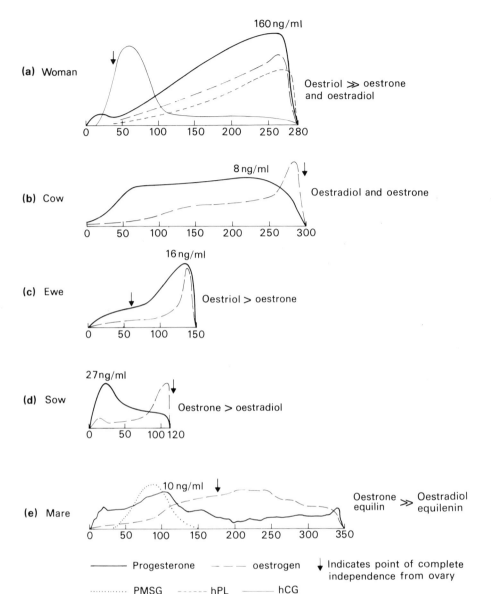

160 ng/ml

**(a)** Woman

Oestriol ≫ oestrone
and oestradiol

0    50    100    150    200    250  280

8 ng/ml

**(b)** Cow

Oestradiol and oestrone

0    50    100    150    200    250    300

16 ng/ml

**(c)** Ewe

Oestriol > oestrone

0    50    100    150

27 ng/ml

**(d)** Sow

Oestrone > oestradiol

0    50    100 120

10 ng/ml

**(e)** Mare

Oestrone  ≫  Oestradiol
equilin        equilenin

0    50    100    150    200    250    300    350

———— Progesterone    – – – oestrogen    ↓ Indicates point of complete
                                            independence from ovary
·········· PMSG    – – – – – hPL    ———— hCG

**Fig. 10.2.** Pattern of plasma hormones during pregnancy in various
species. Maximal progesterone level and principal oestrogens are
recorded. PMSG=pregnant mare serum gonadotrophin; pHL=
human placental lactogen; hCG=human chorionic gonadotrophin.
Pregnancy timed in days.

the maternal blood also rises. These steroid hormones appear
to be important for the maintenance of pregnancy (Fig. 10.2).
Indeed, both oestrogens and progesterone may reach plasma
levels many times greater than those seen in a normal luteal
phase.

In many species the ovarian corpora lutea, which are under
the control of the maternal pituitary, continue to secrete an

**Table 10.2.** Dependence of pregnancy on maternal ovarian and pituitary secretions in various species.

| Species | Duration of pregnancy (days) | Duration of non-pregnant luteal phase (days) | Day of pregnancy when hypophysectomy without effect (days) | Day of pregnancy when ovariectomy without effect (days) |
|---|---|---|---|---|
| *Group A* | | | | |
| Man | 260–270 | 12–14 | ? | 40 |
| Monkey (*M. mulatta*) | 168 | 12–14 | 29 | 21 |
| Sheep | 147–150 | 16–18 | 50 | 55 |
| Guinea-pig | 60 | 16 | 3 | 28 |
| *Group B* | | | | |
| Rat | 22 | 10–12 | 12 | Term |
| Mouse | 20–21 | 10–12 | 11 | Term |
| Cat | 63 | 30–60 | Term? | 50 |
| Horse | 330–340 | 20–21 | ? | 150–200 |
| *Group C★* | | | | |
| Cow | 280–290 | 28–20 | ? | Term |
| Dog | 61 | 61 | Term? | Term? |
| Pig | 115 | 16–18 | Term | Term |
| Rabbit | 31 | 12 | Term | Term |
| Goat | 150 | 16–18 | Term | Term |

★N.B. These results do *not* mean that these species are *solely* dependent on pituitary and ovarian function. It is established that some are partially dependent on support from placental sources which are inadequate when acting alone (see Table 10.4).

essential proportion of these steroids. Removal of either the ovary or pituitary therefore results in abortion (see Table 10.2, group C). In other species (Table 10.2, groups A and B), pregnancy clearly does *not* depend totally on steroids secreted by pituitary–ovarian interactions, since, at varying periods during pregnancy, one or both glands may be removed without inducing abortion. Where does the endocrine support come from in these species? Some of the clearest answers to this question have come from studies on human pregnancy, which shows the least dependence on the maternal ovarian–pituitary axis of any species. We will therefore examine the endocrinology of human pregnancy in some detail and discuss the evidence from other species in relation to the human pattern.

*a Human pregnancy*

We saw in the first section of this chapter that within 2 weeks of fertilization the human conceptus synthesizes and releases the hormone hCG. This hormone then maintains the progestagenic activity of the corpus luteum. Within a further 2–3 weeks, the conceptus is also synthesizing all the steroidal hormones required for pregnancy and although the maternal corpus luteum remains active for the whole of pregnancy, it can be dispensed with after only 4–5 weeks and plays only a trivial role in total progesterone output at later stages. *The human conceptus thus shows a remarkable endocrine emancipation.* It asserts complete control over the mother.

Our understanding of the role of the conceptus in steroidogenesis has come from several sources. For example, much has been learnt from observations on steroid output and interconversions in abnormal pregnancies, and from isolation of tissues from different parts of the normal conceptus and assessment of their steroid biosynthetic capacity *in vitro.* However, by far the most informative and important data have come from *in vivo* studies using the fetuses of induced pregnancy terminations. Direct sampling of umbilical and maternal blood, and the infusions of minute quantities of radiolabelled steroid precursors into the maternal, fetal or placental circulations with subsequent analysis of their interconversions has provided information of great clinical value about the *sites* of steroid synthesis and interconversion in the conceptus (Fig. 10.3).

i Progesterone

The output of progesterone in late human pregnancy exceeds 200 mg/day and blood levels are high. Part of this rise in levels is accounted for by a three-fold increase in transcortin, which increases the proportion of bound progesterone in the blood (Table 2.1). The transcortin rise is stimulated by a direct effect of oestrogens on the liver. Another consequence of the transcortin rise is of course elevated cortisol levels during pregnancy. A side effect of the elevated progesterone levels is a ten-fold increase in aldosterone, an exaggerated version of the luteal aldosterone rise observed in the normal menstrual cycle (Chapter 7). In addition, oestrogens stimulate angiotensinogen output, further stimulating aldosterone output and thereby the $Na^+$ and water retention characteristic of pregnancy.

In pregnancies in which the embryo fails to develop, progesterone output by the extraembryonic components of the conceptus is often only marginally less than in normal pregnancy. For example, *choriocarcinoma* or *hydatidiform mole* (respectively malignant and benign tumours of the chorion) secrete progesterone in the absence of any embryonic tissue. Such observations suggest that the placental trophoblast itself is the principal source of progesterone. *In vitro* studies on the biosynthesis of progesterone by cultures of syncytiotrophoblast confirm this conclusion, and also indicate that the trophoblast can only use cholesterol (not acetate) as a substrate. The cholesterol is usually derived from the maternal rather than the fetal circulation. The rising output of progesterone through pregnancy appears to be completely autonomous; no external controlling mechanism has yet been discovered.

Clinical observations on patients whose ovaries have been removed at various times in early pregnancy indicate that the placenta is capable of synthesizing an adequate, supportive level of progesterone by 5–6 weeks of pregnancy. In the normal pregnant woman there is a plateau, or even a slight fall, in circulating concentrations of blood progesterone between 6 and 9 weeks. The plateau coincides with a marked fall in 17 $\alpha$-hydroxyprogesterone (an ovarian progestagen) and it is probable, therefore, that it is over this period of 6–9 weeks that the placenta normally takes over the major progestagenic support of pregnancy. Progesterone passes from the placenta into both the fetal and maternal circulations, concentrations in umbilical venous blood being high and infused radio-labelled progesterone being distributed widely within the embryo. Its role there, if any, is obscure, although much of it is probably metabolized by the fetal liver. A similar universal distribution of progesterone to maternal tissues is also seen. The principal single excreted metabolite of progesterone is pregnanediol, but it has been found that assays of urinary pregnanediol (some 15% of the progesterone produced) are not a useful indicator of *fetal well-being* and only a crude indication of *placental function*. The failure to correlate progesterone output with fetal well-being is not surprising since the fetus itself plays no part in progesterone synthesis. The poor correlation with placental function is a result mainly of the wide variation in progesterone plasma levels and excretion observed in normal pregnancies.

ii Oestrogens

The principal oestrogen in pregnancy is not oestradiol 17$\beta$ but the less potent oestrogen, *oestriol* (Table 10.3). The conceptus is responsible for the marked increase in secretion of oestrogens during pregnancy (Fig. 10.2), but, unlike progesterone, their secretion is severely impaired in cases of chorio-carcinoma and hydatidiform mole in which a fetus is lacking. Incubation of placental tisssue with radiolabelled cholesterol or pregnenolone yields labelled progesterone but not oes-trogens. This evidence indicates that the placenta alone is inadequate for the synthesis of oestrogens. Similarly, if the radiolabelled pregnenolone is infused into the circulation of isolated perfused human fetuses, little or no labelled oestro-gens are produced. Therefore, the fetus alone is inadequate for the synthesis of oestrogens. Only if labelled precursors are injected into an intact feto-placental unit does complete synthesis of oestrogens occur. We can conclude from these observations that in the human the *fetus and placenta cooperate to produce oestrogens.*

A series of observations on the result of infusing various radiolabelled steroid substrates has shown that the placenta is capable of synthesizing oestrogens from C19 androgens, no-tably DHA, but is not capable of synthesizing its own

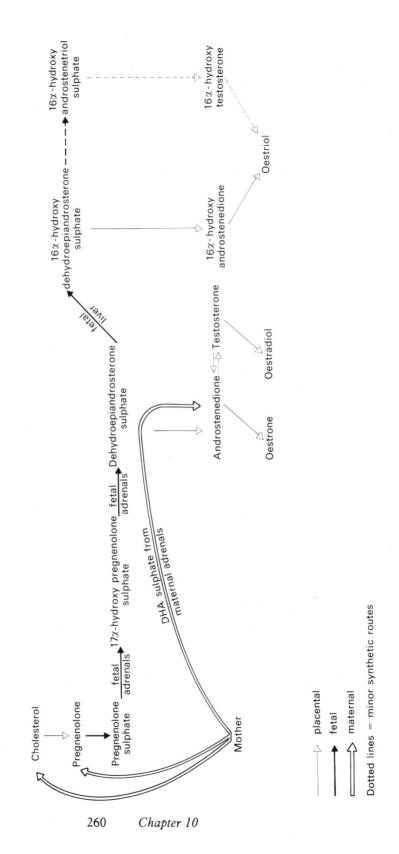

**Fig. 10.3.** Summary of principal routes by which human fetoplacental unit synthesizes oestrogens.

**Table 10.3.** Plasma levels of various steroids in women during late pregnancy.

| Steroid | Values in pregnancy (ng/ml) | | Values in luteal phase of cycle (ng/ml) |
|---|---|---|---|
| Progesterone | 125–200 | | 11 |
| Oestriol | 7 | } 113 | — |
| Oestriol conjugates | 106 | | |
| Oestrone | 7 | } 53 | 0.2 |
| Oestrone conjugates | 46 | | |
| Oestradiol 17β | 10 | } 15 | 0.2 |
| Oestradiol conjugates | 5 | | |

androgens from progestagens. The fetal zone of the adrenal, in contrast, synthesizes the C19 androgens (DHA and some androstenedione) but cannot aromatize them. The C19 steroids therefore pass to the placenta which converts them to oestrogens. The principal pathways involved are summarized in Fig. 10.3.

Two important points should be noted about this cooperation. First, the initial step in the fetal handling of steroids arriving from placenta or mother is conjugation. This conjugation occurs mainly in the fetal liver but also in the adrenals. Moreover, biosynthetic activity of the fetal adrenal involves sulphated intermediates. Conversely, the placenta deconjugates and releases free steroids into the maternal (and fetal) circulation. Thus, the *fetus sulphates, the placenta desulphates.* Conjugated steroids are both more water soluble and biologically inactive, so the conjugation may protect the fetus against untoward steroidal penetration into and activity within fetal tissues (for example, masculinization by high levels of weak androgens, see Chapter 1). This device nonetheless permits the high ambient levels of steroid required by the maternal organism for the maintenance of pregnancy.

Second, the synthesis of oestrone and oestradiol 17β occurs in the placenta from DHA sulphate derived either from the fetal adrenal (40%) or from the maternal adrenal (60%). Oestriol synthesis also occurs in the placenta but the 16α-hydroxylated substrate required comes solely from the fetal liver. It follows therefore that 16α-hydroxylated steroids, such as oestriol, should provide an indicator not only of placental function but also of fetal well-being. In pregnancies at risk, a consistent decline over a period of days of 16α-hydroxylated steroids correlates with fetal distress and may indicate that premature delivery should be induced. A parallel decline in

261     *Maternal Recognition and Support of Pregnancy*

oestradiol $17\beta$ is not seen because 50–60% of its synthesis by the placenta utilizes maternal not fetal DHA sulphate. In anencephalic fetuses, in which fetal adrenal function is impaired, the ratio of oestradiol + oestrone:oestriol is greatly increased, as would be expected. Thus, $16\alpha$-hydroxylated steroids have an important diagnostic value.

iii Corticosteroids

Blood levels of cortisol rise in pregnancy, and the rise can be mimicked by oestrogen injection into non-pregnant women. The elevated levels are partly due to a decrease in the metabolism of free cortisol, but predominantly due to an oestrogen-stimulated synthesis of transcortin from 3.5 mg % to 10 mg %. The transcortin binds both cortisol and progesterone and explains the very high level of total progesterone observed in human pregnancy.

iv Human chorionic gonadotrophin

We saw earlier in this Chapter that hCG was critical to the initiation of pregnancy by extending luteal life from 2 to 6–7 weeks when the placental steroids take over. The rising blood levels and critical role of hCG over this period make it a useful hormone to measure when undertaking pregnancy tests. The action of hCG may be limited to this short-term luteal support, since its blood levels only remain elevated for the first 8 weeks of pregnancy (Fig. 10.2). The fall in blood hCG concentrations correlates with the falling levels of $17\alpha$-hydroxyprogesterone (a luteal product) and the emancipation of the feto-placental unit from ovarian dependence. Low levels of hCG are detectable for the remainder of pregnancy (although in the rhesus monkey rhCG levels are trivial from 40 days onwards.)

v Human placental lactogen (human chorionic somatomammotrophin) and prolactin

As hCG levels decline syncytiotrophoblastic secretion of a second protein hormone rises steadily. Human placental lactogen (hPL) is a polypeptide of 190 amino acids with a molecular weight of 21–23 000. It has both growth hormone and prolactin-like properties. In addition, release of prolactin itself is stimulated by oestrogens (see Chapter 5), and plasma concentrations reach over 200 ng/ml in maternal plasma by the last trimester.

b Pregnancy in other species
i Steroids

Elevated steroid levels are a feature of pregnancy in most mammals studied although it is clear, despite incomplete data, that the patterns and levels of steroids vary considerably from species to species (Fig. 10.2). As we saw in Table 10.2, the degree of independence from the pituitary–ovarian axis varies, but even in species such as the cow and pig which require both pituitary and ovary to be present throughout pregnancy, a fetal steroid contribution also occurs (Table 10.4). Conceptuses of both the horse and sheep become independent of the ovary by about one-third of the way

**Table 10.4.** Major sites of hormone synthesis during established pregnancy.

| Species | Progestagens | Oestrogens | Gonadotrophins |
|---|---|---|---|
| Man | Placenta | Fetal adrenal + placenta | Placenta (hCG and hPL) |
| Horse: early | Corpus luteum | Ovarian follicles | Placenta (PMSG) |
|     mid–late | Placenta | Fetal gonad + placenta | — |
| Sheep: early | Ovary and placenta | Placenta (oestradiol and oestrone sulphate) | — |
|     mid–late | Placenta | Ovary (oestrone), placenta (oestrone sulphate) | Placenta (oPL) |
| Cow | Ovary (and placenta) | Placenta (and ovary) | Pituitary |
| Pig: early | Ovary | Placenta | Pituitary |
|     late | Ovary (and placenta) | Placenta | Pituitary |

through pregnancy. Both are comparable to the human with a fetal source of DHA sulphate being deconjugated and aromatized by the placenta. In the horse, the principal oestrogens formed are oestrone together with two oestrogens found only in equids: equilin and equilenin. The fetal DHA sulphate used for aromatization is derived in the horse not from the fetal adrenal but from the interstitial tissues of the fetal gonads. The spectacular increase in weight of the equine fetal gonads between 100 and 300 days of gestation is due mainly to hypertrophy of interstitial glands. By birth, the gonads have regressed. The sheep fetus, like the human, synthesizes sulphated DHA in the adrenal but maternal sources of DHA may also be available. In addition, the sheep placenta, unlike the human placenta, can undertake conversion of progesterone to oestrogens, and this property is of great importance to parturition (see Chapter 12).

ii Protein hormones

Protein hormones of pregnancy have also been detected in the sheep, goat and horse. The latter half of sheep and goat pregnancies is characterized by the production of ovine and caprine placental lactogens from binucleate cells of the chorion. Between days 40 and 120 of pregnancy in the mare a gonadotrophin with both FSH- and LH-like activity, called pregnant mares' serum gonadotrophin (PMSG), is secreted from the trophoblastic cells of the endometrial cups. Control of its secretion is not known, although the quantity of its secretion is determined by the genetic constitution of the mare. One effect of the PMSG is to promote follicular growth and even secondary ovulation in the maternal ovaries from Day 40 to 120 of pregnancy; as a result, secondary corpora lutea may develop. These remain an active source of progesterone until the decline of PMSG and the take-over by placental progesterone at around 140-150 days.

In a number of species, but notably the guinea-pig and pig, a protein hormone called *relaxin*, of molecular weight 5600–5800 and composed of two polypeptide chains, has been detected in blood at low levels during pregnancy, rising just prior to parturition. This hormone appears to be produced by the corpus luteum of pregnancy, and has also been detected in humans. It has a variety of actions that will be discussed in the context of parturition (see Chapter 12).

*c Summary*

Three comments of general relevance need to be made when comparing the available data on pregnancy hormones. First, the tendency for the feto-placental unit to take over endocrine control from the mother in whole or in part is seen in most species. Second, whilst the levels of plasma oestrogens and progesterone rise, there is great species variation in the absolute level achieved (cf. 160 ng progesterone/ml in man, 8 ng/ml in the cow). Third, the patterns of plasma steroids recorded through pregnancy differ markedly among various species. The large variation, both qualitative and quantitative, in plasma hormone levels is difficult to explain. This difficulty is compounded by the fact that we have very little idea of the functions of the steroid and protein hormones during pregnancy. These problems are addressed in Chapters 11 to 13.

**Further reading**

CIBA Foundation Symposium 62. *Maternal Recognition of Pregnancy*. Excerpta Medica, 1979. (New Series).

Fuchs, F, Klopper A (Eds). *Endocrinology of Pregnancy*. 2nd Edn. Harper Row, 1977.

Greep RO, Kublinsky MA (Eds). *Frontiers in Reproduction and Fertility Control*. Part 2. M.I.T. Press, 1977.

Short RV (Ed.). *British Medical Bulletin on Reproduction*. Medical Department, British Council, 1979.

Steven DH (Ed.). *Comparative Placentation*. Academic Press, 1977.

# Chapter 11
# The Fetus and its Preparations for Birth

The fetus is far from being a quiescent, passively-growing product of conception tucked neatly away in its protected uterine environment. Whilst undoubtedly dependent on the mother's nutrient supplies for its growth and survival, it nonetheless enjoys considerable independence in the regulation of its development. Indeed, the fetus also exerts effects on *maternal* physiology via hormones secreted by the placenta into the maternal circulation and which in part determine the mother's ability to meet the metabolic requirements of pregnancy.

In this chapter we will discuss several topics in fetal and neonatal physiology. Emphasis will be placed on the human and on those developmental processes in the major bodily systems where regulatory mechanisms must be established, in order to maintain the internal environment of the fetus in its transition from a uterine to an external, independent existence at parturition.

## 1 Fetal growth

The pattern of fetal growth is determined primarily by the genome of the fetus but additional fetal and maternal factors modulate the effects of its expression. Insulin-like growth factors (IGFs) produced by a large range of fetal cell types, but particularly by the fetal liver, do seem to provide a major endocrine stimulus to fetal growth. The IGFs are analagous to those produced postnatally, but may not be identical, special fetal IGFs being a possibility. Fetal thyroid hormones also stimulate growth in the latter part of pregnancy. Growth hormone, although produced by the fetus, is not effective in stimulating fetal growth, unlike the situation postnatally.

Maternal nutrition and health are clearly of great significance, the effects of severe malnutrition on fetal well-being and neonatal survival being well known. Additionally, factors such as *parity* (*primiparous* mothers have smaller babies than *multiparous* mothers), maternal size, multiple pregnancies and self-inflicted damage such as smoking (see Chapter 9) may all affect birth weight. It is reasonable to assume that growth is an indicator of fetal well-being. Babies of low birth weight (less than 2500 g) may be so either because they are born prematurely or because of some retarding influence on fetal growth in a baby born full-term. As many as one-third of low birth weight babies come into the latter category and are said to be *'small for dates'* (defined as being of a weight that is two standard deviations below the weight expected of a baby of a particular gestational age).

The rate of fetal growth is relatively slow up to the twentieth week of pregnancy but accelerates to reach a maximum around weeks 30–36, thereafter declining until birth. This is summarized in Fig. 11.1a, where it can also be seen that a postnatal peak in growth velocity occurs during week 8. Protein accumulation occurs early in fetal development to reach its maximum, about 300 g, by week 35 and precedes fat deposition, most of which is subcutaneous, and which only comes to exceed the weight of protein by week 38. By term some three times as much energy is stored as fat rather than protein. The relative amount of water in the fetus (95% in the young fetus) also decreases during development, as the proportion of solids increases. Amniotic fluid also increases in volume until week 34 of pregnancy, after which

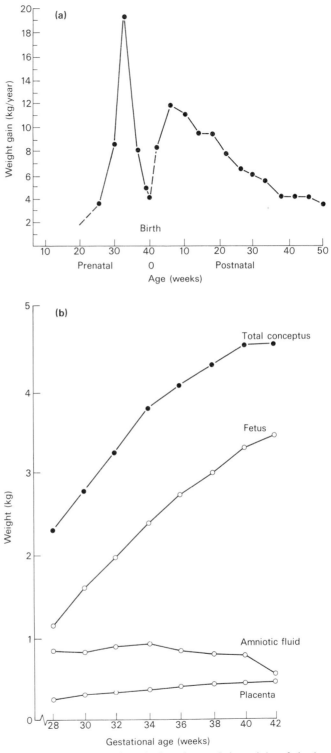

**Fig. 11.1.** (a) Change in velocity of growth in weight of singleton fetuses and children. (b) Weight changes of conceptus, fetus, placenta and amniotic fluid during pregnancy.

*The Fetus and its Preparations for Birth*

time it declines. The placenta increases in size slowly and steadily until birth. However, the rapid growth of the fetus before birth means that the ratio between placental and fetal weight falls markedly during the later stages of pregnancy (Fig. 11.1b). This observation has led to the suggestion that the placenta may limit the transport of nutrients to the fetus late in pregnancy and thereby underly part of the fall in growth velocity after week 36.

The main ingredient of the fetal diet is carbohydrate and about half the calories needed for growth and metabolism come from glucose, the remainder coming equally from amino acids and from lactate formed from glucose in the placenta. The fetus must also, if normal growth and development are to occur, be provided with the basic building materials—essential amino acids, fatty acids, vitamins and minerals. Most, if not all, are transported selectively from mother to fetus across the placenta. Thus, a consideration of metabolic activity in the fetus requires an understanding of transport across the maternal–fetal interface.

## 2 Maternal metabolism, fetal metabolism and placental transport

The discrete nature of the maternal and fetal circulations, separated by cellular and acellular layers, confer on the placenta an important *barrier* property. Simple diffusional exchange between the circulations will only therefore be significant with low molecular weight molecules, such as blood gases, $Na^+$, water, urea, or with non-polar molecules, such as cholesterol and non-conjugated steroids. Hexose sugars, conjugated steroids, amino acids, nucleotides, water-soluble vitamins, plasma proteins and cells will not gain access to the fetal circulation unless either special transport mechanisms exist or the integrity of the barrier is breached. 'Bleeds' across the placenta do occur, but are rare (except at parturition) and probably occur mainly in a feto-maternal direction.

The facility with which *diffusional exchange* occurs varies during pregnancy. In early pregnancy terminal villi in the human placenta are large, with a diameter of 150–200 $\mu$m and the fetal vessel is centrally located beneath a 10 $\mu$m layer of syncytiotrophoblast. Thus, metabolites must diffuse a considerable distance between the two circulations. Furthermore, since the syncytiotrophoblast is itself metabolically active, for example synthesizing chorionic gonadotrophin, it will intercept and utilize some of the maternal metabolites such as oxygen. As pregnancy progresses, the villi thin to 40 $\mu$m diameter, and the fetal vessel occupies a more eccentric position indenting the overlying syncytiotrophoblast to leave only a 1–2 $\mu$m layer separating it from the maternal blood in the intervillous space. This altered anatomical relationship not only increases the capacity for diffusional exchange but also reduces the consumption, by the trophoblast, of oxygen during its diffusional passage. A similar thinning of diffusional barriers also occurs in the epithelio-chorial placenta of the sheep.

Although the interspecies variation in the numbers of layers separating the two circulations may influence the relative efficiency of diffusional exchange (for example, diffusional movement of $Na^+$ across the placenta is considerably faster in the haemochorial primate than the epithelio-chorial ungulate) more freely diffusible molecules such as $O_2$ are much more affected by blood flow than any consideration of 'barrier thickness'. The important point to grasp about the microstructure of the placental interface is that it either permits *adequate* diffusional exchange, i.e. has a large *safety factor*, or employs *special transport systems* to promote selective transport of less freely diffusible but essential substances, such as glucose, fructose, amino acids and various proteins. Both methods of placental transport will be affected critically by circulatory factors that were discussed in Chapter 9. Here we consider the transport mechanisms themselves and more importantly, the changes in maternal and fetal metabolism and physiology that determine the blood levels of metabolites and thereby the gradients across the placental interface.

*a Transport of oxygen and carbon dioxide*

During pregnancy, maternal oxygen consumption is increased compared with the non-pregnant female both at rest and during exercise. However, when allowance is made for the growing tissue mass of the conceptus, the oxygen consumption/lean body mass is not significantly changed. Physical working capacity and the efficiency with which work is performed are not significantly affected by pregnancy. Cardiac output does increase during the first third of pregnancy by about 30% but little more thereafter. Increases in both stroke volume and rate account for this increase. Mean arterial systemic blood pressure increases slightly, but the increase in output is mainly accommodated by a reduction of peripheral resistance by as much as 30% associated with the increasing demands of the conceptus (see also Chapter 9, Section 3c). Blood volume rises by up to 50% near term in humans, partly due to a 20–30% increase in erythrocytes and partly due to increasing plasma volume (up 30–60%).

Oxygen is needed in relatively continuous supply by the fetus because fetal stores of the gas are very small: a 3 kg fetus near term requires 18 ml $O_2$/min, but stores are only 36 ml or 2 minute's worth! As a non-polar molecule, $O_2$ readily diffuses across the placental interface. Likewise $CO_2$, generated by fetal metabolism, has a diffusion constant 20 times higher than $O_2$ and freely diffuses across the placenta. Maternal pulmonary ventilation increases by 40% during pregnancy, possibly due to a direct effect of progesterone on respiratory mechanisms in the brain stem. A decrease of about 25% in

**Table 11.1.** $O_2$ and $CO_2$ composition of human maternal and fetal blood.

| | Maternal blood | | Fetal blood | |
|---|---|---|---|---|
| | Arterial | Venous | Umbilical artery | Umbilical vein |
| 1 Oxygen tension ($pO_2$) mm | 90 | 35 | 15 | 30 |
| 2 % saturation $O_2$ | 95 | 70 | 25 | 65 |
| 3 Oxygen content vol. % | 14 | 10 | 5 | 13 |
| 4 $CO_2$ tension ($pCO_2$) mm | 30 | 35 | 53 | 40 |
| 5 pH | 7.43 | 7.40 | 7.26 | 7.35 |

maternal $pCO_2$ results, with a corresponding fall in bicarbonate concentration and a slight increase in pH. The reduction in bicarbonate buffering leads to more marked changes in pH with exercise than observed in non-pregnant females.

The gradients of the gases at the transplacental interface may be estimated from the figures shown in Table 11.1. Clearly the tension of $O_2$ in the fetal blood is low and that of $CO_2$ high relative to the tension of the gases in maternal blood. Gradients to drive diffusional exchange therefore exist. However, lines 2 and 3 of Table 11.1 reveal that although the $pO_2$ in the oxygenated fetal venous blood leaving the placenta is relatively low, the oxygen saturation and content is not much less than maternal arterial blood (line 3 of Table 11.1). Thus, a much lower oxygen tension leads to a very similar oxygen content. Clearly the 'oxygen trapping' capacity of the fetal blood must be more effective than that of the maternal blood. This *higher affinity* of fetal blood is shown graphically in Fig. 11.2. How is this higher affinity achieved?

The earliest site of *erythropoiesis* in implanting mammalian embryos is the yolk sac mesoderm (Fig. 9.7). The primitive *embryonic erythrocytes* formed are nucleated, and are replaced later during embryogenesis by *fetal erythrocytes* (also nucleated) that are made in the liver. Finally, the spleen and bone marrow take over erythropoiesis as parturition approaches. Embryonic and fetal erythrocytes, like those of the adult, contain haemoglobin as the oxygen-carrying molecule. The embryonic and fetal haemoglobins consist of four globin chains coupled to a haem group. However, they differ from the adult in that the constituent globin chains are *not* two $\alpha$ and two $\beta$ chains. Rather, the embryonic haemoglobin contains two $\zeta$ and two $\varepsilon$ chains and the fetal haemoglobin two $\alpha$ and two $\gamma$ chains. Each globin chain is coded for by a different gene, and during development a programme of gene switching occurs to yield the embryonic, fetal and adult sequence. It is the different globin-chain composition that results in the different oxygen binding curves (Fig. 11.2). In the adult, the highly-charged molecule 2,3-diphosphoglycerate (DPG) tends to drive off oxygen and to stabilize the structure of the

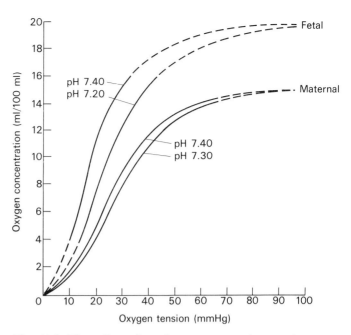

**Fig. 11.2.** The effect of varying oxygen tension on the oxygen content of human maternal and fetal blood under conditions likely to be encountered in the placenta. Note the greater oxygen concentration in fetal blood at any given pO$_2$. Note also that the pH shifts that will be occurring in the placenta (falling pH in maternal blood and rising pH in fetal blood) will further facilitate unloading of oxygen from maternal blood and its uptake by fetal blood. (After Metcalfe *et al. Physiol. Rev.* 1967; **47**: 782.)

deoxyhaemoglobin. It does this by binding to sites on the $\beta$ chain that are exposed only on deoxygenation, and therefore it reduces the opportunity for oxygen binding. A higher pO$_2$ is needed to load oxygen. The embryonic and fetal haemoglobins lack a $\beta$ chain, and the $\beta$ equivalents (e.g. the $\gamma$ and $\varepsilon$ chains) have less binding sites for DPG and so are relatively insensitive to its action. Therefore, at any given pO$_2$ fetal and embryonic haemoglobins will bind more oxygen than will maternal haemoglobin. In some species, the fetal DPG levels may also be lower, which will assist the O$_2$ loading process.

In addition, as Fig. 11.2 shows, a double Bohr effect occurs in the placenta to facilitate exchange of about 10% of the O$_2$, the fall in pH of maternal blood leading to release of O$_2$ and the rise in fetal pH facilitating uptake of O$_2$. Under non-pathological conditions the maternal pO$_2$ and placental perfusion are unlikely to be limiting, and thus the rate of transfer of O$_2$ across the placenta will vary simply with the pO$_2$ of the blood in the umbilical circulation. Thus, the *level of fetal oxygenation will be regulated by the fetal requirement for oxygen.*

*b Transport of water and electrolytes*

During pregnancy, maternal Na$^+$ and water retention is increased for two reasons. First, aldosterone secretion rises ten-fold to compensate for the competitive binding of progesterone to the renal receptor. Second, oestrogens stimulate a four-to-six-fold rise in angiotensinogen output by the liver. Exchange of water between mother and fetus occurs at two main sites, the placenta and the remainder of the non-placental chorion where it abuts the amnion internally (Figs 9.7, 8 and 9). Little is known about the relative quantitative contribution of each route, particularly early in pregnancy, although the placenta is suspected to be the main site of exchange. Most studies in man and in animals indicate that both the amnion and chorion are freely permeable to water molecules. It is not known to what extent water moves by simple diffusion, but there is no evidence to suggest water crosses by active transport or is secreted by the membranes themselves.

At the placenta, intermittent and small osmotic pressure gradients could probably transfer quantities of water in excess of fetal requirements during pregnancy in view of the high permeability of placental membranes. However, the most likely cause of large water movement is a hydrostatic pressure difference between maternal and fetal blood. This would cause water exchange by bulk flow, with more water moving than would be predicted by the laws of diffusion. Present technical limitations have made it extremely difficult to measure hydrostatic pressures in the intervillous space and fetal–placental exchange vessels. However, hydrostatic pressures in maternal and fetal circulations cannot be great since, if large in the maternal–fetal direction, the fetal vessels would collapse and fetal–placental exchange would be impaired, while if large in a fetal–maternal direction an excessive amount of water would move from fetal to maternal blood, dehydrating the fetus. Most likely, then, small or intermittent hydrostatic gradients are responsible for moving the large amounts of water necessary for the fetus.

Large quantities of sodium and other electrolytes clearly cross the placenta, although little is known about net transfer rates *in vivo*. While active pumping of sodium and potassium into and out of trophoblast cells has not been demonstrated directly, it may be that these cells do in fact actively transport these ions. In general, it seems very likely that univalent ions such as sodium and potassium cross the placenta by simple diffusion.

*c Bilirubin metabolism and transport*

Bilirubin is a lipid-soluble product of haemoglobin catabolism and is present in fetal, neonatal and adult plasma in both a free form and bound to serum albumin. Bilirubin passes readily across the placenta in either direction by diffusion. In the mother, bilirubin is transported in the blood to the liver where the enzyme UDP glucuronyl-transferase converts it to *bilirubin glucuronide*, and this polar conjugate cannot cross the

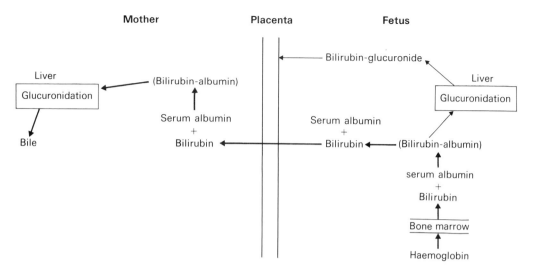

**Fig. 11.3.** The metabolism of bilirubin. The heavy arrows represent the principal route of metabolism during pregnancy. Fetal glucuronidation does not occur significantly until after birth.

placenta but is excreted in the maternal bile (Fig. 11.3). The fetal liver lacks the conjugating enzyme, which is first synthesized late in pregnancy as one of the last clusters of fetal liver enzymes, so preparing the neonate for its own bilirubin excretion. Thus, the combination of a differential diffusion of the non-polar bilirubin compared to its polar conjugate and the differential conjugation capacity of the mother, ensures adequate elimination of bilirubin from the fetus (Fig. 11.3).

Bilirubin transport and metabolism is important clinically. *Hyperbilirubinaemia* is common in the newborn and in its mild form (so-called 'physiological hyperbilirubinaemia') is known as *jaundice* because of the associated yellow coloration of the skin and mucous membranes. A number of factors may be associated with its occurrence such as accelerated red blood cell breakdown in cases where an infant has a large blood volume, as can arise if the umbilical cord is clamped too late, when some 30% of babies may show jaundice. Bilirubin conjugation by the neonate may be depressed as a result of dehydration, low caloric intake or even the actions of steroid hormones, particularly progestagens, which actually inhibit conjugation.

Severe, or *pathological hyperbilirubinaemia* is particularly dangerous since it can result in *encephalopathy* (also known as *kernicterus*). This severe hyperbilirubinaemia may arise as a result of much increased fetal red cell breakdown such as occurs if the mother develops an immune response to fetal erythrocytes (see later in this chapter). Various factors impairing maternal bilirubin conjugation can also give rise to bilirubin toxicity. Damage to the fetal liver as a result of injection of drugs, maternal diabetes, congenital defects and prematurity may all be associated with poor conjugation of bilirubin.

The hyperbilirubinaemia (or its consequences) may be prevented by a number of approaches. Exposure to ultraviolet light (phototherapy) causes the breakdown of bilirubin to a non-toxic product; early feeding prevents hypoglycaemia and dehydration; early clamping of the umbilical cord decreases red cell volume and hence bilirubin plasma levels; exchange transfusion is used in severe cases or if the above fail.

*d Glucose transport and carbohydrate metabolism*

The fetus has little capacity for gluconeogenesis largely because the necessary enzymes, although present, are inactive at low arterial $pO_2$. At birth, when arterial $pO_2$ rises, gluconeogenesis is initiated. The fetus must therefore obtain its glucose from maternal blood, in which the levels depend on both maternal nutritional status and the integrated endocrine control mechanisms which maintain free plasma glucose levels within narrow limits. Thus, the secretion of insulin from the maternal pancreas prevents glucose levels from rising too high by increasing glucose utilization for glycogen and fat synthesis and storage. Conversely, absorption of glucose from the gut and gluconeogenesis (by the utilization of glycogen stores under the action of catecholamines, corticosteroids, glucagon and growth hormone) help prevent maternal glucose levels from falling.

Early on, progesterone apparently increases maternal appetite and stimulates the deposition of glucose in fat stores. Late in pregnancy, human placental lactogen, via its growth hormone-like activity, may assist in the mobilization of fatty acids from these fat 'depots' for use by the mother as an energy source. This source of fatty acids can be important as pregnancy proceeds because maternal tissues become progressively less sensitive to insulin during this time. The cause of the insulin-insensitivity is uncertain, but its consequences are two-fold. First, maternal blood glucose is more available for capture by placental transport mechanisms and thus for transfer to the growing fetus. Second, *latent diabetes mellitus* in women may manifest itself overtly for the first time from mid-pregnancy onwards.

The glucose is conveyed to the fetal circulation by facilitated diffusion. Thus, concentration gradients drive the transfer, but specialized carrier mechanisms carry much of the glucose thereby enhancing the rate of transfer.

The levels of glucose in the fetus will be directly related, then, to those in the mother, and although mechanisms for *regulating* blood glucose concentrations develop in the fetus, they seem not to mature fully until birth. However, the *rate at which glucose is utilized* by growing fetal tissues is probably largely determined by the actions of insulin secreted by the *fetal pancreas*. This is best exemplified by the clinical disorder, diabetes mellitus, in which it is well known that babies born to diabetic mothers are overweight. The poor insulin response of the mother to a glucose load results in elevated plasma glucose concentrations which, as we have seen, cause a parallel increase in fetal blood levels and hence stimulate insulin release by the *fetal* pancreas. In this way fetal growth and fat storage are promoted and an overweight noenate results. It would seem, then, that the *rate of glucose utilization* rather than glucose availability is the primary *determinant of fetal growth* under normal nutritional conditions.

Glucose is stored in two main forms. The storage of glucose as glycogen, particularly in the fetal liver, is important if the metabolic needs of the neonate are to be provided until feeding begins. A progressive increase in the activity of the fetal adrenal cortex near term (see below) is especially important in promoting the deposition of liver glycogen. Indeed, fetal adrenal hypoactivity is associated with major reductions in liver glycogen stores, a situation reversed by corticosteroid replacement, while exogenous ACTH or corticosteroids enhance liver glycogen content in normal fetuses.

The high concentration of glycogen in fetal cardiac muscle probably explains why the heart can maintain its contractile activity in the face of severe hypoxia. The brain, however, since it has no glycogen stores, relies totally for its activity on a supply of glucose from the circulation. Thus, we see why hypoglycaemia can have such deleterious effects on the brain, especially if accompanied by hypoxia, the risk being higher in a fetus with low cardiac glycogen reserves (as occurs through glucoprivation accompanying placental insufficiency). Postnatal hypoglycaemia may also be seen in babies born to a diabetic mother, since the reduction in the maternal supply of glucose at parturition is not immediately accompanied by a corresponding cut in the fetal hypersecretion of insulin. A sharp drop in noenatal blood glucose results. Fortunately, although neonatal hypoglycaemia is a major cause of brain damage, and thereby mental subnormalities, it is highly treatable.

The storage of glucose as fat in the fetus is regulated primarily by insulin. Thus, when glucose levels are maintained optimally, such as occurs when maternal nutrition is good, the glucose available after the requirements for growth are fully met is diverted by fetal insulin into fat stores. In conditions where glucose supplies to the fetus are reduced, the needs of growth have priority over storage and the newborn consequently has an emaciated appearance.

In addition to these storage depots of white fat, *brown adipose tissue* or 'brown fat' is also found in the fetus, newborn and infant. It is deposited in five sites: (i) between the scapulae in a thin diamond shape, (ii) small masses around blood vessels in the neck, (iii) in the axillae, (iv) in the mediastinum between oesophagus and trachea as well as around the internal mammary vessels and (v) a large mass around the kidneys and adrenal glands. It is different in structure to white fat, the lipid being distributed multilocularly with, in close apposition, large numbers of mitochondria bearing prominent cristae. These deposits of brown fat are of immense importance in temperature regulation, having the ability to generate large quantities of heat, independent of other mechanisms such as increased muscular movement and shivering. Indeed, this form of heat production is termed *non-shivering thermogenesis*. The molecular basis of the thermogenic response of brown fat is not entirely clear but seems to be mediated via the sympathetic nervous system. As neonatal development proceeds, brown fat becomes of less importance thermogenically, and regresses. However, since brown fat is a most effective user of calories, there is considerable interest currently in finding ways to 'reactivate' it in the adult as a means of regulating body weight and reducing obesity.

*e Amino acids and urea*

Amino acids in the adult are derived directly from digestion of both dietary and endogenous protein, and in the case of the non-essential amino acids, by interconversion from other amino acids. Deamination of amino acids during catabolism results in release of ammonia, levels of which are kept low by conversion to urea in the liver. Urea constitutes the major source of urinary nitrogen excretion.

Traditionally, protein supplements have been considered an important and desirable feature of pregnancy diets, to cope with the increased protein synthetic demands of the growing conceptus. However, except in cases of extreme malnourishment there is little or no evidence to support this view, and protein supplements appear largely to be used for conversion as energy sources. Since the fetus clearly does grow and thereby increase the total protein content of the pregnant mother, where do the amino acids required come from? There is no evidence for improved maternal digestion of dietary protein (already exceeding 95% in the non-pregnant state). However, the efficiency of the intermediary metabolism of amino acids does appear to increase. Thus, urea excretion falls markedly in pregnancy suggesting that the same intake of dietary amino acids is being utilized more efficiently. A reduced capacity of the maternal liver to deaminate amino acids can be detected during pregnancy, and results largely from the anabolic action of progesterone. Thus, the human conceptus via its production of progesterone regulates maternal metabolism of amino acids such that no extra dietary protein intake is required to support fetal growth. The 'extra' amino acids retained in the mother are transported actively to the fetal circulation, and fetal urea, produced by the limited catabolism of fetal amino acids, diffuses passively into maternal blood with its already lowered endogenous urea levels.

*f Iron, calcium and vitamins*

*Iron* is present in both fetal and maternal blood in both an unbound form and bound to the protein *transferrin*. However, fetal blood contains iron at two to three times the concentration of maternal blood, and since the iron-binding capacity of fetal and maternal blood is similar, the higher fetal serum iron concentration is due to an increased concentration of unbound iron which accumulates through active transport across the placenta. Trophoblast cells contain intracellular iron as a ferritin complex which is involved in iron transport.

In pregnancy, there is a high incidence (40–90%) of maternal iron deficiency based on serum iron levels, although anaemia is less frequent. Additional iron in pregnancy is required to replace an average loss of 300 mg to the fetus, 50 mg to the placenta and 200 mg in blood loss after labour. In addition about 500 mg is required to increase the maternal haemoglobin mass, but this amount is not ultimately lost. The net need is about 550 mg, therefore. However, absorption is enhanced from 10% in the first trimester to 30% or more in the third. A food intake of 12 mg/day and 10% absorption would provide an estimated 335mg and such a diet would be sufficient for the majority for pregnant women, but in general a supplement of 100–150 mg/day iron, especially in the second half of pregnancy, is recommended.

*Folic acid* and *vitamin B12* are two essential compounds obtained from the diet and have very fundamental metabolic actions which make them vital for normal fetal development. They are provided to the fetus at the expense of maternal stores, so making *fetal* deficiency unlikely. However, vitamin deficiencies in the *mother* may affect the fetus indirectly via the resultant maternal metabolic disorders. It is perhaps pertinent to remember here that folic acid and its co-enzyme forms are involved in 1-carbon transfers and thereby in nucleoprotein synthesis and amino acid metabolism—processes imperative for normal fetal development. Vitamin B12 is a co-factor in folate metabolism, in the metabolism of some fatty acids and branched amino acids.

In pregnancy, serum folate levels progressively decrease towards term, and in megaloblastic anaemias of pregnancy they are generally lower than in the folate-deficient anaemias of non-pregnant women. Red cell folate activity accounts for 95% of the blood folate, levels falling progressively from the first to third trimester. The incidence of folate deficiency varies considerably according to the type of population studied, but it is of the order of 2%. The incidence of megaloblastic anaemia is much lower. The precise consequences of folate deficiency in the fetus have not been established definitively, but appear to be associated with

277    *The Fetus and its Preparations for Birth*

prematurity and abortion. Folic acid supplements are generally considered to be desirable in pregnancy as a means of preventing the development of maternal anaemia.

Vitamin B12 is absorbed relatively slowly across the mucosa of the terminal ileum to be transported in blood mainly by a protein, *transcobalamin II*. Serum levels of vitamin B12 decrease during pregnancy, falling to a minimum at 16–20 weeks, although this does not indicate a true deficiency, which is rare in pregnancy; indeed, B12-deficient women are unlikely to become pregnant.

The fetus places a considerable demand for calcium on the mother, largely during ossification in the last trimester. However, the average daily maternal intake of calcium appears to greatly exceed adequacy, there being a positive calcium balance throughout pregnancy partly due to the more efficient absorption of dietary calcium from the intestine. This efficiency increases further during lactation and is a result of higher levels of parathormone stimulating conversion of vitamin D to the active derivative $1\alpha$, 25-dihydroxy-vitamin D3 by the kidney. Lack of vitamin D during pregnancy is more likely to lead to maternal osteomalacia than is dietary deficiency of calcium. The calcium is transferred to the fetal circulation by an active transport mechanism that generates high fetal levels of plasma calcium.

## 3 Amniotic fluid

The composition and turnover of amniotic fluid has become a subject of renewed interest with the advent of the sampling procedure, *amniocentesis*, in the diagnosis of fetal abnormalities.

The volume of amniotic fluid increases during pregnancy from about 15 ml at 8 weeks postconception to 450 ml at week 20, after which time net production declines to reach zero by week 34. The composition of amniotic fluid (see Table 11.2) suggests that it is a dialysate of maternal and/or fetal fluids, save that the concentration of protein is only 5% of that in serum. During the last third of pregnancy, total solute concentration in amniotic fluid falls while the concentrations of urea, uric acid and creatinine increase markedly.

Amniotic fluid is in a dynamic state, complete exchange of its water component occurring every 3 hours or so. There are several routes by which water and solutes enter and leave the amniotic cavity. Exchange with the fetus occurs via its gastrointestinal, urinary and respiratory tracts, and until week 20 or so, the skin of the fetus is not keratinized and so amniotic fluid (osmolarity 260–280 mosmol/l) exchanges freely with fetal extracellular fluid. In addition, the amniotic epithelium, which during the later stages of pregnancy expands to fill the extra-embryonic coelom and becomes apposed to both chorion and umbilical cord (see Fig. 9.8), is a route of exchange.

**Table 11.2.** Compositions of amniotic fluid in early and late pregnancy and the full-term maternal and fetal serum.

| Fluid | Total osmotic pressure (mosmols) | Na (mM) | Cl (mM) | K (mM) | Non-protein nitrogen (mg/100 ml) | Urea (mg/100 ml) | Uric acid (mg/100 ml) | Creatinine (mg/100 ml) | Total protein (gm/100 ml) | Water content (%) |
|---|---|---|---|---|---|---|---|---|---|---|
| Amniotic fluid. First and second trimester | 283 | 134 | 110 | 4.2 | 24 | 25 | 3.2 | 1.23 | 0.28 | 98.7 |
| Amniotic fluid. Third trimester | 262 | 126 | 105 | 4.0 | 27 | 34 | 5.6 | 2.17 | 0.26 | 98.8 |
| Maternal serum. Full term | 289 | 137 | 105 | 3.6 | 22 | 21 | — | 1.55 | 6.5 | 91.6 |
| Fetal serum. Full term | 290 | 140 | 106 | 4.5 | 23 | 25 | 3.6 | 1.02 | 5.5 | — |

Indeed the finding that removal of the fetus does not prevent formation of amniotic fluid in the rhesus monkey emphasizes the capability of the amnion in this context.

It is clear that during early life the fetus swallows from 7 ml amniotic fluid per hour at week 16 to around 120 ml per hour at week 28. Moreover, radiopaque dye injected into amniotic fluid becomes concentrated in the fetal stomach, presumably because after swallowing, most of the water is absorbed by the fetus. Thus, *the fetal gastrointestinal tract is a pathway for removal of amniotic fluid* from the amniotic cavity. The *fetal lungs produce a fluid* which fills the alveoli and also contributes to amniotic fluid since surfactant compounds (see below) are found within it during late pregnancy. Micturition also occurs into the amniotic cavity, and undoubtedly, *the fetal urinary system plays an important role in maintaining amniotic fluid volume.* It has been estimated using ultrasonic techniques that, at 25 weeks, some 3–5 ml per hour of hypotonic urine is formed rising to 26 ml per hour (500–600 ml per day) by week 40, after which time it drops rapidly. The *relative* importance of these pathways for exchange at various times of gestation has not, however, been determined although the fetal kidney would seem to be a principal source of amniotic fluid later in pregnancy. Renal agenesis causes *oligohydramnios* (insufficient amniotic fluid, 'Potter's syndrome') while excessive accumulation of amniotic fluid, *hydramnios,* is associated with impaired or no swallowing, for example in anencephaly or oesophageal atresia.

The diagnostic value of amniotic fluid can be considerable. For example, the glycoprotein *α-fetoprotein* is normally found in very low concentrations in amniotic fluid. However, if the

279     *The Fetus and its Preparations for Birth*

fetus has a defect in neural tube formation, as occurs in *spina bifida* or *anencephaly,* the concentration of α-fetoprotein is elevated markedly. The recovery of *fetal* cells in amniotic fluid makes it possible to scan the karyotype of the fetus for gross chromosomal abnormalities (e.g. Down's syndrome) and assessment of fetal sex (see Chapter 1). It is also now becoming possible to identify various genetic disorders by use of recombinant DNA probes that recognize DNA sequences restricted to chromosomes carrying mutant genes for Duchenne muscular dystrophy, phenylketonuria or haemophilia. The recovery of amniotic fluid at around 16–20 weeks for these diagnostic tests is not without risk of fetal damage, abortion or Rhesus sensitization. However, where an appreciable risk of fetal abnormality exists, as for example in mothers over 39 years of age, a parent with a balanced translocation, exposure of mother to potential teratogens or prior evidence of familial risk, then the offer of amniocentesis with the possibility of therapeutic abortion is becoming routine.

# 4 Development of fetal systems and their maturation for postnatal life
## a The cardiovascular system

The fetal circulation differs from that in the adult because the placenta, not the lung, is the organ of gaseous exchange. The adaptations which achieve this end are ingenious: the two fetal ventricles pump in parallel, not in series as in the adult, and there are a number of vascular shunts which divert the fetal circulation away from the lungs and towards the placenta. These adaptations are exquisitely designed such that conversion to the adult form of circulation is initiated instantaneously at the first breath taken by the newborn.

The fetal circulation is shown diagrammatically in Fig. 11.4a. Oxygenated blood returns from the placenta and is carried into two channels. The larger of these is the *ductus venosus,* a fetal shunt which bypasses the hepatic circulation and delivers blood directly into the inferior vena cava. The smaller channel perfuses the liver and enters the inferior vena cava through the hepatic veins. The inferior vena cava carries blood to the right atrium where it is split into two streams by the *crista dividens,* the free edge of the interatrial septum which projects from the *foramen ovale.* The larger stream passes through the foramen ovale, another fetal shunt, into the left atrium thereby avoiding the pulmonary circulation. The smaller stream continues through the right atrium, as does blood returning from the head region via the superior vena cava and also returning coronary blood. This blood flows into the right ventricle and out through the pulmonary artery. Thereafter it also is split into two channels, the largest passing through yet a third fetal shunt, the *ductus arteriosus,* which carries the blood to the aorta, while the smaller channel conveys blood to the fetal lungs. The small amount of poorly

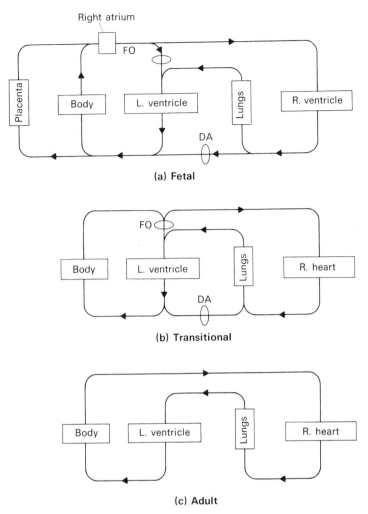

Right atrium

FO

Placenta

Body | L. ventricle | Lungs | R. ventricle

DA

(a) Fetal

FO

Body | L. ventricle | Lungs | R. heart

DA

(b) Transitional

Body | L. ventricle | Lungs | R. heart

(c) Adult

**Fig. 11.4.** Highly schematic representation of the circulations of (a) the fetus, (b) the neonate, and (c) the adult. The transitory form in (b) occurs through the occasional opening of fetal shunts prior to their complete, anatomical closure. DA = ductus arteriosus, FO = foramen ovale.

oxygenated blood passing through the pulmonary circulation returns to the left side of the heart.

The combined cardiac output consists of about one-third from the left and two-thirds from the right ventricle. The three fetal vascular shunts—ductus venosus, foramen ovale and ductus arteriosus—combine in function to ensure the optimal distribution of oxygenated blood to the head and body. It is instructive to examine the quantitative aspects of this process. Blood leaving the placenta in the umbilical vein is 90% saturated with $O_2$. Most of it is shunted past the liver to join with blood only poorly oxygenated (20%) in the inferior vena cava, the resultant mix reaching the heart being 67%

saturated. The blood shunted through the foramen ovale to the left atrium is joined by $O_2$-poor blood returning from the lungs. The result is blood 62% saturated which leaves the heart via the brachiocephalic artery to supply, in large part, the head region. The remainder of the $O_2$-rich blood from the inferior vena cava enters the right atrium to mix with poorly oxygenated blood (31% saturated) returning from the head region via the superior vena cava. The blood thereby entering the pulmonary artery via the right ventricle is 52% saturated with $O_2$ and the largest proportion of it is shunted through the ductus arteriosus to join oxygen-rich blood in the aorta to yield an $O_2$ saturation of 58% in the descending aorta. The effectiveness of the ductus arteriosus shunt results largely from the high pulmonary vascular resistance due to constriction of pulmonary arterioles in response to the low fetal oxygen tension (20–25 mmHg compared to 80–100 mmHg in the adult).

The changes in the fetal circulation at birth involve closure of the three fetal shunts, thereby replacing the placental circulation with a pulmonary circulation (Fig. 11.4b,c). With the obliteration of the umbilical circulation, the ductus venosus ceases to carry blood to the heart. At the same time there is a dramatic fall in pulmonary vascular resistance due to inflation of the lungs with the first breath and the rise in pulmonary $pO_2$. Thus, there is overall a net *drop* in pressure on the right side of the heart (with loss of umbilical input and rise in pulmonary outflow) and a *rise* in pressure on the left side (with a return of pulmonary venous blood). This pressure imbalance leads to a brief reversal of blood flow through the ductus arteriosus, the muscular wall of which responds to the elevated $pO_2$ of neonatal blood by contracting. The foramen ovale has a flap valve over it in the left atrial chamber. In the fetus, this is maintained open by the stream of blood from the right atrium, but with the reversal of interatrial pressure, the flap is pressed against the interatrial wall, thereby separating the two sides of the heart to yield two pumps working in series.

There is a recognizable, transitory form of circulation in the newborn which is the result of functional, rather than anatomical, closure of the foramen ovale and ductus since these shunts are able to reopen from time to time. The ductus venosus is closed permanently in most individuals within 3 months of birth, the ductus arteriosus by 1 year and the foramen ovale obliterates very slowly and not in all individuals—in 10% of adults a probe may be passed through it.

*b The respiratory system*

It is clear that the fetus spends at least 1–4 hours each day making rapid respiratory movements which are irregular in amplitude and frequency and generate negative pressures of 25 mmHg or more in the chest. Interestingly, these movements occur in episodes of up to 30 minutes during rapid eye movement sleep (see section (e) below) but not during wakefulness or slow wave sleep. The breathing movements are purely diaphragmatic and move amniotic fluid in and out of the lungs (see 3 above). The functions of these breathing movements probably include an element of 'practice' of the reflex neuromuscular activities to be initiated in breathing at birth in order to fill the lungs with air, and also the promotion of the growth which follows lung distension. Prevention of fetal breathing, as occurs in congenital disorders of the nervous system or diaphragm, retards lung development to the extent that they may be incapable of supporting extra-uterine life.

The fetal lungs undergo major structural changes during pregnancy, especially as parturition approaches. Primitive air sacs are apparent in the lung mesenchyme from about week 20 and blood vessels appear soon afterwards at around week 28. The pressure required to expand the fetal lung decreases as the time of birth approaches and this is in large measure the result of the appearance of a surface active agent, or *surfactant,* which reduces the surface tension of pulmonary fluid thereby aiding lung expansion. The surfactant is a phospholipid (a disaturated lecithin, mainly dipalmitoyl lecithin, that is attached to an apoprotein). The enzymes required for its synthesis are found in the human fetal lung from weeks 18 to 20 onwards, but increase enormously in concentration in the 2 months before birth.

One of the major findings of recent years is that synthesis of surfactant is promoted by fetal corticosteroids particularly during their preterm rise (Fig. 11.5). Failure to produce sufficient surfactant has serious consequences for lung expansion, as is seen in so-called *idiopathic respiratory distress syndrome (hyaline membrane disease).* Lung maturation can be accelerated by injecting ACTH to stimulate fetal adrenal activity, or by administering potent corticosteroids to the mother. Such treatment is highly effective for infants about to be born prematurely.

To initiate normal, continuous breathing at birth, the first breath must overcome the viscosity and surface tension of fluid in the airways and also the resistance of lung tissues. Removal of fluids in respiratory pathways occurs through the mouth during vaginal delivery due to the rise in intrathoracic pressure, but this does not occur in babies born by Caesarian section. Remaining fluid is resorbed through the agency of pulmonary lymphatics and capillaries. Prenatal episodic breathing is replaced rapidly with normal, continuous postnatal breathing and the gaseous exchange function of the lungs is established quite rapidly, within 15 minutes or so of birth.

The mechanisms bringing about the pronounced inspiratory effort at birth are varied. Cold exposure, tactile, gravi-

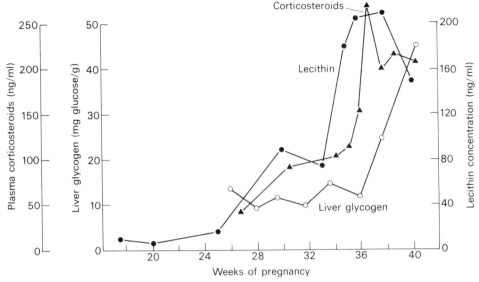

**Fig. 11.5.** Time course of concentrations of corticosteroids in umbilical cord plasma, lecithin in amniotic fluid and glycogen in fetal liver. Lecithin may be used as an indicator of surfactant production. The rise of lecithin content of human amniotic fluid just before birth is reflected in the lecithin: sphingomyelin ratio test for monitoring fetal development and well-being. (Sphingomyelin is a phospholipid, the concentration of which *does not* change near term and serves, therefore, as a baseline against which to measure lecithin.) A ratio > 2 indicates surfactant production is normal, < 2 indicates the fetus may have hyaline membrane disease (see text). The close temporal relationship between corticosteroid concentration in blood and amniotic lecithin levels is suggestive of a causal relationship between the two (see text). The increase in liver glycogen near term is also the result of the rise in corticosteroids which induces the late cluster of fetal liver enzymes, including those required for glycogen synthesis.

tational, auditory and noxious stimuli may all enhance respiration at birth, but are not absolutely necessary for the initiation of continuous postnatal breathing. Whichever factors are important, they operate on a newborn in which the neuromuscular activities of respiration and swallowing (see below), including the ejection of foreign bodies from the pharynx and trachea, have been rehearsed, and medullary respiratory rhythmicity has been established. Pulmonary stretch receptors and arterial and central chemoreceptor mechanisms are also functional by this time.

*c The gastrointestinal system*

By using intra-amniotic injections of inert substances which may subsequently be identified in maternal urine, or of opaque media which are visualized by subsequent X-ray, it has been demonstrated that near term the human fetus swallows about 500 ml amniotic fluid daily. The opaque medium is transferred rapidly through the stomach to the large bowel where it remains for the duration of pregnancy. Water, which forms the major part of the swallowed fluid, is absorbed readily through the small bowel as are electrolytes and other small molecules, such as glucose. Debris from fetal skin found in amniotic fluid, as well as larger molecules, accumulate in the large bowel together with sloughed cells from the small intestine and bile pigments, to form a green faecal mass, *meconium*. Defaecation *in utero* does not normally occur.

Shortly before birth there is a major and rapid increase in glycogen concentration in the *liver,* associated with the development of a cluster of liver enzymes (the *neonatal cluster*) apparently induced by corticosteroids. Enzymes involved in gluconeogenesis are also induced at this time (see Section 2d). Fetal corticoids also appear to regulate maturation of pancreatic islets of Langerhans. We have discussed already that the concentration of glucose in fetal blood is fairly constant coming, as it does, from a relatively constant maternal pool. The $\beta$-cell response to increases in circulating levels of glucose is very weak—even absent—for much of pregnancy unless hyperglycaemia is sustained, as in the case of maternal diabetes. Thus, enhanced secretion of fetal adrenal corticosteroids near term has a major, orchestrating role in the maturation of pancreatic and liver functions essential for postnatal survival.

## d The renal system

Although the placenta is the main organ of excretion, the fetal kidneys are functional during pregnancy and produce substantial quantities of hypotonic urine, the tubules being inefficient at $Na^+$ reabsorption. Fetal urine contributes to total amniotic fluid volume to the extent of half a litre or so per day. Renal agenesis (Potter's syndrome) is associated with a marked reduction of amniotic fluid volume, but such fetuses survive to term and are born alive although with various growth and developmental defects.

At parturition, renal function must undergo a quite radical alteration, since the constant supply of water, sodium and other electrolytes through the placenta will be lost. Soon after birth, urine flow occurs at a high rate while sodium reabsorption is rather low. Over the following hours, urine flow is reduced rapidly but rises again at the end of the first week. Glomerular filtration rate is relatively low in the newborn, perhaps only a third of what would be estimated in relation to body size. Indeed, a mature rate of glomerular filtration is not achieved until 1.5–2 years of age. The newborn is in danger of hyponatraemia since the ability to retain sodium is poor. Prematurity represents a marked risk in this context, since the kidney leaks up to three times more sodium than that of babies born at term.

The developmental changes in the urogenital tract during early pregnancy have been discussed in Chapter 1, with special reference to the role of sex steroids in the differentiation of internal and external genitalia.

*e The nervous system*

Little is known about the development of function in the *fetal brain*, although it may reasonably be assumed that its general course is genetically determined and relatively independent of environmental events. This is not to say, of course, that intrauterine sensory stimulation has no effect on the developing brain but information on this point is very sparse. Fetal hormones, particularly sex steroids and thyroxine (see Chapter 1 and below) also have major effects on neural development.

The fetus is certainly capable of responding to extraneous stimuli. Loud noises and intense light, noxious stimulation of the skin and rapid decreases in the temperature of its fluid environment will result not only in movement but also autonomic responses, such as acceleration of heart rate.

Presumably the level of stimulation in these sensory modalities is normally rather low and unvarying in the fluid-cushioned, constant-temperature, light- and sound-attenuated chamber in which the fetus grows.

Fetal movements occur early in pregnancy and these can be felt readily by the mother by week 14. The function of the movements are uncertain, but 'exercise' which will contribute to muscle growth must be among them. During the long, human gestation period, the innervation of muscles by motor nerves and partial maturation of both ascending sensory and descending motor systems in the CNS means that some purposeful and well coordinated movements become possible late in pregnancy. A number of simple postural and other stereotyped reflexes are apparent in the fetus from a relatively young gestational age.

The fetus shows periods of slow-wave (SW) and rapid eye movement (REM) sleep between periods of wakefulness. The neonate sleeps for about 16 hours each day, with REM and SW sleep in roughly equal proportion. This sleeping time progressively decreases over the first 2 years of life to about 12 hours each day and the proportion of REM sleep decreases to about one-quarter compared to SW sleep.

Of tremendous clinical and sociological significance is the impact of drugs from the maternal circulation on the developing brain of the fetus. Many are lipid-soluble and have no barrier to their free diffusion both across the placenta and, therefore, into the brain as well. The immature status of the blood–brain barrier also ensures that other chemical agents may gain access to the fetal brain in a way not seen in the adult. Addictive drugs such as the opiates (heroin, morphine) taken by pregnant women produce withdrawal symptoms, and, indeed, dependence in their babies. Pain-killing drugs which are sometimes given during labour may depress the behavioural repertoire of the newborn;

for example, sucking reflexes may be impaired. This may have adverse consequences both for lactation and mother–infant interaction (see Chapter 13).

*f Summary*

In this section we have examined the ways in which selected fetal systems operate during intra-uterine life and the developmental changes which occur towards the time of parturition that are essential for extra-uterine survival. The mechanisms involved vary from the instantaneous (in the cardiovascular system) to the gradual (after birth in the urinary system) with some occurring late in pregnancy (gastrointestinal and respiratory systems) and regulated like parturition itself (see Chapter 12) by endocrine events involving, in particular, the fetal adrenal gland.

## 5 Fetal and neonatal endocrinology

It has already been emphasized that the fetal endocrine system exerts marked influences on maternal physiology, especially through the hormones of the feto-placental unit (Chapter 10). In contrast, the maternal endocrine system does not influence the fetus directly, except in pathological circumstances, and few maternal hormones other than unconjugated steroids cross the placenta. In general terms, the fetal endocrine organs function by the end of the first quarter of pregnancy, and from this time the fetus is autonomous in its endocrine requirements and its hormones do not cross the placenta to the maternal circulation. Furthermore, the fetal endocrine system has a number of unique functions—for example, in the differentiation of the reproductive tract, the lungs, gastrointestinal system, even the brain itself—that are not apparent in the adult.

*a Anterior pituitary hormones*

The adenohypophysis is functional through most of fetal life, as evidenced, for example, by early-occurring thyroid, gonadal and adrenal activity. Thyroid hormone production under TSH control is discussed below in 5b. The gonadotrophins FSH and LH are essential for the androgenic output of the testis. The importance of ACTH is discussed in Section 5c (below). Growth hormone is produced, but its role, if any, in the fetus is not clear. Prolactin is present in amniotic fluid in concentrations 100-fold greater than those in either maternal or fetal circulations. Since only very small amounts of the maternal hormone cross the placenta, it seems reasonable to assume that amniotic prolactin is of fetal origin. After week 30 or so, the low levels of prolactin in fetal plasma rise markedly until term, but decline in the neonate after a brief postpartum rise. The functions of fetal prolactin are unknown, although some role in regulating the permeability of the chorion and amnion to water, and hence amniotic fluid production, has been suggested.

**Table 11.3.** Functions of the fetal adrenal cortex.

| Function | Mechanism and/or relevant section |
|---|---|
| 1 Lung maturation | Induces enzymes necessary for surfactant synthesis (Ch. 11.4b) |
| 2 Parturition | Induces placental oestrogen-synthesizing enzymes. Increases oestradiol precursor (DHA) concentrations (function of fetal zone) (Ch. 12) |
| 3 Glucose storage and gluconeogenesis | Induction of enzyme systems in liver and myocardium (Ch. 11.2d; 11.4c) |
| 4 Insulin secretion | Regulates maturation of fetal Islets (Ch. 11.2d) |
| 5 Lactogenesis | Ductal-lobule-alveolar growth in pregnancy (Ch. 13) |
| 6 Synthesis of adrenaline | Induction of phenylethanolamine-$N$-methyl transferase in adrenal medulla |
| 7 Production of thyroxine | May promote conversion of $T_3 \rightarrow T_4$ |
| 8 Haemoglobin formation | May promote 'switch' in production of fetal to adult haemoglobin |
| 9 Activation of ANP | Maturation of salt:water regulation? |

*b The thyroid gland*

Thyroxine ($T_4$) is essential for the normal development of the fetus. Fetal hypothyroidism is associated with a bone age far behind chronological age, deficiency in body hair and, most important, behavioural retardation. $T_4$ is essential for the normal differentiation of the central nervous system. Maternal thyroxine does have limited access to the fetal circulation since it crosses the placenta, particularly during the second half of pregnancy, although not in amounts sufficient to meet fetal needs. The secretion of $T_4$ from the fetal thyroid increases, under the influence of TSH, from about week 20 of gestation such that its circulating levels may even exceed those in the maternal circulation at term.

*c The adrenal gland*

We have already encountered the importance of adrenal cortical activity for the fetus (in this chapter) and the endocrine function of the placenta (Chapter 10), and will do so again in Chapter 12 in the context of parturition. There is some evidence that ACTH from the anterior pituitary has an inductive role in the growth and development of the adrenal cortex. Fetal serum ACTH concentrations are quite high during weeks 12–19 of gestation and gradually decline by around week 40. Very high levels are found in the fetus at term, probably reflecting a stress response to parturition.

Cortisol itself is found in fetal blood by week 10 and levels increase as gestation proceeds, to become especially high during labour. After birth, the adrenal shows major structural changes. The *fetal zone* which is responsible for the synthesis of DHA (Chapter 10) regresses by the end of the first month, its job having been done, while the cortex proper differentiates into the distinctive three zones characteristic of the adult gland. Table 11.3 summarizes the actions of fetal corticosteroids as parturition approaches. Aldosterone levels rise late in pregnancy, but appear not to become responsive to a reduction in blood volume or nephrectomy until after birth. Atrial natriuretic factor (ANF) is also produced prenatally and does regulate $Na^+$ excretion. The production of ANF may be regulated by adrenal corticosteroids.

*d The parathyroid glands and calcium-regulating hormones*

The parathyroid glands are capable of secreting parathormone by about week 12, although plasma concentrations are low until 2 or 3 days after parturition. This relative suppression of parathormone secretion is largely a reflection of the rather high circulating levels of maternally-derived calcium (see Section 2f). Thyrocalcitonin secreted from thyroid parafollicular cells, on the other hand, is at relatively high concentration in fetal plasma and declines slowly after birth. The hormone is able to exert its hypocalcaemic effects and abnormally high circulating levels may cause a severe hypocalcaemia in the fetus.

*e Glucagon and Insulin*

The endocrine pancreas is active early in pregnancy, glucagon ($\alpha$ cells) and somatostatin ($\delta$ cells) being the predominant hormones at first, followed by insulin ($\beta$ cells) clearly present by 10 weeks in man. Development of $\beta$-cell function appears to depend upon the activity of the anterior pituitary, as a result of GH and ACTH activity. Initially, each type of cell is clustered separately, and only later do $\beta$ cells become surrounded by $\alpha$ and $\delta$ cells. Abnormalities in early pancreatic development tend to affect $\alpha$ and $\delta$ cells preferentially, leading to relatively uncontrolled $\beta$-cell activity, hyperinsulinaemia and hypoglycaemia (nesidioblastosis syndrome). Insulin secretion from mid-pregnancy onwards responds positively to amino acids, glucose and short-chain fatty acids, and negatively to catecholamines.

Details of how insulin acts in the fetus are given in Section 11.2d.

*f Summary*

Some components of the endocrine system of the fetus achieve autonomy very early during pregnancy. With the exception of adrenal cortical, gonadal and thyroid secretions, the essential functions of fetal hormones have yet to be determined, although experiments on other species, such as the sheep, are beginning to give valuable information.

For example, prolactin may be involved in the developing responsiveness of the adrenal cortex to ACTH and hence, indirectly, in the timing of parturition itself.

## 6 Immunological aspects of pregnancy

On the first page of this book we discuss the particular value of sexual reproduction in generating genetic diversity. The whole edifice of sex differentiation, reproductive cyclicity and pregnancy, with their social ramifications, is constructed upon the biological advantages conferred by producing *a genetically distinct individual.* Yet we have now come full circle. For the genetically distinct individual will also be *phenotypically unique.* A component of this unique phenotype is the array of cell-surface glycoproteins that constitute the *system of histocompatibility antigens.* Thus, the biological advantages of genetic variation appear to confront and conflict with those of viviparity. The problem may be illustrated dramatically. If the skin of a newborn child is grafted to its mother—she rejects it. Why then does she not reject the whole fetus? Several mechanisms have been proposed to explain the survival of the fetus *in utero.* None is in itself adequate.

### a Fetal antigenicity

The developing fetus does not lack target antigens. The major histocompatibility antigens appear on embryonic cells around the time of implantation and, although present in smaller amounts than in the adult, are detectable throughout pregnancy. Thus, grafting of fetal tissues to another individual (or to the mother) results in their rejection.

### b Maternal immune responsiveness

The pregnant mother *is* competent to respond immunologically to the fetus. Indeed, examination of maternal blood in late pregnancy indicates that a regular feature of pregnancy is an immunological reaction against the conceptus. Thus both antibodies and lymphocytes that react against antigenic specificities present on fetal cells can be detected. Furthermore, active presensitization of the mother, by injecting or grafting paternal tissues, does not prejudice the establishment and maintenance of subsequent pregnancies by the same father. Thus, neither a generalized nor a specific depression of maternal immune responsiveness can adequately explain fetal survival.

Indeed recently it has become clear that an immune reaction of the mother against paternal histocompatibility antigens may be an *essential or highly desirable requirement* for pregnancy to succeed. The basis for such a requirement is not at present clear but would result of course in forcing increased heterozygosity in the population at large.

There is some evidence that during pregnancy the *quality* of the maternal immune response is different, perhaps modified by the high levels of pregnancy hormones. Helper T cells decline relative to supressor cells, and the classes of immuno-globulin produced change their balance. It is possible that these qualitative changes contribute to the survival of the fetal 'graft'.

*c Immunological filter*

The fetus and its circulating blood are separated from the mother by the investments of fetal membranes (Fig. 9.8). These membranes, the outermost of which is usually chorionic trophoblast, are part of the conceptus and therefore are also genetically alien. However, if the chorionic trophoblast was itself able to resist maternal rejection, and also effectively preclude maternal antibodies and lymphocytes from entering the fetal circulation, then protection for the fetus would be achieved.

A number of studies on syncytiotrophoblast antigenicity have revealed convincing evidence to suggest that either it *lacks* any histocompatibility antigens on its surface, or, if present, they are in some way *masked in vivo* so as to render the cells insusceptible to immune damage. The masking could be provided by a surface coat of glycoprotein molecules. While cytotrophoblast does express a unique class 1 antigen, it is not clear whether this antigen is a target for cytotoxic action. Thus, it appears that the trophoblast layer is presented to the mother as being effectively 'antigenically neutral'. In addition, the placenta is bathed in fluids containing high levels of progesterone, corticosteroids and chorionic gonadotrophin. There is evidence that these hormones may act as *local immunosuppressants* to reduce the cytotoxic effectiveness of immune cells locally.

We saw earlier that the discrete nature of the fetal and maternal circulations prevented appreciable passage of maternal cells to the fetus. In many species, including the pig, sheep, cow and horse, antibody is also excluded. However, in some species, such as man, IgG antibodies normally pass across the placenta into the fetal circulation via a special transport mechanism. The maternal antibodies will include some directed against prevalent bacteria and viruses and thus will confer *passive immunity* on the fetus and afford the child temporary protection for a few weeks postnatally. (In the farm animals, a similar protection is afforded by antibodies transmitted postnatally in the milk; see Chapter 13). However, in addition to antibodies reactive to bacteria, IgG antibodies directed against fetal antigens presumably will also be transferred. That such transfer does indeed occur is seen in cases of sensitization to the major blood group antigen called *Rhesus*. A woman may lack the Rhesus antigen (Rhesus negative) and if she carries at fetus possessing it (Rhesus positive), she may mount an IgG immune response to the antigen, particularly at parturition when extensive fetal bleeding into the mother may occur. In subsequent pregnancies the IgG antibody is transferred across the placenta and destroys the fetal erythrocytes. Here then is a clear example of the mother rejecting her fetus immunologically. But the Rhesus antigen is only one of many antigens by which mother and fetus may differ. Can we gain any clues from the Rhesus example as to why fetuses are not normally rejected?

The Rhesus antigen differs in two ways from most of the other cellular antigens expressed on fetal cells. First, the Rhesus antigen is present only on red blood cells. Most of the other important histocompatibility and blood group antigens are also present on several other types of fetal cell and are thus widely distributed amongst the tissues of the fetus. Second, the Rhesus antigen exists only as a structural component of the cell membrane. Other antigens, such as ABO blood group and major histocompatibility antigens seem to be present not only as structural membrane components but also in solution in the fluids of the fetus such as the blood and amniotic fluid. Thus, if an antibody, or indeed the odd lymphocyte, directed against these other antigens should enter the fetal circulation, it will first be 'mopped up' harmlessly by free soluble antigen. Any remaining antibody will be distributed amongst a wide range of cell types and so will effectively be 'diluted-out'. Any given single cell will be unlikely to bind a large number of antibody molecules and since only the binding of a large number of antibodies will debilitate the cell, gross tissue damage is avoided. In the case of the Rhesus antigen, 'mopping up' and 'diluting-out' cannot occur and so the chance of tissue damage increases considerably with obvious pathological consequences.

Thus, the protection of the fetus from the immune response of the mother appears to depend upon:

1 an antigenically inert trophoblast forming the front-line defences possibly in association with local, endocrinologically-mediated depression of immune reactivity;

2 a complete (or in man highly selective) barrier to the transmission of immune cells or antibodies from mother to fetus;

3 properties of fetal antigens that mean that such aggressive immune cells or antibodies as do get across the placenta are mopped-up and diluted out before they can cause extensive tissue damage.

## 7 Summary

Viviparity provides the developing embryos with the optimal environment for growth. The uterus may be viewed as the ultimate 'nest', temperature-controlled and a continuous supply of food and protection from predators. This environment is created at the mother's expense and her metabolism is brought, to varying degrees, under the control of the fetus which to a large extent functions autonomously within its protected environment. It should, however, be clear from this account that we still have much to learn about fetal and maternal function in pregnancy and its control. Central to the maintenance of pregnancy is the trophoblast of the placenta. This remarkable tissue elaborates and secretes the steroid and protein hormones, is involved in the transplacental passage of metabolism both by diffusion and active transport, acts as a selective barrier between the two circulations and presents an antigenically inert front to the mother's immune system. The role of the trophoblast ends with delivery of the new infant but the mother still has a major and crucial role to play postnatally. This subject is discussed in the following chapters.

**Further reading**

Biggers JJ. Fetal and Neonatal Physiology. In *Medical Physiology* (Ed. Mountcastle V) pp. 1947–1982. Mosby, 1979.

Ciba Foundation Symposium 86. *The Fetus and Independent Life.* 1981.

Dancis J. Feto–maternal interaction. In *Neonatology* (Ed. Avery GB) J.B. Lipincott Co., 1975.

Garel JM. Hormonal control of calcium metabolism during the reproductive cycle in mammals. *Physiol Rev* 1987; **67**: 1–66.

Gluckman, PD. The role of the pituitary hormones, growth factors and insulin in the regulation of fetal growth. In *Oxford Reviews of Reproductive Biology* Vol. 8, pp.1–60. Oxford University Press, 1986.

Jones CT, Rolph, TP. Metabolism during fetal life: a functional assessment of metabolic development. *Physiol Rev* 1986; **65**: 357–430.

Loke, YW. Female reproductive immunology. In *Scientific Foundations of Obstetrics and Gynaecology* (Eds Philipp E E, Barnes J, Newman M) pp. 66–77. Heinemann, 1986.

Philip EE, Barnes J, Newton M (Eds). *Scientific Foundations of Obstetrics and Gynaecology.* William Heinemann Medical Books, 1986.

Rogers Brambell FW. *The Transmission of Passive Immunity from Mother to Young.* North-Holland, 1970.

Scott JS, Jones WR (Eds). *Immunology of Human Reproduction.* Academic Press, 1976.

Shearman RP (Ed). *Human Reproductive Physiology.* Blackwell Scientific Publications, 1979.

# Chapter 12
# Parturition

In the previous eleven chapters we have described the events leading to the development of a mature fetus. In this chapter we consider *parturition,* the process by which the uterus expels the products of conception usually at a time when the newborn child can exist semi-independently of the mother.

Factors affecting the onset of parturition are poorly understood in the human female and we must therefore rely on experimental data largely from the sheep and the goat which, although having markedly different endocrine strategies underlying their pregnancies (see Chapter 10), nevertheless show considerable similarity in the mechanisms and timing of onset of parturition. This fact has undoubtedly encouraged workers in their attempts to unravel the mysteries of birth in women in whom, as we shall see, the ultimate *mechanism* of parturition is probably the same as in goats, sheep and possibly all species even if the *means* by which it is activated shows considerable variation.

The fetus lies within its fetal membranes in the uterus and is retained there by the cervix (Fig. 12.1). Expulsion of the fetus is achieved by coordinated contractions of the *myometrium* (assisted later in labour by involuntary contractions of striated muscle in the abdominal wall and elsewhere). However, the *cervix* will resist these contractions, and the large increases in intra-uterine pressure which accompany them,

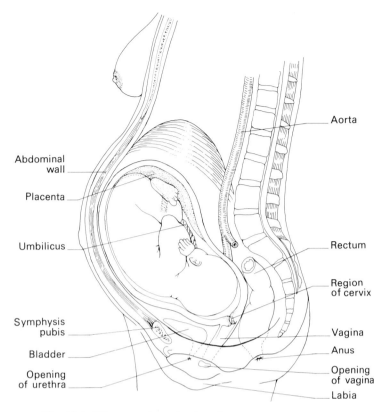

**Fig. 12.1.** Diagram of a sagittal section through a pregnant woman. Notice, particularly, the *size* and *position* of the uterus, and how much of the abdomen it occupies.

unless the cervical structure changes to become soft and compliant. Thus, expulsion of the fetus requires both *myometrial contractions* and *cervical softening*. Before examining the processes underlying these changes, we will first consider briefly the nature of the myometrium and cervix.

**1 The myometrium**

The myometrium consists of bundles of non-striated muscle fibres, intermixed with areolar tissue, blood and lymph vessels and nerves. During pregnancy myometrial bulk increases enormously, due primarily to an increase in muscle cell *size* from about 50 to 500 μm (hypertrophy). *Functionally*, this system of muscle cells behaves as a *syncytium* being electrically coupled via specialized regions of contact—the so-called *gap junctions* or *nexuses*—which allow the flow of electrical current and thereby coordination of the spread of contraction through the myometrium. A *uterine contraction* requires more or less simultaneous activation of all smooth muscle cells in the uterus. It is not altogether clear how this happens, but in an attempt to understand it we will consider the mechanisms controlling activity in one cell, and then the way in which contractions might be propagated from one group of cells to others. Both these events are affected by steroids and other hormones.

The contraction of myometrial cells depends on the movement of calcium ions ($Ca^{2+}$), both by liberation from intracellular binding sites and by entry into the muscle cell from the extracellular fluid. Calcium binds to regulatory sites on the contractile proteins, actin and myosin, to allow expression of ATPase activity, and hence contraction. Release of calcium is stimulated by the presence of *action potentials* within the muscle cell. In relaxed, oestrogen-primed uterine muscle, spontaneous depolarizing *pacemaker potentials* occur. If the magnitude of such potentials exceed a critical threshold, a burst of action potentials is superimposed on the pacemaker potentials. This event is associated with a sharp increase in intracellular $Ca^{2+}$ and subsequent contraction. During relaxation, $Ca^{2+}$ is both removed from the cell and rebounds intracellularly.

Understanding the mechanisms which regulate myometrial activity at parturition therefore involves elucidating factors which regulate the intracellular $Ca^{2+}$ concentration and its relationship to electrical potential. Two hormones are of direct importance. *Prostaglandins* act mainly by liberating $Ca^{2+}$ from intracellular binding sites, and oxytocin has a direct action to increase the rate of $Ca^{2+}$ influx and also lowers the excitation threshold of the muscle cell.

## 2 The uterine cervix

The cervix is of major importance in retaining the fetus in the uterus, and this function is a reflection of its high connective-tissue content, which helps resist stretch. The connective tissue is derived from collagen-fibre bundles embedded in a proteoglycan matrix. In order that the fetus can move from the uterus to the outside world, the *nature* of the pregnancy cervix must change—soften or '*ripen*'.

The process by which this occurs appears to involve two changes in the intercellular matrix: a *loss of collagen* and a marked *increase in glycosaminoglycans* (GAGs), which are important in determining the degree of aggregation of the collagen fibres. Of the GAGs in the human cervix, *keratan sulphate* seems to be the one which increases most postpartum. This compound does not bind at all to collagen, and its increased proportion relative to *dermatan sulphate*, which binds very tightly, may explain the 'loosening' of collagen bundles seen at parturition.

Prostaglandins, in addition to their role in inducing myometrial contractions, also seem to be important factors controlling cervical ripening. A number of clinical trials have demonstrated that $PGE_2$ in particular but also $PGF_{2a}$ given intravaginally or intracervically, increase compliance of the cervix. They are able, therefore, to induce abortion or birth at a stage of pregnancy when massive doses of oxytocin alone are relatively ineffective despite causing uterine contractions.

From this account of the changes in the myometrium and uterine cervix during late pregnancy, it is clear that oxytocin and prostaglandins play a key role in bringing them about and, thereby, are essential for the process of parturition itself. Before considering the mechanisms underlying parturition, we will briefly examine the biochemistry of prostaglandins and the neural mechanisms regulating oxytocin secretion.

## 3 Prostaglandins

Prostaglandins (PGs) are biologically-active lipids probably synthesized in every tissue of the body, including the brain. They are essentially *local hormones* acting at, or near, their site of synthesis and are inactivated in the lung during one circulation in the bloodstream. At parturition, the endometrium is probably the most important site of PG synthesis. In Chapter 10 we saw that, in some species, presence of the conceptus in the luteal phase of the cycle inhibits the synthesis and release of $PGF_{2\alpha}$ by the endometrium, and so prevents luteolysis. This condition endures until parturition when, probably in all species, something must happen to disinhibit or promote PG synthesis and release. It is probable that the myometrium, cervix, placenta and fetal membranes also synthesize PGs.

The pathways of PG biosynthesis are shown in Fig. 12.2. *Arachidonic acid* is the common precursor of $PGF_{2\alpha}$ and $PGE_2$ which are of main importance in the context of parturition. The availability of arachidonic acid is probably rate-limiting, rather than the activity of PG *synthetase*. Glycerophospholipids are the most important source of arachidonic acid, and their liberation depends on the activity of an acyl hydrolase, *phospholipase* $A_2$. This enzyme is found largely in an inactive form, membrane-bound in lysosomes of the decidua and fetal membranes. Factors which *increase* PG synthesis are widely believed to act primarily by altering the stability of membranes binding phospholipase $A_2$, to cause liberation of the active enzyme from the lysosomes. Those factors which *decrease* PG synthesis probably do the converse, i.e. stabilize the lysosomal membrane. It is interesting to note, therefore, that steroid hormones which may change quite markedly at parturition (see below) exert opposing effects on phospholipase $A_2$. Some ultrastructural evidence suggests that oestrogens *labilize* lysosomes, while progesterone *stabilizes* them. Thus, a *rise* in the *oestrogen:progesterone ratio* would result in increased production of arachidonic acid and hence PG synthesis (Fig. 12.2). The oestrogen:progesterone ratio affects not only *synthesis* of PGs but also their *release*. It does

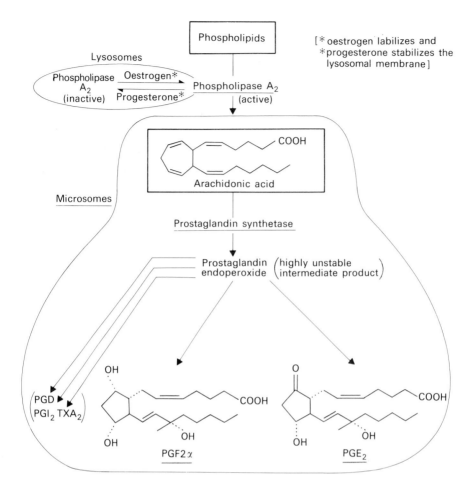

[*oestrogen labilizes and
*progesterone stabilizes the
lysosomal membrane]

**Fig. 12.2.** Biosynthesis of prostaglandins. The most important rate-limiting factor is the availability of arachidonic acid. Factors, for example steroids, which affect the activity of phospholipase $A_2$ are therefore critical determinants of the rate of PG synthesis. $TXA_2$=thromboxane A2; $PGI_2$=prostacyclin.

this via effects on oxytocin action. Oxytocin stimulates the release of $PGF_{2\alpha}$ directly from the uterus. In the sheep, oestradiol has been shown to enhance this effect by increasing the number of oxytocin receptors in the endometrium, while progesterone has the reverse effect. This means that *a rising oestradiol:progesterone ratio can affect PG release via an oxytocin-dependent mechanism* without any alteration in circulating oxytocin levels. Receptor regulation may represent an important means of hormone action at parturition (at other times, see Chapter 2) which makes the absence of changes in the plasma concentrations of hormones potentially misleading.

Thus, two routes to increased PG availability at parturition exist, and both can be induced by an increase in the oestrogen:progesterone ratio. In some species such a change in the

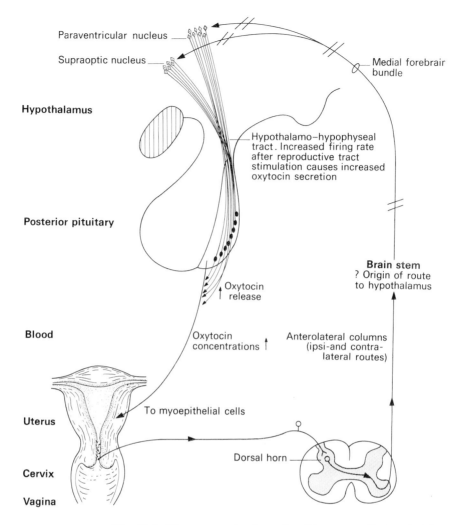

Paraventricular nucleus

Supraoptic nucleus

Medial forebrain bundle

Hypothalamus

Hypothalamo–hypophyseal tract. Increased firing rate after reproductive tract stimulation causes increased oxytocin secretion

Posterior pituitary

**Brain stem**
? Origin of route to hypothalamus

↑ Oxytocin release

Oxytocin concentrations ↑

Anterolateral columns (ipsi- and contra-lateral routes)

Blood

To myoepithelial cells

Uterus

Dorsal horn

Cervix

Vagina

**Fig. 12.3.** The neuroendocrine reflex (Ferguson reflex) underlying oxytocin synthesis and secretion.

oestrogen:progesterone ratio has been shown to occur at parturition.

**4 Oxytocin**

As we saw in Chapter 5, oxytocin is a nonapeptide synthesized by magnocellular neurons in the supraoptic and paraventricular nuclei of the hypothalamus. It is transported along their axons to the terminals to be stored for release into the circulation from the posterior lobe of the pituitary.

Oxytocin is released in response to tactile stimulation (usually) of the reproductive tract, particularly the uterine cervix. This *neuroendocine reflex* has, as its *afferent limb* (i) the sensory nerves from the vagina and cervix (ii) the ascending somatosensory pathways in the spinal cord (the anterolateral columns), (iii) an as yet unspecified series of projections

299      *Parturition*

through the brain stem, via the medial forebrain bundle, which courses through the lateral hypothalamus ultimately to reach the hypothalamic magnocellular nuclei. The *efferent limb* of the reflex is the blood-borne carriage of oxytocin to the uterus, where the hormone exerts its actions on myometrial contractility in interaction with both steroid hormones and prostaglandins. This reflex, often referred to as the 'Ferguson reflex', is illustrated in Fig. 12.3. Significantly for our discussion here, this reflex is greatly facilitated in the presence of a high plasma oestrogen:progesterone ratio.

The reflex release of oxytocin, which occurs in all species, is very similar to the coitus-induced release of prolactin seen in the rat which we discussed in Chapter 5. In the next chapter, we will see that stimulation of the nipple during suckling also causes the release of both oxytocin and prolactin, events vital for lactation and milk let-down.

## 5 Hormonal changes and timing of the onset of parturition

The medical literature contains a confusing array of ideas relating endocrinology to parturition. Current hypotheses place the prostaglandins at the centre of events. In the section which follows, therefore, we will attempt to answer the question *'How is PG synthesis and release increased at parturition?'* We will see that different species achieve this in different ways, but that in all species fetal and maternal endocrine mechanisms are critical in bringing it about. But even more important, we must consider *'What times these changes at the onset of parturition?'* The answer appears in many species to be *only* the *fetus*. Maturational changes in the fetal hypothalamo-pituitary-adrenal axis provide the key to the way in which this amazing and important task is achieved.

### a Goat

The goat is representative of those species that are dependent on progesterone secreted from the corpus luteum for the maintenance of pregnancy (see Chapter 10). There is no doubt that parturition in the goat is initiated by the sharp *fall in circulating concentrations of plasma progesterone* (occurring during the last 24 hours before delivery) which is the consequence of luteal regression. The factor responsible for luteal regression is $PGF_{2\alpha}$ and there are marked increases in its concentration in uterine venous blood about 48 hours before parturition, preceding the decrease in plasma progesterone levels by about 20 hours. This increased production of $PGF_{2\alpha}$ originates in the placenta and is dependent in turn on an increased output of *glucocorticoids* by the *fetal* adrenal cortex. Indeed, parturition can be induced prematurely by injections of ACTH into the fetus. These fetal corticoids induce placental aromatizing enzymes which act on DHA and DHA sulphate substrates (also derived from the fetal adrenal) to boost the synthesis of oestrogens. The increased oestrogens enhance the synthesis and release of $PGF_{2\alpha}$ in the placenta. The $PGF_{2\alpha}$ causes the corpus luteum to regress, progesterone plasma concentrations plummet thus removing the 'block' on the myometrium where additional $PGF_{2\alpha}$ synthesis and release is enhanced and myometrial contractions begin. The consequent increased oestrogen:progesterone ratio also facilitates oxytocin release from the posterior pituitary—a phenomenon that is reinforced directly via the neurally-derived input from uterine contractions and cervical dilation as parturition proceeds. The oxytocin, as we saw above, stimulates further release of PGs, and so a positive feedback system is established.

Rising fetal corticoid levels may also close down production of placental lactogen (caprine placental lactogen, cPL) during the last 15 days of pregnancy, thereby removing an important luteotrophic stimulus. How the corticoids shut off cPL synthesis and how important this shut off is to parturition in relation to pituitary luteotrophic support in the goat is at present unclear.

Cortisol clearly plays a central role in initiating caprine parturition. The mechanism underlying the enhanced output of cortisol by the fetal adrenal will be discussed in Section b below.

*b Sheep*

Although closely related to the goat and with a gestation period of similar length, the sheep is dependent on the *placenta* and not the corpus luteum for steroid-hormone production in the later stages of pregnancy (Chapter 10). The fetal pituitary-adrenal axis, as in the goat, has a dominant role in timing the onset of parturition. Thus, *fetal* hypophysectomy, stalk section and bilateral adrenalectomy all indefinitely *prolong* pregnancy, while administration in sufficient quantities of ACTH or dexamethazone (a synthetic glucocorticoid) to the *fetus* induces parturition prematurely. It is not surprising, therefore, to discover that normal parturition is preceded by a steadily rising output of fetal ACTH (probably driven by hypothalamic corticotrophin releasing hormone secretion) and this is followed by a marked *increase* in circulating cortisol concentrations in the *fetal lamb*. An equally marked *decrease* in progesterone in the *maternal* circulation follows. However, the relationship between the rise in corticosteroids and the fall in progesterone is very different from that in the goat.

The elevated levels of cortisol in the fetal circulation of the sheep induce activity of the *placental enzymes* 17$\alpha$-hydroxylase, steroid $C_{(17)}$-$C_{(20)}$-lyase and probably also aromatases (see Fig. 12.4 and Chapter 2). The effect of this activation is to *divert placental progesterone into the synthesis of oestrogens* and, consequently, progesterone secretion falls while oestrogen secretion rises. The increase in the oestrogen:progesterone

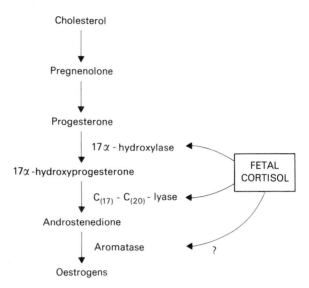

**Fig. 12.4.** Pattern of oestrogen synthesis in the placenta of the sheep. Cortisol secreted from the fetal adrenal enhances conversion of progesterone to oestrogens by activating the enzymes $17\alpha$-hydroxylase, $C_{17}$-$C_{20}$-lyase and, possibly, aromatases.

ratio in the maternal circulation which follows leads to *stimulation of the synthesis and release of PGF$_{2\alpha}$* in the decidua and fetal membranes (see Section 3 above). Thus, uterine myometrial contractions begin and these induce oxytocin release from the maternal neurohypophysis which in turn reinforces PGF$_{2\alpha}$ release and myometrial contractility. That PGF$_{2\alpha}$ is the final link in this chain of events resulting in myometrial contractions is emphasized by the finding that indomethacin, or other PG inhibitors, block naturally occurring and ACTH-, cortisol-, or dexamethasone-induced parturition in the sheep (and goat).

In the goat and the sheep, regardless of the source of hormones in the maintenance of pregnancy, therefore, the *fetus times the onset of parturition* by increasing the output of adrenal cortisol. The mechanisms which time this are at the moment not understood. Although the relationship between rising fetal cortisol and falling maternal plasma concentrations of progesterone are different in these species, the ultimate result is enhanced PGF$_{2\alpha}$ synthesis and release. This appears, then, as a common mechanism of parturition—not only in these two but in many species. Is it also true in the human female?

*c Human*

Despite the background of research summarized above, which points very clearly to the likely mechanisms surrounding parturition in the human female, little is known of the timing mechanism and its relation to alterations in PG synthesis and release at the onset of parturition.

Anencephaly, with attendant absence of corticotrophin releasing hormone in the human fetus, in pregnancies not further complicated by polyhydramnios (excess fluid in the amniotic cavity), is not reliably associated with prolonged gestation. Fetal adrenal hypoplasia may be associated with postmaturity, but not necessarily so—normal gestation and parturition have been seen in fetuses with congenital absence of adrenal glands. Infusions of ACTH, or synthetic glucocorticoids, do not apparently induce parturition in the human. Even so, plasma cortisol concentrations in the fetus and in the amniotic fluid have been seen to rise during the last few weeks of normal pregnancy. This phenomenon is suggestive of a role of the human fetus, via its adrenal cortex, in timing the onset of parturition. However, fetal cortisol cannot divert placental progesterone into the secretion of oestrogens because the enzyme $17\alpha$-hydroxylase does not exist in the human placenta (Chapter 10). Thus, maternal plasma progesterone concentrations do not fall at parturition. Instead, the fetal adrenal influences placental oestrogen synthesis by providing androgen precursors, notably DHA and DHAS. These androgens are secreted from the specialized *fetal zone* of the adrenal (which regresses soon after birth: see Chapter 11) and they form 80–90% of the substrate for placental oestrogen synthesis late in pregnancy. However, treatment of women throughout pregnancy with corticoids that markedly depress fetoplacental androgen and oestrogen production does not delay or otherwise alter the onset of labour. The onset of labour does not appear to be critically dependent, therefore, on the secretory activity of the fetal adrenal and, reflecting this, there are no clear and consistent changes at term in the placental secretion of oestrogens (an indicator of fetal steroidogenic capacity, see Chapter 10).

There is some evidence, in carefully selected subjects, that maternal plasma oestrogen concentrations rise from week 36 of gestation while plasma progesterone concentrations fall. These changes are neither abrupt nor universally seen, and many women enter labour without any obvious changes in circulating sex steroid concentrations. Of course, alterations in the binding of these steroids or local changes in their concentration in uterine tissues (Chapter 2) may bring about effective local alterations in the oestrogen:progesterone ratio without detectable changes in their circulating plasma levels. Evidence for either of the possibilities is, however, generally lacking at present.

Controversy also surrounds possible changes in PGs in the human *before* parturition. Many studies have failed to reveal increased $PGF_{2\alpha}$ concentrations in peripheral blood or amniotic fluid until labour has begun. But recent data have demonstrated an increase of $PGF_{2\alpha}$ in amniotic fluid *before* the onset of labour, and levels progressively increase as parturition proceeds and cervical dilation increases. The cause of the increase in PG levels, as will be clear from the above account, is not immediately apparent. Surprisingly, experiments in non-human primates, notably the rhesus monkey, have not particularly clarified the mechanisms of parturition occurring in the human. The situation is just as complex and difficult to unravel there! Thus infusion of progesterone, oestradiol and dexamethasone into mother or fetus, or of ACTH into the fetus has no effect on gestation. But, as in women, cortisol and $PGF_{2\alpha}$ concentrations in amniotic fluid increase prior to normal parturition which is itself blocked by PG inhibitors.

*d Summary*

In a number of mammalian species (not just those exemplified here) the *onset of parturition is timed primarily by the fetus* via secretions of the adrenal cortex. Although the *exact* consequences of this increased secretion of fetal cortisol vary, the *general* result is an increased oestrogen:progesterone ratio, which stimulates the synthesis and release of $PGF_{2\alpha}$ in the uterus, the common result of the hormonal changes at term. $PGF_{2\alpha}$ is the activator of the mechanical events at parturition, i.e. myometrial contractions and cervical ripening. Oxytocin may further enhance the synthesis and release of $PGF_{2\alpha}$ as parturition proceeds. Although in the human the beginning and ending of this sequence appear similar, i.e. increased fetal cortisol plasma concentrations and increased levels of $PGF_{2\alpha}$ in maternal peripheral blood or amniotic fluid, the intervening changes in the oestrogen:progesterone ratio have not been measured consistently. This may mean that such changes occur at a local or cellular level, but evidence in favour of this is sparse.

It has not been stated explicitly here, but it is generally regarded that maternal mechanisms may contribute to or modulate the timing of parturition controlled largely by the fetus. The nature of these modulating influences which may, for example, restrict births to certain hours of the day or night, are largely unknown, but might be neural rather than hormonal in origin.

**6 Relaxin, pregnancy and parturition**

The existence of relaxin was first postulated in the 1920s to explain the phenomenon of prenatal separation of the maternal pubic symphysis in some species. The separation process which aids parturition is caused by relaxation of the interpubic ligament; hence the name *relaxin* and its implied role as an aid to parturition.

Relaxin is a basic polypeptide consisting of two subunits of similar size linked by disulphide bonds. The A chain consists of 22 and the B chain 26 residues, and its molecular weight is 5600–5800. Relaxin bears some similarities in structure to insulin and may perhaps be cleaved from a larger 'prorelaxin' form of molecular weight 42 000.

Measurement of relaxin has been largely by bioassay, the most common approach being to measure the relative ability of relaxin preparations to induce separation and stimulate elongation of the interpubic ligament in the guinea-pig or mouse. Such a bioassay is laborious, of low sensitivity and the endpoint is not particularly precise. More recently, radio-immunoassays of relaxin have been developed and are the methods of choice.

*a Source and secretion of relaxin*

The major source of relaxin is apparently the corpus luteum—at least in the sow where ultrastructural, immuno-cytochemical and radioimmunoassay data are in good accord. It is stored in cytoplasmic granules of the granulosa lutein cells and secreted in relatively small amounts during most of pregnancy. However, immediately before parturition, the hormone is rapidly released in large amounts and this is accompanied by degranulation of luteal cells. There are similarly good data for luteal localization of relaxin in the rat and mouse.

In pregnant women, the hormone has been measured in blood using both bio- and radioimmunoassay. Maximum plasma concentrations are seen during weeks 38–42 but relaxin is detectable as early as weeks 7–10. The corpus luteum again seems the most likely source, since venous blood from ovaries bearing corpora lutea contains higher plasma concentrations than that from ovaries lacking corpora lutea. Moreover, removal of the corpus luteum causes systemic levels of relaxin to fall. Plasma relaxin levels are apparently not elevated in women during labour induced with either $PGF_{2\alpha}$ or oxytocin.

*b The actions of relaxin*

Separation of the innominate bones occurs in many species at term and is a mechanism for removing the obvious physical obstacle to parturition that the pelvic girdle presents. Other ligaments associated with the pelvis also soften as parturition approaches, the end result enabling distention of the girdle during passage of the fetus. A role for relaxin prior to *pubic separation,* although not necessarily in ligamentous relax-ation, is well established in a wide variety of species, but particularly rodents.

In pregnant women, there are a number of reports demon-strating relaxation of pelvic ligaments associated with separ-ation of the pubic bones at term (by 10 mm or so) and closure by 5 months postpartum. However, it has not been established directly that relaxin, even though elevated in blood at this time, is itself responsible for comparable changes in the pelvis. Relaxin also has a more rapid action on the myometrium to cause its quiescence. This occurs through a different mechanism to that underlying the remodelling of connective tissue and involves cAMP-dependent alterations in the phosphorylation of myosin light-chain kinase. This action, like that on the pelvic ligaments, results from the interaction of relaxin with its specific receptors. Relaxin may also inhibit oxytocin secretion and facilitate cervical softening. The hormone may thus be seen as assisting the maintenance of pregnancy and paving the way for parturition.

## 7 Labour

In the above account we have discussed those factors that determine the onset of *labour,* which is the process by which the mother expels the fetus. *Premature* labour is said to occur after 27 weeks but before 37 weeks of human gestation. Labour at *term* occurs between 37 and 43 weeks and after this time, it is called *post-term.* Labour is usually taken to consist of three stages: *Stage 1* begins with the onset of labour and ends when the uterine cervix is fully dilated. It may be further divided into latent and active phases; *Stage 2* begins with full dilation of the cervix and ends with complete expulsion of the fetus; *Stage 3* begins with completion of fetal expulsion and ends with expulsion of the placenta.

With the onset of labour, as we have seen, large contractions of the uterine musculature occur. These are regular, occur at shorter and shorter intervals and result in intra-uterine pressures of between 50 and 100 mmHg compared with about 10 mmHg between contractions. One of the tasks of these contractions is to retract the lower uterine segment and cervix upwards to allow expulsion of the fetus. The phenomenon of *brachystasis* reflects the property of myometrial cells which helps to achieve this end. Thus, shortening of each muscle cell during contraction is followed, during relaxation, by failure to regain its initial length. With each subsequent contraction further shortening of the cell occurs, and so eventually, each myometrial cell becomes shorter and broader, the fundal musculature becomes thicker and uterine volume decreases. The lower uterine segment does not take part in these contractions and remains quite passive during labour. As a result of the muscular phenomenon described above, therefore, the lower segment moves upwards, i.e. is retracted. This event may even be palpated abdominally because the junction between the two segments (the *physiological retraction ring,* marked because of the contrast between thick myometrium above and thin lower uterine segment below) gradually moves upwards.

During this time the cervix has softened, and when *fully* dilated, can no longer be pulled upwards because of its attachment to uterine and uterosacral ligaments and pubocervical fascia. Moving into Stage 2 of labour, the fully dilated cervix is at about the level of the pelvic inlet. Subsequent

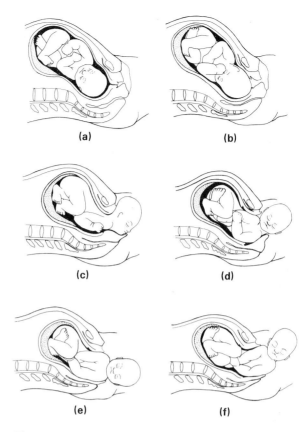

**Fig. 12.5.** Normal labour showing: (a) engagement and flexion of the head; (b) internal rotation; (c) delivery by extension of the head after dilation of the cervix; (d–f) sequential delivery of the shoulders.

uterine contractions, and the resultant decrease in uterine volume, push the fetus downwards and through the pelvis (see Fig. 12.5 for summary). This whole process of labour takes from about 8 hours (in multipara) to 14 hours (in primigravida), the first stage occupying much of this time and the second stage only 1–5 hours.

A few minutes after delivery of the fetus and clamping the umbilical cord, the placenta becomes detached from the uterine wall as the result of a myometrial contraction. Within a short time the placenta will be completely extruded by uterine contractions, but this is often aided by the obstetrician abdominally massaging the fundus and accelerating the first contraction.

## 8 Fetal monitoring and Caesarian section

Parturition is a time of vulnerability for the fetus as it undergoes the transition to neonatality. The necessary adjustments that it must make to its own cardiovascular and respiratory systems (Chapter 11) are preceded by a period of dwindling maternal support as labour progresses. If this

307　　*Parturition*

period is unduly protracted, the effectiveness of metabolic exchange can decline such as to cause fetal distress with increased heart rate and asphyxia. The traditional obstetric approach to this problem is to monitor the fetal heart rate by intermittent auscultation with a stethoscope, and to resort to Caesarian section if necessary. More recently, continuous monitoring of fetal heart rate and/or fetal blood pH has replaced this approach in some centres. This technique is costly and invasive, involving sampling from the fetal scalp, and therefore increasing the risk of postpartum infection. It has been questioned whether any advantage this approach might bring—in terms of a reduced rate of intrapartum still-births for a few mothers—is more than offset by the invasive procedure used. The balance of evidence is not yet decisive. There clearly is an improved survival rate in obstetrically difficult cases, but whether this is entirely due to monitoring itself or due also to the increased general attention given to monitored patients is not clear. In addition, infection rates are higher and many studies report a tendency to resort to Caesarian section more readily, thus increasing the use of this delivery procedure disproportionately. However, the likely increase in the use of monitoring procedures in the future, together with advances in understanding the fetal events underlying the traces, may mean that obstetric practice will be greatly benefitted in due course.

## 9 Summary

Once born, inspected, and sexed, the fruit of the past eleven chapters enters into a period of prolonged parental care during which, early on, its nutritional requirements are usually provided by the lactating mother. These topics are the subject of the next chapter.

## Further reading

Challis JRG, Lye SJ. Parturition. In *Oxford Reviews of Reproductive Biology*, Vol. 8, pp. 61–129. Oxford University Press, 1986.

Ciba Foundation Symposium. *Foetal Autonomy*. Churchill Livingstone, 1969.

Ciba Foundation Symposium 47. *The Fetus and Birth*. Elsevier, 1977 (New Series).

Liggins GC (Ed.). *Parturition Seminars in Perinatology*. Vol. II, No. 3. Grune and Stratten, 1978.

Nathanielsz PW. *Fetal Endocrinology—an experimental approach*. Monographs in Fetal Physiology, Vol. 1. North-Holland, 1976.

Nathanielsz PW. *Ann Rev Physiol* 1978; **40**: 411-445.

Niswander KR. *Obstetrics. Essentials of Clinical Practice*. Little, Brown and Co., 1976.

Porter DG. Relaxin: old hormone, new prospect. In *Oxford Reviews of Reproductive Biology*, Vol. 1, pp. 1–57 Oxford University Press, 1979.

Short RV (Ed.). Reproduction. *Br Med Bull*; 1979.

Wynn, RM (Ed.). *Biology of the Uterus*. Plenum Press, 1977.

# Chapter 13
# Lactation and Maternal Behaviour

In this chapter, we will consider the early postnatal events which are designed to ensure the survival of the newborn baby. These are the provision of food through the process of lactation and associated nursing, and the relatively extended period of maternal care which provides a protective environment in which the young can grow and gradually attain independence.

## 1 Lactation

Among the many changes occurring in the mother during pregnancy are those that involve the breast. In most species, this process is as vital to the success of reproduction as gamete production and fertilization, since failure to lactate will result in early postnatal death of the young. Breast-feeding in Western society, however, is by no means the inevitable sequel to pregnancy and birth, largely because bottle-feeding with milk substitutes represents an alternative. It is not our purpose here to enter the contemporary and important debate concerning the relative values for mother and infant of breast- or bottle-feeding. Instead we shall describe the factors which

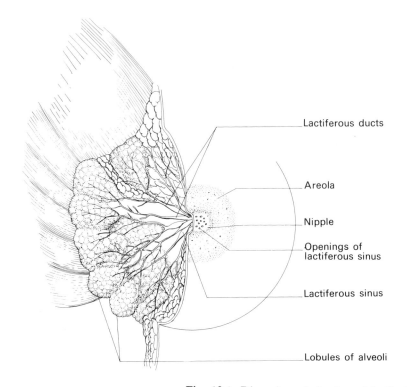

Lactiferous ducts

Areola

Nipple

Openings of
lactiferous sinus

Lactiferous sinus

Lobules of alveoli

**Fig. 13.1.** Dissection of the lateral half of the right breast of a pregnant woman. When fully developed the lobules consist of clusters of rounded *alveoli* which open into the smallest branches of milk-collecting ducts. These, in turn, unite to form larger, *lactiferous ducts*, each draining a lobe of the gland. The lactiferous ducts converge towards the *areola* beneath which they form dilatations, or *lactiferous sinuses*, that serve as small reservoirs for milk. After narrowing in diameter each lactiferous sinus runs separately up through the *nipple* (or papilla) to open directly upon its surface. The skin of the areola and nipple is pigmented brown and wrinkled. Also opening onto the peripheral area of the areola are small ducts from the *Montgomery glands*, which are large sebaceous glands whose secretions probably have a lubricative function during suckling.

induce, control and regulate breast development, lactation and milk removal by the young, primarily in the human female. We will refer to other species only when data in the human are lacking, or when important differences between species are apparent.

*a Anatomy of the breast*

The mammary gland forms a lobulated mass comprised of glandular (or parenchymatous) tissue, fibrous tissue connecting the lobes and adipose tissue between them. The lobes are 15 to 20 in number in the human breast and themselves consist of *lobules* interconnected by areolar tissue, blood vessels and ducts.

The basic pattern of breast structure shown in Fig. 13.1 is common to all species, even though the number of mammary

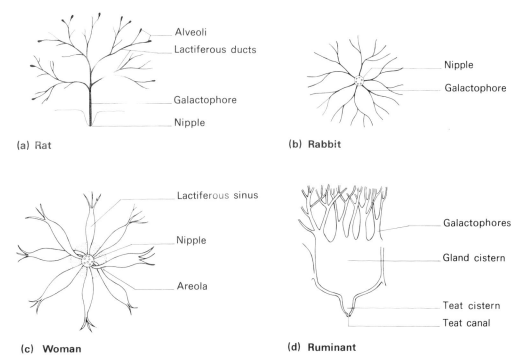

(a) Rat

(b) Rabbit

(c) Woman

(d) Ruminant

**Fig. 13.2.** Different patterns of the ductular system in the mammary glands of four mammals. (a) The rat in which the lactiferous ducts unite to form a single galactophore, which opens at the nipple. (b) The rabbit in which a number of lactiferous ducts unite to form several galactophores. (c) The human female where one lactiferous duct drains each of 15 to 20 mammary lobes, dilating as a lactiferous sinus before emerging at the nipple. (d) The ruminant udder in which the galactophores open into a large reservoir or *gland cistern*, which opens into a smaller teat cistern and thence to the surface via a teat canal.

glands, their size, location and shape vary greatly. The sow, for example, has up to 18 mammary glands (nine pairs) while in the cow and goat their pairs of glands (two and one respectively) are closely apposed in an abdominal *udder*. In addition, there is some variation in the pattern of the duct system (Fig. 13.2).

Microscopically, the alveolar walls are formed by a single layer of cuboidal to columnar epithelial cells, their shape depending on the fullness or emptiness, respectively, of the alveolar lumen (Fig. 13.3). It is these cells which are responsible for milk synthesis and secretion during lactation. The *myoepithelial cells* situated between the epithelial cells and the basement membrane have a contractile function, and are important for moving milk from the alveoli into the ducts prior to milk ejection (Section g).

*b Effects of hormones on the developing breast*

At birth the mammary gland consists almost entirely of lactiferous ducts; few, if any, alveoli are present. Apart from a

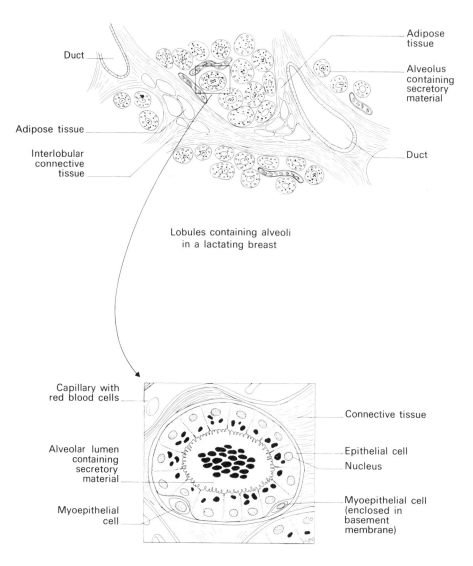

Duct

Adipose
tissue

Alveolus
containing
secretory
material

Adipose tissue

Interlobular
connective
tissue

Duct

Lobules containing alveoli
in a lactating breast

Capillary with
red blood cells

Connective tissue

Epithelial cell

Nucleus

Alveolar lumen
containing
secretory
material

Myoepithelial
cell

Myoepithelial cell
(enclosed in
basement
membrane)

An alveolus

**Fig. 13.3.** Microscopic structure of (above) lobules in a lactating mammary gland and (below) a high-power view of an alveolus. Note the rich vascular supply from which the single layer of secretory epithelial cells draws precursors used in the synthesis of milk. The myoepithelial cells situated between the basement membrane and epithelial cells form a contractile basket around each alveolus.

little branching, the breast remains in this state until puberty (see Chapter 1). At this time, and under the action primarily of oestrogens, the lactiferous ducts sprout and branch and their ends form small, solid, spheroidal masses of granular polyhedral cells, which later develop into true alveoli. As menstrual cycles establish themselves, successive exposure of mammary tissue to oestrogen and progesterone induces additional, if limited, ductal–lobular–alveolar growth, and the

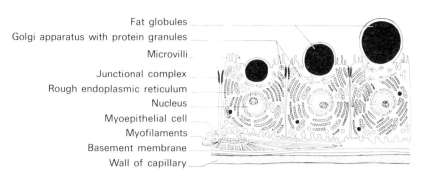

Fat globules
Golgi apparatus with protein granules
Microvilli
Junctional complex
Rough endoplasmic reticulum
Nucleus
Myoepithelial cell
Myofilaments
Basement membrane
Wall of capillary

**Fig. 13.4.** Schematic drawing of the ultrastructure of secretory alveolar epithelial cells. The position of the myoepithelial cell can clearly be seen (cf. Fig. 13.3). Adjoining alveolar cells are connected by junctional complexes near their luminal surfaces which also bear numerous microvilli. The cell cytoplasm is rich in rough endoplasmic reticulum, particularly in the basal part of the cell. Many mitochondria are present and the large Golgi apparatus is situated nearer the luminal surface and close to the cell nucleus. The fat globules and protein granules are the cellular secretory products destined for the alveolar lumen. Their manner of extrusion from the cell is also indicated.

breasts increase in size due to the deposition of fat and growth of connective tissue. Adrenal corticosteroids may also contribute to duct development at this time. Cyclic changes in the breast occur in the non-pregnant woman, and these are most marked premenstrually when there may be an appreciable increase in breast volume (Chapter 7). In addition, some secretory activity may occur in the alveoli and small amounts of secretory material which result from this activity can be expressed from the non-pregnant breast during the premenstrual period. The degree of mammary development in non-pregnant women, therefore, is considerable—particularly when compared with other mammals, including non-human primates, in whom appreciable mammary growth is not achieved until the middle or end of pregnancy.

During early pregnancy, and under the influence of oestradiol, progesterone and possibly insulin and prolactin, the previously developed ductular–lobular–alveolar system undergoes considerable hypertrophy. Prominent lobules form in the breast, and the lumina of the alveoli become dilated. Differentiation of the alveolar cells to the form shown in Fig. 13.4 occurs during mid-pregnancy at a time when duct and lobule proliferation has largely ended. The epithelial cells contain substantial amounts of secretory material (Fig. 13.4) from the end of the fourth month of human pregnancy, and the mammary gland is fully developed for lactation, awaiting only the endocrine changes described below for full activation.

The hormonal determinants of mammogenesis in women

differ quite considerably from the situation in other species, notably rodents, in which a complex of sex steroids, adrenal steroids, growth hormone and prolactin combine to induce mammary growth. This increased hormonal requirement probably reflects the relatively poor mammary development which has occurred before pregnancy in these species compared with humans. It is doubtful whether prolactin or placental lactogen (despite its name) are absolutely necessary for induction of full breast development in women. Similarly, growth hormone, an essential factor in rodents, appears unnecessary in humans since even with congenital growth hormone deficiency, mammary growth and lactation can still occur.

*c Cellular mechanisms of milk secretion and the biochemistry of milk*

It is now well recognized that *milk fat* (lipid) is synthesized in the endoplasmic reticulum of the alveolar epithelial cells. The membrane-bound lipid droplets then move towards the luminal surface of the cell, increasing in size as they do so (Fig. 13.4). The droplet then pushes against the cell membrane, causing it to bulge and lose its microvilli. Gradually the cell membrane constricts behind the lipid droplet to form a 'neck' of cytoplasm which ultimately pinches off to release it, membrane-enclosed, into the alveolar lumen (Fig. 13.4). The *milk protein* seems to be formed within the Golgi apparatus, where it eventually appears as granular material within vacuoles, which gradually move to the luminal surface of the cell to release their contents by exocytosis (Fig. 13.4). These processes are much affected by prolactin, and the alveolar cells bear receptors for the hormone on their plasma membrane.

The composition of milk varies from the first week postpartum, when a yellowish, sticky secretion (*colostrum*) is produced, through a transitional phase of 2–3 weeks which leads to the production of *mature milk*. Colostrum is secreted in amounts up to 40 ml day during the first week postpartum, and, compared with mature milk, contains less water-soluble vitamins (B complex, C), fat and lactose, but greater amounts of proteins, some minerals and fat-soluble vitamins (A, D, E, K). Colostrum also contains immunoglobulins (IgG). Mature milk is produced after a transitional period during which the concentration of immunoglobulins and total proteins declines, while lactose, fat and the total calorific value of the breast milk increase. A summary of the constituents of mature milk are listed in Table 13.1. One or two features of human milk will be emphasized here. The main energy source in this milk is *fat*, which is almost completely digestible, partly due to the fact that it is present as well-emulsified, small fat globules. Milk fat is also an important carrier for vitamins A and D. *Lactose* (milk sugar) is the predominant carbohydrate in milk. It is less sweet than common sugars and is important for

**Table 13.1.** Some contents of human mature milk.

| | |
|---|---|
| Water | approx. 90 gm% |
| Lactose | approx. 7 gm% |
| Fat | inc. essential fatty acids |
| | saturated fatty acids |
| | unsaturated fatty acids |
| Amino acids | inc. essential amino acids |
| Protein | inc. lactalbumin and lactoglobulin |
| Minerals | inc. calcium, iron, magnesium, |
| | potassium, sodium, phosphorus, |
| | sulphur |
| Vitamins | inc. A, $B_1$, $B_2$, $B_{12}$, C, D, E, K |
| pH | 7.0 |
| Energy value | |
| (kcal 100 ml) | 650 |

promoting intestinal growth of *Lactobacillus bifidus* flora (lactic acid-producing) as well as providing an essential component (galactose) for myelin formation in nervous tissue. Lactose is formed within the Golgi apparatus of alveolar cells and is dependent on the prior synthesis of *α-lactalbumin* (the whey protein) which is itself formed in the endoplasmic reticulum. The *α*-lactalbumin is passed to the Golgi apparatus where a second enzyme involved in lactose synthesis, *galactosyltransferase,* is situated. The combination of these two factors (as *lactose synthetase*) then permits completion of lactose synthesis and the sugar passes to the alveolar lumen with the protein granules. This enzyme system is stimulated by prolactin (see below).

*d Initiation of milk secretion (lactogenesis)*

From the account so far it should be clear that the breasts are sufficiently developed, and the alveoli adequately differentiated, to begin milk secretion by 4 months of pregnancy in women. But they do not usually do so until after parturition. Only at this time does the copious milk secretion essential for the initiation of full lactation occur. It is now widely agreed that the *disappearance of oestrogen and progesterone from the maternal circulation,* which does not occur until or soon after parturition in women, holds the key to the initiation of lactation. Prolactin and placental lactogen increase in plasma concentration throughout pregnancy and reach their maxima at term (Fig. 13.5), but the *breast remains unresponsive to them* until steroid levels fall after delivery of the placenta. The steroids appear to inhibit milk secretion by acting directly on mammary tissue, probably on the alveolar cells. After parturition, placental lactogen levels fall abruptly while prolactin levels fall more slowly, remaining high for the duration of lactation. Bursts of prolactin secretion occur during suckling episodes, and this is important for the maintenance of lactation (see below).

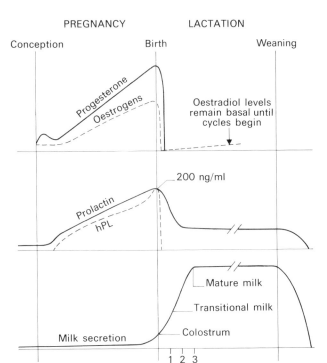

PREGNANCY          LACTATION

Conception          Birth         Weaning

Progesterone

Oestrogens

Oestradiol levels
remain basal until
cycles begin

200 ng/ml

Prolactin

hPL

Mature milk

Transitional milk

Colostrum

Milk secretion

1 2 3
weeks post-partum

**Fig. 13.5.** The sequence of hormone changes in the maternal circulation which underly the onset of lactation in women. Withdrawal of oestrogen and progesterone is critical and removes a block to prolactin-induced milk secretion in the gland. Withdrawal of hPL may have a similar function but this is less certain (see text for details).

Milk secretion will occur after parturition and last, if scantily, for 3 or 4 weeks even in the absence of suckling. This period is one in which blood prolactin concentrations remain well above normal, which presumably explains this temporary phase of lactation. But if full lactation is to continue with copious milk secretion, suckling and attendant nipple stimulation are essential. This brings us to the question 'What maintains milk secretion once it is initiated by sequential steroid hormone and prolactin exposure?'

*e Maintenance of milk secretion (galactopoiesis)*

The critical hormonal requirement for the maintenance of established lactation is prolactin, at least in women. The ovarian steroids, oestradiol and progesterone, seem of little importance and ovariectomy has no effect on lactation. This lack of importance of ovarian steroids is further emphasized by the demonstration that lactation can be induced and maintained in postmenopausal women by repeated exposure to suckling stimuli alone. Frequent reference to the *suckling stimulus* or *nipple stimulation* in the above account emphasizes the important function it fulfills in galactopoiesis. It *induces the* reflex release of prolactin from the anterior lobe of the pituitary, providing another example of a neuroendocrine reflex such as encountered in Chapter 12. The afferent limb consists of neural pathways conveying sensory information from the nipples to the spinal cord and up, eventually, to the hypothalamus. Denervation of the nipple prevents prolactin release in response to nipple stimulation, indicating the important neural afferent component of the reflex. The links within the CNS between somatosensory inputs, which ascend the spinal cord to the brain stem, and altered output of prolactin by the adenohypophysis are unclear at present. Suppression of hypothalamic tubero–infundibular dopamine neuron activity, and hence decreased dopamine (PIF) secretion into the portal vessels, although it occurs, is probably not sufficient to explain the increased secretion of prolactin. It is now clear from experiments in the rat that, during suckling, there is a marked increase in the secretion of VIP into the portal vessels (see Chapter 5). In addition, increased amounts of VIP mRNA are demonstrable, using in situ hybridization techniques, in the neurons of the medial parvocellular paraventricular nucleus whose axons terminate in the palisade zone of the median eminence in association with portal capillaries. This increased secretion of a potent prolactin releasing factor is probably responsible, in large part, for the enhanced secretion of prolactin from the anterior pituitary. It has also been demonstrated that the reduction in TIDA neuron activity and hence dopamine release during suckling sensitizes the lactotrophs to VIP, indicating that the two mechanisms interact to control prolactin secretion during suckling. Plasma-borne prolactin represents, therefore, the efferent limb of this neuroendocrine reflex and ultimately it reaches the mammary alveoli (Fig. 13.6). The amount of prolactin released is determined by the strength and duration of nipple stimulation during suckling. Suckling at both breasts simultaneously, when feeding twins for example, induces a greater release of prolactin than occurs during stimulation of one breast.

As lactation progresses, the amount of prolactin released at each suckling bout may decline, but basal plasma concentrations often remain elevated, although how this is achieved is uncertain (see also discussion of delayed implantation, Chapter 9). The circulating plasma concentration of prolactin during lactation appears, more than any other factor, to be most important in determining the amount of milk secreted. This in its turn is regulated by the frequency or duration or intensity (or all three) of the suckling stimulus. That prolactin is of major importance is made clear by the demonstration that declining milk secretion in women can be boosted, with consequent breast engorgement, by a TRH treatment schedule which releases prolactin. Administration of VIP would no doubt be even more effective.

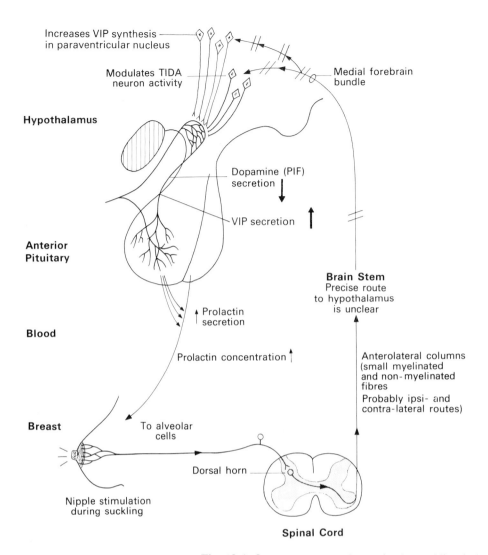

Increases VIP synthesis
in paraventricular nucleus

Modulates TIDA
neuron activity

Medial forebrain
bundle

**Hypothalamus**

Dopamine (PIF)
secretion

VIP secretion

**Anterior
Pituitary**

**Brain Stem**
Precise route
to hypothalamus
is unclear

↑ Prolactin
secretion

**Blood**

Prolactin concentration ↑

Anterolateral columns
(small myelinated
and non-myelinated
fibres
Probably ipsi- and
contra-lateral routes)

**Breast**

To alveolar
cells

Dorsal horn

Nipple stimulation
during suckling

**Spinal Cord**

**Fig. 13.6.** Somatosensory pathways in the suckling-induced reflex release of prolactin. The exact route taken by sensory information between brain stem and hypothalamus is speculative. Although tubero-infundibular dopamine neuron (TIDA) activity is modulated as a result of the arrival of this somatosensory derived input, the increased secretory activity of VIP-containing neurons in the paraventricular nucleus is probably also of crucial importance in driving prolactin secretion during suckling.

Although in other species additional hormones may be essential for the successful maintenance of lactation (growth hormone, insulin and adrenal steroids among them) *prolactin alone is essential in women*. Its action seems to be exerted directly on the alveolar cells of the breast to stimulate the synthesis and secretion of milk proteins (casein, lactalbumin and lactoglobulin), lactose and lipids. Prolactin, having been bound to its receptors on the cell, stimulates RNA-dependent

318     *Chapter 13*

**Table 13.2.** Summary of events leading to full lactation in the human.

| Hormone or activity | Effect |
| --- | --- |
| *Early to mid-pregnancy* | |
| Oestrogen, progesterone and corticosteroids | Ductular–lobular–alveolar growth. Considerable branching of the duct systems up until mid-pregnancy. Followed by considerable differentiation of epithelial stem cells into true alveolar secretory epithelium. |
| Oestrogen and progesterone, ? hPL | Little or no milk secretion occurs due to the inhibitory effects of these hormones on prolactin stimulation of alveolar cells. Continues until . . . |
| *Late pregnancy and term* | |
| Oestrogen and progesterone | Pronounced alveolar epithelial cell differentiation. |
| Steroid and prolactin levels high. Steroids begin to fall | Colostrum secretion. |
| *Parturition* | |
| Oestrogen and progesterone fall precipitously. Prolactin levels decline but basal concentrations remain high | Stimulation of active secretion of colostrum and, over 20 days or so, secretion of mature milk. Full lactation initiated. |
| Suckling | Induces episodic prolactin release at each feed. Maintains milk secretion by promoting synthesis of lipids, milk (particularly $\alpha$-lactalbumin) and lactose. |

protein synthesis. Stimulation of $\alpha$-lactalbumin synthesis is, as discussed above, a critical step in the biosynthesis of lactose, and is also inhibited by progesterone. This step may represent, therefore, one site at which progesterone–prolactin antagonism within mammary tissue may occur prior to parturition. The fact that nipple stimulation *during* a feed (suckling) induces prolactin release which *subsequently* (i.e. after removal of the milk) induces further milk secretion suggests that the baby actually orders its next meal during its current meal.

*f Summary*

We have seen, then, that *development* of the mammary gland's ductular–alveolar system and milk-secreting capacity during pregnancy is under the influence of adrenal and ovarian/ placental steroids. The *initiation of milk secretion* depends on the presence of high levels of prolactin and withdrawal of oestrogen, progesterone and possibly placental lactogen. The *maintenance* of milk secretion depends, in the human female, solely on the continued secretion of prolactin, maintained by nipple stimulation during suckling. These events are summarized in Table 13.2. In the next section we will examine how the suckling infant removes milk from the breast, and the milk ejection reflex which subserves this task.

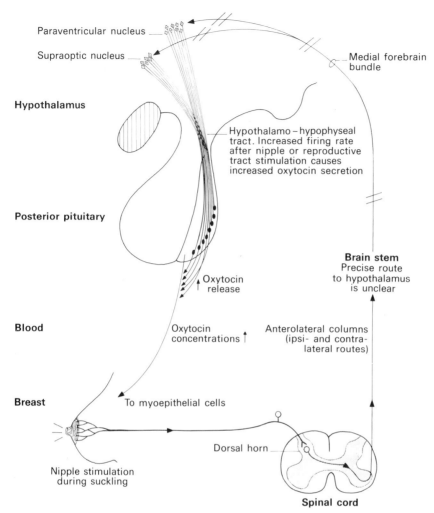

**Fig. 13.7.** Somatosensory pathways in the suckling-induced reflex release of oxytocin. Doubts exist about the brain stem–hypothalamic route taken by the sensory information, but it probably involves the medial forebrain bundle.

*g Milk removal and the milk ejection reflex (MER)*

Milk removal involves transport of milk from the alveolar lumina to the nipple (or teat) where it becomes available to, and is removed by, the suckling infant. The milk ejection reflex (MER), which underlies this function, has much in common with the induced release of prolactin described above.

i MER and its neuroendocrine mechanisms

*Stimulation of the nipple* during suckling probably represents the most potent stimulus to milk ejection. The sensory information so generated travels via the spinal cord and brain stem and ultimately activates oxytocin neurons in the paraventricular and supraoptic nuclei in the hypothalamus (Fig. 13.7). This boosts not only the *synthesis of oxytocin*, but also its *release* from the posterior pituitary into the bloodstream. On reaching the mammary gland, oxytocin causes *contraction of the myoepithelial cells* which surround the alveoli, to induce expulsion of milk into the ducts (milk *let-down or 'draught'*), with a consequent build up of intramammary pressure which may cause milk to spurt from the nipple or teat. Although touch and pressure at the nipple are very potent stimuli to oxytocin release and milk let-down, the MER can be conditioned to occur in response to stimuli other than these, such as a baby's hungry cry or, in cows, the rattling of a milking bucket. Such a conditioned reflex release does not occur in the case of prolactin, where nipple stimulation seems to be the only effective inducer of its release. The fact that the cow's udder contains, in the gland cistern, all the milk obtainable during milking emphasizes that let-down is due to the increase in mammary pressure resulting from the expulsion of milk from alveoli to fine ducts, rather than any sudden increase in milk secretion as was once thought. This is also the case in the human female.

In Chapter 12, we described how stimulation of the female's reproductive tract, particularly the vagina and cervix, also induces the release of oxytocin. The oxytocin release so induced explains the phenomenon of milk ejection during coitus in lactating women and the rather ancient, enduring and erstwhile puzzling practice of blowing air into the vaginae of cattle to induce milk draught!

The MER is particularly susceptible to inhibition by physical and psychological stresses; discomfort immediately after parturition or worry and uncertainty about breastfeeding are potentially important inhibitors of the successful initiation and early maintenance of lactation. The way in which 'stress' inhibits the MER is not clear, but may involve inhibition of oxytocin release, release of catecholamines, adrenaline for example, or activation of the sympathetic nervous system, or all three. Constriction of mammary blood vessels induced by the latter two events might limit access of oxytocin to the myoepithelial cells.

The afferent route taken by sensory information arising at the nipple (or reproductive tract) is quite well established from research on rodents. It involves the peripheral sensory nerves, which enter the spinal cord via the dorsal roots, and a number of synaptic relays in the dorsal horn before transmission up the anterolateral columns. These pathways contain fibres destined for various sites in the thalamus, spinal cord and, in particular, the brain stem reticular formation. From this point, the route to the hypothalamic, magnocellular paraventricular and supraoptic nuclei (Fig. 5.2) has not been precisely defined at present, but the midbrain peripeduncular nucleus and medial forebrain bundle, which courses through the lateral hypothalamus, appear to be particularly important.

321     *Lactation and Maternal Behaviour*

## ii Suckling

There has, for a number of years, been considerable debate as to the mechanics of suckling. Many workers hold that infants obtain milk from the breast by actively *sucking,* and that an airtight seal between lips and breast is essential for the negative pressure-dependent transfer of milk. However, X-ray cinematographic evidence suggests this may not be the case and that milk is obtained by *expressing* it from the nipple or teat, sucking merely aiding the process. Thus, in the human, the nipple and areola are drawn out to form a teat which is compressed between the infant's tongue and hard palate. The milk is then stripped out of this 'teat' by the tongue compressing the nipple from base to apex against the hard palate. Pressure on the base of the teat is then released, allowing its rapid refillment with milk because of the oxytocin-induced increase in intramammary pressure that subserves let-down. There are undoubtedly species differences in this process, and young also readily adapt to alternative means of obtaining milk. Bottle-feeding using stiff teats, for example, requires more sucking than stripping, and calves or human infants learn this skill rapidly.

## iii Summary

Removal of milk from the breast is dependent on the *suckling-induced milk ejection reflex.* By this means, stimulation of the nipple and other cues associated with nursing induce the release of *oxytocin* from the neurohypophysis. This hormone *stimulates contraction of the myoepithelial cells* which surround each alveolus, forcing milk out of the alveoli into the smaller lactiferous ducts. The resulting *increase in intramammary pressure* results in *milk let-down,* causing milk to spurt from the nipple and to be removed from the breast by the suckling infant. The neural pathways mediating the MER are not understood completely at present, but are likely to be complicated. This is particularly the case when psychological factors associated with the induction or inhibition of the MER are considered.

## h Involution of the breast

After the cessation of lactation, the process of involution of the mammary gland takes about 3 months. When nursing is discontinued, milk accumulates in the alveoli and small lactiferous ducts causing distention and mechanical atrophy of the epithelial structures, rupture of the alveolar walls and the attendant formation of large hollow spaces in the mammary tissue. The alveolar distention also results in alveolar hypoxia and reduction in nutrient supply because mammary capillaries are compressed. Milk secretion is, therefore, greatly suppressed not so much by the fall in plasma prolactin (as suckling frequency diminishes), but as a consequence of the effects of local mechanical factors. As desquamated alveolar cells and glandular debris become phagocytosed, the lobular-acinar structures become smaller and fewer, so the ductular system in the breast again begins to predominate. The alveolar lumina decrease in size, may eventually disappear, and their lining changes from the secretory single-layered to a non-secretory, two-layered type of epithelium, previously seen prior to pregnancy.

This whole involutional process is more intense if nursing is stopped rapidly than if it continues at reduced frequency during a more gradual weaning. Although these changes in the mammary gland are pronounced, they are markedly different to those occurring in the postmenopausal woman. In the latter, there is clear structural atrophy of the breast rather than a transition to a period of inactivity, such as occurs postlactation. The breasts invariably remain larger after lactation than they were prior to pregnancy because of the increased deposition of fat and connective tissue which occurs between these two time points.

*i Suppression of lactation*

It may be necessary to suppress lactation in women for a variety of reasons. Abortion after 4 months of pregnancy, and stillbirth, will be followed by unwanted and unnecessary lactation. Nursing may be contra-indicated for clinical reasons or the mother may simply prefer not to breast-feed her baby. A number of methods for suppressing lactation have been used in the past and may still be employed, if less regularly, today. They include binding the breasts and applying ice packs, after which lactation often ceases within a week or so; administering diuretics—although this is less favourably viewed at present; treatment with sex steroids just before or soon after parturition, often using a mixture of a long-acting androgen (testosterone enanthate) and oestradiol valerate given by intramuscular injection. The high plasma levels of steroids which result antagonize the effects of prolactin on the breast (as occurs naturally by high circulating levels of oestrogens and progesterone during pregnancy). Reference to Chapter 5, however, will remove any surprise that the most widely used lactation suppressant today is *bromocryptine*, a dopamine receptor agonist which markedly depresses prolactin, and hence milk secretion. This drug is, of course, used to reinstate menstrual cycles in women with hyperprolactinaemia, who often wish to become pregnant. This must, therefore, be considered in puerpereal women in whom lactation is suppressed by giving them bromocryptine, since they are likely to be fertile after parturition more rapidly than would otherwise occur.

*j Return of menstrual cycles and fertility during lactation*

Menstruation and ovulation return more slowly in lactating than non-lactating women postpartum—normal reproductive function usually re-establishes itself by 3–6 months. However, menstruation itself is a poor *indicator* of fertility during this period, as many women bearing successive children within a year can testify! Conception often occurs in lactating women without an intervening menstruation. Neither ovulation nor menstruation normally occur before 6 weeks postpartum, but approximately half of all contraceptively unprotected, nursing mothers become pregnant during 9 months of lactation.

The reasons for lactational amenorrhoea and anovulation are at present unclear. The most popular hypotheses favour a critical role for prolactin in suppressing the initiation of cyclic release of gonadotrophins. These are discussed more fully in the context of hyperprolactinaemia (which also characterizes lactation) in Chapter 5 and of facultative delayed implantation (Chapter 9).

## 2 Maternal behaviour

In Chapter 1, we gained some insight into the important influences of behavioural interaction between a growing infant and its parents in terms of the progressive development of sexual identity. This is, of course, but one example of the vital role of parents in ensuring not only the normal development but also the very survival of their child. Early on, the mother is of special importance and she must display a whole range of interrelated patterns of *maternal behaviour* in order that her offspring is given protection, warmth, food and affection. The infant, however, is not merely the passive receiver of all this attention, but an active participant in a two-way interaction, eliciting by its own actions appropriate responses from the mother.

A moment's reflection on these facts makes it clear that the nature of the determinants of efficient maternal, and indeed paternal, care are of great importance. Our understanding of them, particularly in the human, is at a most rudimentary stage, but we will examine them in a comparative setting in the hope that some general principles will emerge.

### a General considerations

Maternal behaviour may be considered to occur in three sequential stages : behaviour *preparatory* for the arrival of young displayed during gestation, for example nest building. Behaviour concerned with the *care and protection* of the young early after parturition and associated with *lactation*. Behaviour associated with the progressive independence of the young which are associated with *weaning*.

During gestation many mammals enter a phase of *nest-building activity* which is highly characteristic for each species. Very often at this time, the female may show a marked increase in aggression and defend the area in which the nest has been made. Females of few species, and human females may be among them, will accept the sexual advances of males as gestation proceeds.

After birth the females of all mammalian species are critical to the survival of young, because they provide the essential food supply during lactation. The role of the mother during the early postnatal period varies from species to species depending on the stage of development of the young at birth.

Those species that give premature birth to their young, notably the marsupials, show little change in their behaviour at parturition. The young crawls into the mother's pouch, attaches to a teat, and the mother, apart from cleaning her pouch more often, shows little additional maternal behaviour, apparently not recognizing her own as distinct from other young at this time.

Young who are born naked and blind, so-called *altricial*, need the essential requirements of both food and warmth to be provided by their mothers. Baby mice, rats, rabbits, ferrets and bears, among others, fall into this category, and they are generally attended by their mothers in *nests,* where they keep warm and suckle. Retrieval behaviour forms an important element of maternal behaviour in these species, and misplaced young often make ultrasonic calls to direct and elicit it.

*Semi-altricial* young are those born with hair and sight, but have poorly developed motor skills and may need to be carried by their parents. The carnivores fall into this category, and the parents have nests or dens in which the female stays with the young most of the time. Among the canids, the males return periodically to the nest and regurgitate food, initially for the non-excursive mothers, but later for the young as well. Primates do not normally leave their young in nests, but carry them around—provided the infants have well-developed clinging reflexes! Males often help in this task.

Ungulates, hystricomorph rodents and aquatic mammals give birth to *precocious young* which can move about well, and, to some extent, fend for themselves. The mother–infant bond in these species seems to reflect more a need for contact rather than nourishment, and mothers appear to recognise their own young quickly, and thereafter will feed only them.

As infants grow there is a gradual change in both their behaviour and that of their mothers, which ultimately results in the attainment of independence. Infants, during this time, tend to move away from the mother more often and to greater distances. Mothers, on the other hand, retrieve them less and encourage this exploration by rejecting them more often. Suckling occurs less frequently, lactation ceases and the infant is *weaned.* In solitary species, the young move away from their mother quite rapidly, the females stop maintaining the nest and the temporary family aggregation disintegrates. In more social, group-living species, the young become independent of, and are rejected by, their mothers, but are progressively integrated within the group.

Maternal behaviour comprises, therefore, a complex and variable pattern of interaction between mother and young, beautifully adapted to the social and environmental context in which it occurs. However, little is known in detail of the mechanisms underlying the onset and maintenance of the mother–infant bond in many species.

b Factors affecting the onset and maintenance of maternal behaviour in non-primates

A prevailing belief in studies of maternal behaviour has been that the distinctive profile of pregnancy hormones, progesterone, oestradiol and prolactin must in some way be involved in the appearance of nest building prior to, and maternal behaviours promptly after, birth. The presence of the newborn must also contribute to the maintenance of the postnatal period of care until weaning and independent life is possible.

A huge experimental literature, largely involving the rat, has generally been consistent with this dogma but, until recently, there has been no adequate explanation for the remarkable *speed* with which maternal behaviours appear at parturition, nor their *selectivity* in that within a very short time afterwards, a mother will generally only nurse her own young. Some of the relevant recent data on these points will be summarized.

Firstly, *hormones are not an essential requirement for the display of maternal behaviour.* Ovariectomized, hypophysectomized female rats, castrate and intact males, will all display elements of the maternal behaviour pattern when *exposed to pups for a sufficient length of time.* This process is often called '*sensitization*' and is not, apparently, mediated by any hormonal factor. It is also generally accepted that maternal behaviour in the normal, postpartum lactating female is *not dependent* on hormones, since ovariectomy and hypophysectomy are not followed by any decline in behaviour. (Lactation will, of course, be greatly affected by hypophysectomy, and this must, therefore, be taken into account when performing such experiments, for example by providing foster mothers to nurse the pups.)

Thus, so-called *pup-stimulation,* i.e. attributes of the pups evoking behavioural changes in the mother, can both *induce* and *maintain* behaviour in virgin and postpartum rats. In rats the *sensitization* period necessary to elicit such behaviour is 4–7 days. From about the tenth day of exposure to pups onwards *both* sensitized and lactating mothers show a decline in their maternal care, avoid nursing pups and increase their rejection of, and withdrawal from, them. There is little difference between the two groups of mothers in the timing of the rate at which they show these changes in behaviour.

It would be quite *wrong,* however, to suppose that hormones have no influence on elements of maternal behaviour. Non-postpartum females (or males) *require several days* exposure to pups before they display such behaviour. However, *postpartum females display all elements of maternal behaviour immediately* after birth of their first litter, and do not require a long period of sensitization. This finding suggests that the hormones of gestation, and particularly the changes occurring prior to parturition, may be important determinants of the prompt onset of maternal behaviour. Thus, sequences of injections of oestradiol, progesterone and prolactin, which mimic the changes in these hormones during pregnancy, induce a more rapid display of maternal behaviour in virgin females when first presented with pups. Additional experiments suggest that oestradiol is the most important hormone in this regard, although very large doses are required to elicit maternal responsiveness.

The apparent transition from prepartum *hormonal* to postpartum *infant* determinants of maternal care seems relatively securely established. However, the *immediacy* of maternal care following normal delivery has never been adequately mimicked by any of these manipulations. Clearly, there is something special about parturition itself which renders the mother uniquely sensitive to the newborn. Recent experiments on the sheep have shown this indeed to be the case. Non-parturient ewes or parturient ewes 2 hours or more after having given birth, will not accept or nurse an orphaned lamb even if oestrogen primed. They can be induced to do so with 50% or so success by being made temporarily anosmic by use of a nasal spray, or alternatively by draping the pelt of the ewe's own dead lamb over an orphaned lamb that requires fostering. Olfactory cues from the lamb apparently, then, prevent it from being nursed by any mother, other than its own. The parturient ewe *will* accept an alien lamb and nurse it along with its own, provided it is presented within *2 hours or so of the ewe having herself given birth*. The reason for this seems to be the ewe's altered *olfactory* responsiveness to the lamb during the immediate postpartum period. There are two possible mechanisms: first, the ewe may selectively recognize the odour of her *own* lamb and respond only to it; secondly, the ewe may be functionally *anosmic* in the immediate postpartum period and, therefore, does not respond negatively to an alien lamb by rejecting it. The latter explanation seems more likely, but in either case, how is this olfactory mechanism brought into play? In an ingenious experiment, stimulation of the cervix and distension of the vagina (using a vibrator and bladder of a rugby football, respectively!) in oestrogen-primed non-parturient ewes caused the *immediate* (within minutes) display of maternal behaviour towards alien lambs (Fig. 13.8). The same immediate response has now been seen to occur in the rat i.e. cervical stimulation of a virgin, oestrogen-treated female causes maternal behaviour to be shown within minutes of exposure to pups, whereas many hours are required without cervical stimulation.

*c Summary*

The sequence of events which determines the display of maternal behaviour would appear to be: (i) exposure to

327  *Lactation and Maternal Behaviour*

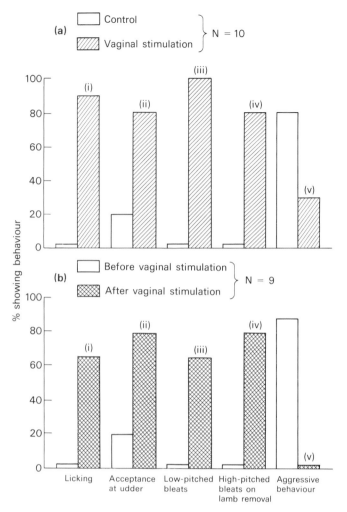

**Fig. 13.8.** (a) The effects of vaginal stimulation on the maternal behaviour of non-pregnant ewes. After stimulation, ewes lick the alien lambs (i), allow them to suckle at the udder (ii), they emit low-pitched bleats characteristic of a maternal ewe (iii), and high-pitched ones—a form of distress—if the lamb is removed (iv), and exhibit a marked decrease in aggression towards the lamb (v). (b) Here, the controls in (a) who showed little or no maternal behaviour were subjected to vaginal stimulation at the end of the observation period. As can be seen, their change in behaviour towards the alien lamb immediately afterwards is dramatic.

hormones, particularly oestradiol, during late pregnancy; (ii) parturition or cervical stimulation itself; (iii) continuing exposure to the newborn and growing infant. Maternal behaviour will be displayed in the absence of either (i) or (ii), but with reduced success. The realization that cervical stimulation has such dramatic effects will have great impact in the world of farming.

Perhaps it will provide a new stimulus to studies of maternal behaviour in women as well. As we will see, mother–infant bonding after Caesarian delivery may be different to that seen after normal delivery. Whether or not this is related to the absence of cervico-vaginal stimulation in the former has not been directly studied. In man and other primates, little is known of hormonal or other determinants of maternal behaviour, but some work, particularly on the rhesus monkey, has revealed a number of factors which contribute to the development of the mother–infant bond and its subsequent severance. These findings may have particular relevance to our understanding of human maternal behaviour. Although species differences should never be minimized, comparative studies of maternal behaviour may serve to focus our attention on critical, and often analogous, elements of behaviour, and their phases of development in man.

*d Maternal behaviour in non-human primates and its relationship to human maternal behaviour*

Baby monkeys spend the majority of their first few months of life (apes, their early years of life) clinging to their mothers, and occasionally fathers, aunts and juveniles, in characteristic ventro-ventral, back-riding or arm-cradled positions. During this period, the mother gives relatively little active physical support to her young, particularly if they cling ventro-ventrally, and so the infant must cling tenaciously with its hands and feet to the mother's abdomen, and often with its mouth to her nipple. Thus, the newly-born monkey's motor capabilities are essential to successful, early contact with its mother. Similarly a well-developed 'rooting reflex', also seen in human babies, by which the head turns towards a tactile stimulus around the mouth, particularly the cheeks, ensures the infant gains the nipple in order to suckle. These clinging and rooting reflexes are present immediately after birth, and are primarily elicited by cues from the mother, largely because she is the individual most likely to be in proximity with the infant. 'Contact comfort' also seems to be an important determinant of the infant's early goal-directed movements, as experiments using surrogate mothers have shown. Thus, a towelling-covered, and therefore 'cuddly', wire model is much preferred by a baby monkey to a bare wire surrogate, even if the latter can provide milk. Similarly, a warm, moving, milk-providing surrogate will be preferred to one without these attributes. Although it is difficult to say why babies prefer, and derive pleasure from, such cues, it presumably involves feelings of security and comfort. There are many analogous instances when the behaviour of human infants towards a warm, rocking mother or soft blanket is considered.

Clinging, contact comfort and rooting behaviour serve to emphasize that baby monkeys contribute to the interaction between themselves and their mothers at the earliest moments after birth, and this is likely to be important in establishing the bond between them. The infant is able, therefore, to secure the immediate needs of warmth, contact and food from its mother with only minimal help from her. In return, apart from actually providing milk and some physical support, the mother must keep her infant clean and protect it. These behaviours are clearly elicited quite specifically by the infant, which emphasizes that the close contact at suckling forms the focal point around which subsequent patterns of maternal behaviour develop. Although the baby accounts in large part for the contact with its mother by clinging, it is clear that the mother derives considerable rewards from close contact with her young, as evidenced by the fact that a mother will carry her dead baby (who, of course, cannot cling) around for some days and show distress when it is removed. It is extremely difficult to assess what the nature of this reward is.

In social primates, females have plenty of opportunity to learn and 'rehearse' skills they will subsequently need as mothers. They will watch other monkeys, particularly their mothers, holding infants and may even 'practice' by holding siblings themselves. Experiments demonstrating that females reared in a socially deprived environment make poor, aggressive and rejecting mothers are well known. There may also be important parallels in disturbances of parental care in man.

Initially, the infant does not respond to its mother as an individual, and its filial responses can be elicited by a wide range of stimuli (fur, nipples, etc.) generally associated with it's own mother's body, but other bodies elicit the response as well. Eventually the range of stimuli eliciting responses in the baby become narrowed, through the developing process of maternal recognition, to those from its mother alone. Similar processes occur in human babies, although when and how a baby, monkey or human, recognizes its mother as distinct from other individuals is, at present, not clear, except that it seems to take weeks, and may involve olfactory as well as facial characteristics. The point at which a mother recognizes her own, as opposed to other babies, appears to be after several days.

Communication between mother and infant takes several forms, and these reinforce both mutual recognition and the mother–infant bond. Thus, we have already seen that the mother gains some positive stimulation from carrying her infant, and qualities of the latter's coat, its size, shape and smell may contribute to its 'attractiveness' in this context. Baby monkeys also make characteristic vocalizations, for example 'whoos' when separated from their mothers and 'geckers' when frightened or denied access to the nipple. Physical contact following such vocalizations (the mother usually responds rapidly to them) has an immediate soothing effect. There are obvious parallels with human babies who cry when in pain, frightened, hungry or in a temper, and who are soothed by close contact with their mothers. Furthermore, mothers can distinguish the different types of crying and learn to respond appropriately to their babies' needs. Undoubtedly, there are many more subtle and, as yet, poorly defined means of communication between mother and infant—facial expressions, e.g. 'grins' in monkeys and smiles in human babies, which convey different things at different times, also probably contribute in an important way to the mother–infant bond.

As the infant grows and begins to move away from its mother, it becomes increasingly essential that it understands and complies with its mother's signals concerning potential hazards, for example proximity of a predator. Thus, communication between mother and infant changes in time, as does the bond between them, in parallel with and often as a consequence of the infant's cognitive and motor development. This, of course, one knows intuitively to be the case, but it cannot be over-emphasized since the very gradual process of gaining independence from the mother and exploring the environment is vital, in its turn, for cognitive development. Extreme fear of strangers and novel surroundings could easily result in the infant never leaving its mother, and thus failing to gain experience of the wider world in which it must move. Indeed, monkeys reared in isolation show great fear of novel stimuli when released, because their cognitive development has been restricted and impaired. It is in this context, then, that the delicate balance between a mother's rejection of the infant and the latter's curiosity and exploratory tendencies (to be discussed below) are of extreme importance. The mother's proximity and availability allow the infant to resolve the conflict of explore–retreat tendencies and so increase its familiarity with strange objects while assessing their safety or hostility.

Elegant experiments on rhesus monkeys, summarized in Fig. 13.9, have thrown considerable light on the nature of the changing mother–infant relationship and, in particular, who contributes when to the gradual evolution of the infant's independence. Initially, the infant stays on the mother all the time, clasping her ventro-ventrally. Gradually, it spends more time off her, both at hand and also out of arm's reach (Fig. 13.9). Early on the mother is primarily responsible for the close contact, restricting the infant's sorties by hanging on to a tail or foot. During this time, therefore, the infant is responsible for breaking, and the mother for making, contact (Fig. 13.9). Later, however, the infant becomes primarily responsible for making contact, since the mother rejects its approaches more often and initiates contact less often.

*e Summary*

Clearly, there are dynamic changes in the interaction between a mother and her infant. During the first weeks of life the mother is virtually her child's only social companion. It clings to her and suckles all day and sleeps on her at night. After a few months, it spends much more time away from its mother playing with peers (an important period of development), but

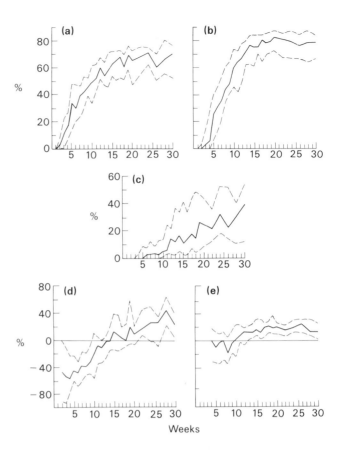

**Fig. 13.9.** The course of mother–infant interaction in small captive groups of rhesus monkeys. (a) Total time infant spends off mother as % of total time watched. (b) Time spent out of arm's reach (>60 cm) of mother. (c) Relative frequency of rejections (*ratio* of numbers of occasions on which infant attempted to gain ventro-ventral contact and was rejected to number of occasions on which it made contact on mother's initiative, on its own initiative or attempted unsuccessfully to gain contact). (d) Infant's role in ventro-ventral contacts (number of contacts *made* on infant's initiative, as a % of total number made *minus* number of contacts *broken* by infant as a % of total number broken). (e) Infant's role in the maintenance of proximity. For descriptions see text. Weeks = first 30 weeks of life.

the mother exists as a secure base from which excursions can occur and to which the infant will return for food and contact. If she disappears it makes loud and persistent distress calls to bring her back and continues, for a year or so, to sleep close to, or on her. During this time more prolonged separation of the infant from its mother, by taking her away, has devastating effects. The baby shows considerable distress, withdrawal— adopting a hunched, depressed posture—and a decrease in motor activity. Reuniting the pair is followed by a period of intense contact, but with eventual and gradual rejection of the infant by the mother to re-establish preseparation patterns and levels of interaction. The longer the separation the more severe the effects, and the mother's behaviour is critical in restoring the infant's security, just as it is during normal mother–infant bonding. The fact that, under some circumstances, a period of separation can have long-term effects on behaviour, particularly in terms of fear responses to strange objects in an unfamiliar environment, has led to the belief that such events can occur in human babies. It is important to emphasize, however, that because mother–infant separation *can* have long-term behavioural effects it *need not necessarily* have them. In fact, explicit statements about which factors underlie normal mother–infant bonding and patterns of interaction cannot be made because we know too little about them. It follows, then, that we must know even less about factors which determine disturbances in this interaction. But the comparative method may be particularly valuable in trying to understand patterns of normal and deficient behaviour in man.

*f Some data on human mother–infant interaction*

Studies of human mother–infant interaction are virtually still at a descriptive stage, and it is beyond the scope of this book to do credit to the vast wealth of fascinating information on this subject. Instead, one or two facets of contemporary work will be presented, particularly since these show striking parallels with and amplify the data described above.

Despite the almost folklorish assertion that newborn babies cannot see, it is now quite clear that they can, and, furthermore, that they can selectively direct their attentive responses to specific visual stimuli. In particular, stimuli associated with the human face seem particularly important. Indeed, there is some evidence for facial mimicry in babies just a few hours old! The fact that mothers tend to look intently at their baby's face from the earliest moments after birth therefore indicates a reciprocity in mother–infant interaction of considerable importance in establishing the bond between them. This will also lead subsequently to their mutual recognition. Observations of early maternal behaviour after home deliveries in California show that, immediately postpartum, mothers pick up their infant, stroke its face and start breast-feeding while gazing intently into its face. In hospital deliveries the pattern differs only slightly, the mothers particularly exploring their baby's extremities, but still with considerable emphasis on eye-to-eye contact. It is the contention of many workers that this early period of intense eye-to-eye contact and physical exploration, which often forms quite a consistent sequence, may be very important in mediating the early formation of the mother–infant bond. But the interaction is not one way, the baby emits signals to the mother which evoke her maternal responses and willingness to nurse.

Reference was made above to the effects of early separation on subsequent mother–infant interaction in rhesus monkeys. Some evidence indicates that enforced early separation of women from their newborn for periods up to 3 weeks, for example as might occur after premature delivery, can be associated with differences in attachment behaviour (bonding) when compared with mothers similarly separated from their babies, but allowed additional contact during the first few days after birth. Modern paediatric practice takes account of such findings, and, where possible, ensures as much contact as is practicable between mothers and their young. Again, it must be emphasized that although postnatal separation of mother and baby *can* have delayed and long-lasting effects, and these have been reported to include child abuse, it need not necessarily do so. Many other factors may intervene to alter the predicted outcome.

In addition to the stimuli associated with the mother's face being important for eventual maternal recognition by the infant, there are data to suggest that olfaction can be used by neonates to differentiate between their own and another mother. In experiments in which an infant was presented with breast pads (which had absorbed milk) from its own mother or from an 'alien' lactating female, significantly more time was spent, by 6 days of age, turning towards its own mother's pad. Babies, particularly when several weeks old, attune to their mother's facial expressions when they talk and coo. In one study, a 4-week old infant with a blind mother who displayed a mask-like face during speech (she had never been sighted) tended to avert his eyes and face from the mother when she leaned over to talk to him. When interacting with other, sighted, individuals, however, this was not the case. The normal interaction had, therefore, been distorted—but not completely so, since other modes of communication (verbal–auditory, in particular) were successfully used to overcome this interaction deficit.

The essential contribution of the baby to the bond between it and its mother is highlighted when babies display behaviour which disrupts the relationship. A baby who cries and shows avoidance responses when picked up may very easily induce feelings of frustration, confusion and anxiety in the parents. They may, in fact, feel rejected by the infant—quite the opposite to what is usually encountered when examining the occurrence of rejection in a mother–infant diad. The behaviours subsequently displayed by the 'rejected' parents will affect the infant's developing behaviour and so the path is potentially set for an unsatisfactory and enduring pattern of interaction between them.

*g Summary*

It is not really possible to summarize the data on human parental care since far too little is known about the mechanisms regulating its onset, course and maintenance. The brief account above, which cannot do justice to the rich literature, should not be taken as dogma. It is not intended to represent the 'way' a mother or father should behave towards their baby, nor that adverse consequences will result if they do not behave in this way. However, by studying the behaviour of both human and non-human primate mother–infant pairs, we may begin to understand what factors contribute to the success and richness of the mother–infant bond. Equally, we may discover what contributes to its breakdown, and how failure to establish an adequate bond at an appropriate time leads to disturbed behaviour in the parents, or the infant, or both, later on.

Given the pervasive effects of good or bad parental care on subsequent social and, indeed, parental behaviour, the importance of such research is made apparent.

**Further reading**

Bowlby J. *Attachment and Loss.* Vol. 1, Attachment Vol. 2, Separation. Hogarth, 1969, 1973.

Ciba Foundation Symposium 33. *Parent–infant interaction.* Elsevier, 1975 (New Series).

Everitt, BJ, Keverne EB. Reproduction. In *Neuroendocrinology* (Eds Lightman SL, Everitt BJ) pp. 472–537. Blackwell Scientific Publications, 1986.

Harlow HF, Suomi SJ. Nature of love—simplified. *Am J Psychol* 1970; **25**: 161–168.

Harlow HF, Zimmerman RR. Affectional responses in the infant monkey. *Science* 1959; **130**: 421–432.

Hinde RA. *Biological Bases of Human Social Behaviour.* McGraw-Hill, 1974.

Keverne EB, Levy F, Poindson P, Lindsay DR. Vaginal stimulation: an important determinant of maternal bonding in sheep. *Science* 1983; **219**: 81–83.

Larson BL, Smith VR (Eds). *Lactation. A Comprehensive Treatise.* Vols I and IV. Academic Press, 1974, 1978.

McNeilly AS. *Journal of Biosocial Science (Suppl)* 1977; **4**: 5–21.

McNeilly AS. Reproductive disorders. In *Neuroendocrinology* (Eds Lightman SL, Everitt BJ) pp. 563–588. Blackwell Scientific Publications, 1986.

Mepham TB. *The Secretion of Milk.* Camelot Press, 1978.

Noirot E. The onset of maternal behaviour in rats, hamsters and mice: a selective review. In: *Advances in the Study of Animal Behaviour.* Vol. 4 (Eds Lehrman DS, Hinde RA and Shaw E) pp. 107–145. Academic Press, 1972.

Rutter M. *Maternal Deprivation Reassessed.* Penguin, 1977

Shulman FG. *Hypothalamic Control of Lactation.* Springer, 1970.

Trivers RL. Parental investment and sexual selection. In *Selection and the Descent of Man* (Ed. Campbell B) pp. 136–179. Aldine Publ. Co., 1972.

Walser ES. Maternal behaviour in mammals. *Symp Zool Soc Lond* 1977; **41**: 313–331.

Wilson EO. *Sociobiology. The New Synthesis.* Belknap, 1975.

Yen SSC, Jaffe RB (Eds). *Reproductive Endocrinology.* Saunders, 1978.

# Chapter 14
## Fertility

In the preceding chapters we have attempted to describe the important mechanisms that underlie the establishment of sex, the attainment of sexual maturity, the production and successful interaction of male and female gametes, the initiation and maintenance of pregnancy and the production and care of the newborn. These processes absorb much of our physical, physiological and behavioural energies, reproduction and its associated activities permeating all aspects of our lives. This ramification of sex in turn makes *it* highly susceptible to social and environmental influences, and these influences can be of adaptive value as we have seen at several points in the book. However, whilst the adaptive value of sensitivity to environmental and social signals is undeniable for the reproductive efficiency of the species as a whole, it is not necessarily so for the individual. In contemporary human society, the increasing emphasis on individual rights and welfare has prompted a more sympathetic attitude towards the fertility of the *individual*, both its limitation and its encouragement. Moreover, sexual interaction among humans has ceased to be related directly and exclusively to fertility, but rather has itself assumed a wider social role. Sexuality and its expression has, in its own right, become an important element within human society. In this final chapter, therefore, we stand back from a detailed consideration of the mechanisms of reproduction and look instead at human fertility patterns, and at the factors, both natural and artificial, that may influence them. Furthermore, we draw heavily on examples of the reproductive mechanisms described earlier to illustrate our points, and so the chapter cross-refers extensively.

The fertility of a population or an individual is expressed in terms of the number of children born. In no human society does the fertility of women (the 'limiting' sex reproductively) reach its maximum theoretical limits. Clearly, therefore, constraints are placed on fertility. It is the nature of these constraints that concerns us here.

## 1 Fertility and age

The fertility of a woman varies with her age, and earlier puberty coupled with delayed menopause and longer life spans have expanded the average fertile period in most human societies. Menstrual cycles begin during puberty, and menarche marks the earliest expression of the potential fertility of the female with its implication that a full ovarian and uterine cycle has been achieved. If menarche fails to occur, a diagnosis of *primary amenorrhoea* is recorded. This may result from a failure of normal maturation of the underlying neuroendocrine mechanism (Chapter 6), from primary defects in the gonad, such as dysgenesis or agenesis due to chromosomal abnormality (e.g. Turner's syndrome and true hermaphroditism, Chapter 1), or from primary defects in the genital tract, such as lack of patency between uterus and vagina (*cryptomenorrhoea* or hidden menses) or indeed the absence of internal genitalia as seen in the testicular feminization syndrome (Chapters 1 and 2). However, even in normal females, early menstrual cycles are rarely regular. Some cycles are *anovulatory* and may lack, or have abbreviated, luteal phases (Fig. 14.1). In general the follicular phase tends to become shorter with age and the luteal phase to lengthen but the reason for this is not at all clear.

Fertility appears to be highest in women in their early twenties and declines thereafter, but data on changing human fertility patterns with age are notoriously difficult to interpret, since older women may have a reduced chance of, or inclination for, unprotected intercourse. It does seem clear

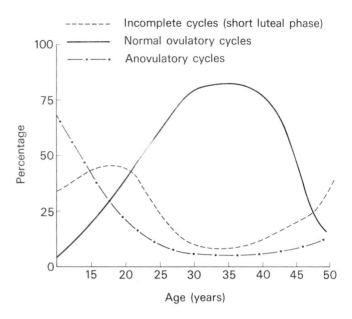

**Fig. 14.1.** Relative incidence of three types of menstrual cycle with age of woman.

that whether or not the incidence of fertilization declines with age, the ability to sustain a pregnancy through to successful parturition declines slowly to age 35 years and more rapidly thereafter when an increasing frequency of perinatal or neonatal mortality, low birth weights, maternal hypertension and congenital malformations especially but not exclusively due to fetal chromosomal imbalance are encountered. Thus, estimates of the percentage of infertile women aged 20–25 are around 3–6% whereas at 35–40 years of age the proportion of women that are infertile is around 20%.

Irregular menstrual cycles begin to reappear in women in their early forties onwards and mark the onset of the *menopause*. This period of change may last for 10 years or more and is known as the *climacteric*. It results from a decline in both the number of ovarian follicles and in their responsiveness to gonadotrophins, and thus *cessation of reproductive life is a function of ovarian failure*. Secretion of follicular inhibin and oestrogen declines as the follicles fail, but androgen output tends to rise (Table 14.1). The fall in negative feedback stimulus leads to elevation of gonadotrophin levels, that of FSH preceding that of LH (see Fig. 5.9). Associated with these ovarian changes are corresponding physical, functional and emotional changes. Oestrogen withdrawal seems to be responsible for vasomotor changes such as 'hot flushes' and 'night sweats' and the increased risk of coronary thrombosis, as well as for involution and fibrosis of the uterus and breasts, a reduction in vaginal lubrication and a rise in the pH of vaginal

**Table 14.1.** Steroidogenesis in the postmenopausal woman.

| Hormone | Change from resting plasma concentrations (%) | |
|---|---|---|
| | Dexamethasone suppression (% decrease) | hCG stimulation (% increase) |
| Oestradiol | >50 | 0 |
| Oestrone | >75 | 10 |
| Progesterone | >75 | 10 |
| 17 α-OH Progesterone | 53 | 42* |
| Dehydroepiandrosterone | 65 | 60* |
| Androstenedione | 70 | 10 |
| Testosterone | 40 | 60* |
| Dihydrotestosterone | >30 | — |

Dexamethasone inhibits adrenal cortical steroidogenesis by depressing ACTH output. The degree of *reduction* in plasma steroid levels shown suggests the degree of adrenal contribution. hCG provides a gonadotrophic stimulus to the ovaries. The degree of *increase* in plasma steroid levels shown suggests the degree to which the *ovaries* are able normally to secrete the hormones. Astersks denote the only 3 significant changes in column 2.

fluids, in consequence of which discomfort during intercourse (*dyspareunia*) and *vaginitis* may occur. The anti-parathormone activity of oestrogen is also lost at this time and as a result increased bone catabolism occurs with the resultant oesteoporosis, witness the Dowager's hump. All these symptoms of the menopause may be prevented or arrested by oestrogen, although there is not uniform agreement about the safety and advisability of using oestrogen therapy.

A number of behavioural changes occur during the climacteric, for example depression, tension, anxiety and mental confusion. However, these changes may not be related directly to steroid withdrawal but rather may result secondarily from difficulties in psychological adjustment to a changing role and status, and in part from insomnia due to night sweats. Loss of libido is rather common, but may also be due to worries about sexual attractiveness or even due to pain at coitus as a result of poor vaginal lubrication.

Men, in contrast to women, experience a less dramatic fall off in fertility, producing spermatozoa well beyond 40 years of age. However, loss of libido, impotence and failure to achieve orgasm occur with higher frequency from 30–40 years onwards.

Puberty and menopause may mark the limits of the period of fertility but within this period most women are not continuously reproducing. Traditionally three factors have limited fertility, namely: social or religious practices and traditions, natural infertility or subfertility, use of contraception and abortion. Until this century, the first of these constituted the single most important means of regulating fertility in the world as a whole.

## 2 Social constraints on fertility

Amongst the important social (in which we include religious) variables that influence fertility are :

1 the accepted social roles of men and women;
2 the age of women at marriage and at birth of the first child;
3 the accepted size of the family and the preferred sex ratio;
4 the desirability of spacing children; the anticipated mortality pattern;
5 the perceived economic advantages of a given family size;
6 the permitted or expected frequency of intercourse in relation to the point in the menstrual cycle, the time of year or religious calendar, age of partners, the delivery or suckling of children;
7 the extent to which maternal lactation (with its consequent hyperprolactinaemia and depression of fertility— see Chapters 5 and 13) is replaced by use of milk substitutes or surrogate mothers;
8 the acceptability of sexual interactions outside (or even to the exclusion of) the usual framework in which successful pregnancy might result (e.g. prostitution, extramarital sex, homosexuality);
9 the presence of celibacy;
10 the incidence of divorce and the attendant delay before remarriage.

Each of these factors will affect to varying degrees the overall fertility of the society, and moreover the various factors will interact with each other. For example, high economic expectations coupled with low infant mortality tends to reduce fertility both by delaying birth of the first child and reducing overall family size.

Perhaps the single largest change that has occurred in Western society over the past century, and is occurring now in developing countries, is the fall in expected infant mortality. This fall has led to a corresponding rise in the numbers of females surviving beyond puberty. The time lag between the decline in infant mortality and the decline in fertility has resulted in a continuing population explosion in the developing world that we witness today. The other compensatory social factors listed above that regulate fertility have not come into play rapidly enough to counter the increased fecundity of the population as a whole caused by the abrupt introduction of antibiotics, improved diet and hygiene, and prophylactic immunization. Indeed, it is not clear that social adjustment alone is adequate to correct this imbalance. The increased use of artificial constraints on fertility in various societies has become an essential part of the social response to limit

population growth and to regulate individual fertility according to desired social patterns.

## 3 Artificial control of fertility

The three socially accepted artificial controls over fertility
available to varying extents in different societies are contraception, induced abortion and sterilization. Not all of these
controls are 100% effective and thus they should be seen as
regulating fertility by delaying births or increasing the birth
interval. Their relative cost-effectiveness in terms both of the
population as a whole and the individual in particular is, of
course, of the greatest importance.

### a Sterilization

Sterilization should be 100% effective and should be entered
into as an irreversible procedure. It is therefore the method of
fertility limitation selected by individuals or couples that have
achieved their desired family size or that for eugenic or health
reasons must or should avoid reproduction.

Sterilization of the male involves *vasoligation* or *vasectomy*
(ligation, or removal of part, of the vas deferens). The
operation can be done under local anaesthetic as an outpatient, is quick, and leads to an aspermic ejaculate within 2–3
months. Performed competently, vasectomy provides an
efficient method for limiting fertility. However, the obstruction caused by the ligation of the vas can lead frequently to a
local granuloma due to the build up of spermatozoa and the
small volume of fluid that continues to pass out of the
epididymides. Certain consequences follow: first, there may
be chronic or intermittent tenderness in the scrotal region;
second, leakage of large numbers of spermatozoa into the
systemic circulation from the site of inflammation induces in
many men an immune response to their own spermatozoa;
third, in several species of experimental animal a progressive
decline in spermatogenic output occurs—it is not clear
whether this is accompanied by *orchitis* (inflammation of the
testis), nor whether it occurs in men. Some success with
reversal of vasectomy has been achieved. Psychological factors, particularly arising from erroneous equation of fertility
with potency, may result in anxiety that leads to impotence.
This fear also prevents many men from accepting the procedure. Sexual arousal and ejaculation are, of course, independent of sperm release.

In the female, sterilization involves ligation, electro-
coagulation or removal of all (*salpingectomy*) or a section
(*fimbriectomy*) of the oviducts. A general anaesthetic is usually
required although regional anaesthetic can be used. The
surgical approach is generally made with the use of a
laparoscope or through the abdominal wall by laparotomy,
more rarely a transvaginal approach is used. Although the
procedures require surgical skill the patient can be brought in

**Table 14.2.** Effectiveness of various methods of contraception expressed in terms of the number of pregnancies encountered per 100 woman years (= 10 women for 10 fertile years, or 100 women for 1 fertile year, etc.).

| Method | Pregnancy rate (%) | Estimated usage by married women in the U.K. (%) |
|---|---|---|
| No protection (in fertile marriages) | 60–75 | 14 |
| Douche alone | 20–40 | 1 |
| Coitus interruptus | 12–40 | 20 |
| Foams, jellies, creams | 5–40 | 2 |
| Rhythm* | 0–40 | 7 |
| Condom | 5–30 | 29 |
| Diaphragm + spermicide | 5–30 | 4 |
| Intra-uterine device | 2–4 | 3 |
| Steroidal contraceptives | | 20 |
|    Combined | <1 | |
|    Sequential | 1–2 | |
|    Progestagens—oral | 5–10 | |
|    Progestagens—depot | 0.5–1.5 | |
|    Postcoital—oestrogen | 0–1.5 | |
|    Postcoital—progestagen | 5–30 | |

*Clearly this method will be effective if coitus is strictly limited. Pregnancy rates for the other methods do not include any restriction on coitus during the cycle.

as a day patient. The operation has a very low failure rate (less than 1%) and cannot be reversed easily. Sterilization may also be accomplished by hysterectomy, especially in premenopausal women troubled by irregular, painful or heavy menses. The side effects of sterilization include postsurgical infection or obstruction, development of fistulae, the very slight risk of mortality attendant on use of anaesthetics and possible psychological disturbance associated with loss of fecundity.

*b Contraception*

Contraception differs from sterilization only in its potential or actual ease of reversibility and thereby in the control that the individual exerts over its use. It is thus the artificial control of choice for those who wish to delay the expression of fertility or to exercise it with discrimination. The methods currently available are listed in Table 14.2, together with estimates of their efficiency in terms of their impact on pregnancy rates. The reported rates vary widely depending upon the education, motivation and experience of the users, and perhaps the most important figures are the upper limits.

i Natural methods

Contraceptive approaches that rely on coital technique, rather than the use of technical or pharmaceutical aids, straddle the boundary between social and artificial approaches to fertility control. In consequence, such approaches seem to pose particular problems for theologians and lawyers when prescribing or proscribing sexual behaviours. For example, of the techniques listed below only the rhythm method is sanctioned by the Catholic church on the grounds that the other approaches are unnatural. We use the expression 'natural' here in a biological and not a theological sense.

*The rhythm method*: For many centuries common belief, reinforced more recently by misinformed religious approval, equated menstruation with the period of maximum fertility. In fact, of course, the period of abstinence for avoidance of maximum fertility is roughly mid-cycle, as was first discovered in the 1920s.

Given the potential life of spermatozoa in the cervix of several days, the possibility (albeit remote) of coitally induced ovulation (Chapter 5), the variability in the length of the cycle and in particular its susceptibility to emotional or stressful disturbances, and the requirement for discipline by the sexual partners, the success of the rhythm method used alone in reducing fertility is often not high, being some 20-fold less effective than oral contraceptives at their worst (Table 14.2). The devising of a simple test for incipient ovulation is unlikely to improve the success of the method, given the nature of its limitations outlined above. Additionally, there is evidence that a significantly increased frequency of genetic abnormalities in embryos derived by fertilization of aged eggs occurs in couples using this method of fertility control (see Chapter 8 and Section 4c of this chapter).

*Coitus interruptus*: The withdrawal of the penis prior to ejaculation has been, and remains, one of the most frequently used forms of contraception in Europe, and the technique is credited (together with abortion) as being responsible for much of the decline in birth rate at the time of the industrial revolution. Failures will occur due both to lack of adequate control on the part of the male partner and possibly the leakage of semen containing spermatozoa prior to ejaculation itself.

*Masturbation and other forms of sexual interaction*: Masturbation is undoubtedly the commonest means of achieving sexual gratification prior to establishing a sexual relationship. Thereafter, mutual masturbation is one commonly employed means of reducing the incidence of pregnancy, and is highly effective as such. Oral and anal sex are commonly used in many societies.

ii Condoms

Probably the commonest used form of mechanical contraceptive throughout the world; the improved strength, lubrication and design of modern condoms has considerably enhanced their efficiency and durability. Condoms are cheap, readily available, relatively easy to use and may also give some protection against venereal diseases particularly gonorrhoea and AIDS.

**iii Caps and spermicidal foams, jellies and creams**

The combination of a physical barrier sealing off the cervix and a spermicide at the site of seminal deposition has been used for over a century. The method was popular since, for the first time, it put the responsibility for contraception on the woman. However, the method has decreased in popularity with the development of more modern methods of contraception for females. More recently, publicity surrounding side effects of the pill and the IUCD has led to an increased use of the diaphragm.

**iv Intra-uterine contraceptive devices (IUCD)**

Modern IUCDs are made of plastic and may also contain copper. They function as foreign bodies within the uterus to produce a low-grade, local, chronic inflammatory response, the composition of the uterine luminal fluid resembling a serum transudate containing large numbers of invading leucocytes. There is a broad correlation between the capacity of the IUCD to induce inflammation and its efficiency as a contraceptive. Such a uterine environment does not appear to inhibit fertilization but is cytotoxic to the developing pre- or post-implantation conceptus and may inhibit or impair decidualization. In addition, luteal life is abbreviated perhaps due to premature prostaglandin release (see Chapter 9). Not surprisingly, in light of its proposed mechanism of action, the IUCD must either be of a shape that extends throughout the uterus or be placed high in the uterus, presumably to ensure that the conceptuses entering from the oviduct are exposed to local cytotoxic effects. The recent use of copper in IUCDs enhances their contraceptive effectiveness perhaps simply via the inflammatory effect of the metal, although a specific embryotoxic effect of copper has also been proposed.

For obvious reasons, contra-indications to IUCD insertion include pelvic inflammatory disease, endometriosis, uterine bleeding and recent septic abortion. Insertion into nulliparous young women who intend to have a family subsequently is not recommended. Complications of IUCD insertion are heavy menstrual or irregular uterine blood loss, uterine pain or muscular spasm, uterine perforation and an increased tendency to pelvic sepsis especially in non-copper-containing devices. Unnoticed expulsion of the IUCD can also occur, as can pregnancies with the IUCD still in place. The skills required for insertion, and the desirability of regular monitoring of patients, mean that trained and available medical or paramedical staff are needed. Despite these disadvantages, the IUCD is a widely used and very effective contraceptive device, and may be particularly suited to young parous women.

**v Steroidal contraceptives**

Since the initial development of 'the pill' in the 1960s, a range of steroid-based contraceptives has been developed, and must now account for a substantial part of the reduction in fertility. Synthetic steroids as described in Chapter 2 are used in these preparations as their half life in the body is longer and their effect therefore sustained.

*Combined oral contraceptives*: Combined preparations contain both oestrogens and progestagens and are taken for 21 days with a 7 day break to induce a withdrawal bleed simulating menstruation. The maintained level of *both* steroids throughout the 21 day period functions to depress, via negative feedback effects, gonadotrophin output rather as we saw occurs during the luteal phase of the normal cycle (Chapter 5). In this way endogenous follicular maturation and increasing steroid output is suppressed along with the oestrogen surge, LH surge and ovulation. The exogenous steroids develop and maintain the uterus and other target organs. The level and balance of oestrogens and progestagens varies in different preparations, thereby offering a range of contraceptives to suit women with different clinical histories or levels of motivation.

*Sequential oral contraceptives*: These preparations use *oestrogens only* for 7–14 days and a *combined oestrogen/ progestagen* mix for 14 or 7 days followed by 7 pill-free days. The same contraceptive principle applies, but higher doses of oestrogen are required to counter the absence of progesterone in the first half of the 'cycle'. Since the hormonal regime is closer, at least in sequence if not in dose, to the natural cycle, the side effects tend to be less, but so does the contraceptive efficiency. For this reason sequentials are now no longer used.

*Biphasic and triphasic oral contraceptives*: The principle here is to combine the mimicry of a normal cycle attempted in sequential preparations with the contraceptive efficiency of the traditional combined preparations. Thus, in biphasic preparations both oestrogen and progestagen are given for the first half of the cycle, but the progestagen dose is stepped up at mid-cycle. The contraceptive efficiency is as good as in the conventional combined tablets but the doses of steroid used are lower. In triphasic preparations, lower steroid doses still are used in the sequence: 5 or 6 days with low oestrogen + low progestagen, 6 or 5 days with both oestrogen and progestagen slightly elevated, 10 days with low oestrogen and doubled progestagen. *These preparations place increasing emphasis on the capacity of progestagens to block the positive feedback effect of oestrogen* and less on direct negative feedback effects (see Chapter 5). In consequence, oestrogen levels are lower, with fewer side effects, and a more normal 'cycle' is achieved.

*Long-acting oral contraceptives*: a high-dose, potent, long-acting oestrogen is combined with a progestagen and taken once a month. Despite falling levels, the steroids nonetheless retain sufficient negative feedback action to suppress gonadotrophin output and thereby follicular growth. These pills are suitable for highly motivated women with good memories or women who need to take contraceptive precautions under supervision of a nurse or health worker. However, the contraceptive efficiency is lower than that of triphasic preparations, and the high doses give cause for worry about side effects.

*Low dose progestagenic contraceptives*: progestagens may be taken daily by an oral route or given by depot injection subcutaneously for slow continuous release over a period of 1 month to 5 years (Norplant). The suppression of gonadotrophin secretion, and therefore of ovulation, is not so effective with these preparations, but additional contraceptive effects on cervical mucus, spermatozoal transport and capacitation, and tubal transport of the embryo are present (Chapter 7). Side-effects are relatively few, but menstrual irregularity including intermenstrual spotting may occur. A more effective way of delivering progesterone to the female genital tract is to place silastic capsules directly *in utero* or to use progestagen-impregnated IUCDs or vaginal rings. Here the effect is exclusively local, no effect on ovulation being possible, but the contraceptive efficiency is nonetheless, extremely high.

*Postcoital oral contraceptives*: Oestrogens given in high doses within 72 hours of unprotected intercourse ('morning-after pills') interfere with transport of the conceptus and implantation (Chapter 7). The high doses given may cause nausea and vomiting, and failure of contraceptive action can lead to an increased incidence of ectopic pregnancy (Chapter 7). A range of high dose progestagen, postcoital contraceptives are being tried, but success rates are variable.

*Steroidal contraceptives for men*: the principle here is to suppress endogenous gonadotrophin, particularly LH, with exogenous progestagenic steroids, thereby depressing testicular androgen synthesis and release. Local androgen levels in the testis thus fall leading to aspermatogenesis. Restoration of systemic androgens with an exogenous preparation would then overcome any tendency to demasculinization. The development of a contraceptive along those lines appears to be highly acceptable, and controlled trials have clearly shown a depression of spermatogenesis. However, in those same trials the exogenous androgens tested have not adequately compensated for loss of testicular androgens, with resultant adverse effects on libido, erections and the general quality of sexual interaction.

*Side effects*: the possible side effects of steroidal contraceptives are the subject of widespread concern and discussion. There is evidence that the higher doses of oestrogen in long-acting, sequential and some combined oral contraceptives increase blood coagulability (see Chapter 7). Although the evidence relating some steroidal contraceptives *causally* to increased risks of thrombo-embolic disorder is only suggestive, nonetheless, any other conditions that tend to have a similar action, such as heavy smoking, obesity, age, varicose veins and prospective or recent surgery, contra-indicate steroidal contraceptives containing oestrogens. Between 1 and 5% of normotensive women become slightly hypertensive when taking a combined preparation and this phenomenon may be related to the increased angiotensin levels in response to oestrogens (Chapter 7). Oestrogen on its own at high levels is associated with an increased incidence of endometrial cancer, but this is not a problem with most currently available oral contraceptives. The latent diabetes revealed during pregnancy (see Chapter 11) can also become evident in users of steroidal contraceptives, who should therefore only use low dose oestrogen preparations under careful supervision.

Very recently, it has been reported that oral contraceptive preparations with a high progestagenic potency may be associated with an increased risk of breast cancer. This link appears particularly strong in women having taken such oral contraceptives for several years before the age of 25 and may be related to the effects of progestagens on cell division in the postpubertal, developing breast. A contemporaneous study also shows an increased incidence of cervical cancer related to duration of oral contraceptive use (regardless of type, other than low dose progestagen-only pills) although a causal link was not established. These findings, which result from controlled studies, suggest not only the importance of monitoring patterns of oral contraceptive use but also the value of regular cervical smear and breast examinations. Clearly, careful notice of the above associations should be taken by prescribing practitioners. Progesterone-only contraceptives are also associated with mild transvaginal bleeding that some women find unacceptable.

Altered liver function is detected in women using steroidal contraceptives, although the changes are slight and not obviously of any importance. These also are induced mainly by oestrogens. A history of jaundice or liver malfunction therefore contra-indicates use of oral contraceptives. The sodium and water retention observed in a normal cycle (Chapter 7) may be exacerbated by steroidal contraceptives in some women, and behavioural effects have been reported (see Chapter 7) in women as well as men. There have been suggestions that restoration of fertility may be delayed after termination of steroid contraceptive use, but the evidence for this is somewhat ambiguous.

vi Summary

A range of variably effective sexual techniques, devices and pharmaceutical preparations exist to control fertility contraceptively. It is unlikely, given the restrictions on clinical trials imposed in recent years, that a totally new range of contraceptive approaches will be available in the near future. Some progress has been made with immunological approaches. For example, immunity to the specific $\beta$-chain terminal sequence of hCG neutralizes the embryonic hormone but leaves LH unaffected, thereby giving normal (or slightly lengthened) cycles but protection against pregnancy (see Chapter 10). Results of preliminary trials using this approach have been hopeful and more extensive trials are planned. No side effects have been noted, and the immunity declines after about 6 months in the absence of a booster injection, so the approach may be reversible. Additionally, immunity to antigens on spermatozoa and on the zona pellucida has been associated with infertility. However, there is a reasonable reluctance to tinker with the body's immune system until more is understood about its natural regulation, in case uncontrollable side effects, such as a wider autoimmune response, develop after use of a 'contraceptive vaccine', particularly one that utilizes cellular or cell-associated antigens.

The recent progress in characterizing the actions of inhibin on FSH secretion has prompted speculation about its use contraceptively. In addition, analogues of GnRH have been synthesized that bind to GnRH receptors but provide poor stimulation of gonadotrophin release. Their use, in conjunction with exogenous oestrogen and progesterone to maintain peripheral steroid-dependent functions, might provide a controlled fertility.

*c Abortion*

Induced abortion, whether performed legally or illegally, has been, and remains, a major approach to fertility control in all communities. However, the importance of its role varies. Thus, in Western Europe where contraception is readily available abortion tends to be used when failure occurs, and only more rarely as *an alternative* to contraception by, for example, poorly educated or inexperienced women. In some Eastern European countries, abortion is used as a principal method of fertility control. In countries where contraception is proscribed on religious grounds or is not easily available, abortion is frequently resorted to; for example, it provides the major artificial control on fertility in South America, many Islamic countries, Japan and Spain.

In the first *trimester* (or 3 months) of pregnancy, abortion is usually performed by either *dilating* the cervix with metal sounds and scraping out the conceptus with a *curette,* or more frequently now in the UK by use of *vacuum aspiration* with or without attendant curettage. The use of vacuum aspiration alone is particularly useful in the first month or so of conception and is frequently resorted to after a *missed period* when it is euphemized as *menstrual regulation.* The euphemism is useful, since legal or religious proscription of abortion often applies only if a firm diagnosis of pregnancy is made. A delayed period can therefore be treated therapeutically as such and remain within the law and ethical teaching; hence the

348    *Chapter 14*

extensive use of this approach in South America. The failure rate using these approaches is low (1–1.5%) and complications such as major blood loss, incomplete aspiration or damage to the cervix are about 2–3%. Side effects of this approach to abortion include post-operative infection, damage to the uterus or cervix that may affect subsequent fertility, and haemorrhage. The possibility of conducting outpatient abortions without the need for surgery under aseptic conditions has been realized with the development of orally active anti-progestins (such as RU 384 86), which compete for progesterone receptors. They appear very effective at first trimester terminations, especially if combined with a low dose injection of prostaglandins. Such anti-progestins may also be useful as 'once a month' pills to terminate the luteal phase.

Later in pregnancy, in mid-trimester, premature delivery was induced traditionally by cervical injection of hypertonic saline intra-amniotically, or of soaps, pastes or saline extra-amniotically. These approaches are little used now in the U.K. More recently, controlled prostaglandin infusions into the cervical region, intra- or extra-amniotically or intravenously to induce both cervical softening and myometrial contractility (see Chapter 12) provide an alternative method of termination. However, side effects include vomiting, diarrhoea and nausea.

*d Risks versus effectiveness in the artificial control of fertility*

The individual requires control over his or her fertility to be 100% effective at each sexual encounter, with zero risk of side effects. In practice this is very difficult to achieve for conventional coital techniques with existing methods of fertility control. Thus, resort to multiple, or hierarchical, levels of fertility control tends to occur. Social regulation of sexual encounters is followed by use of natural techniques or caps and condoms. Failure of these approaches leads to use of the postcoital 'morning-after' pill or to menstrual regulation. There should be little need for unwanted pregnancy in more sophisticated societies given the cumulative contraceptive efficiency of these various approaches.

With the use of many contraceptive approaches there may be attendant risks to health, well-being or even life. However, whilst clearly these risks should not be minimized, their impact is often somewhat exaggerated. For example, many of the contra-indications to use of steroidal contraceptives such as thrombo–embolic episodes, cardiovascular problems or latent diabetes, also contra-indicate pregnancy or the surgical procedures required for sterilization. Among groups not at risk from steroidal contraceptives, pregnancy itself and abortion probably constitute greater risks to health and life. Thus, the balance of risks must be evaluated for each individual. The optimal situation is a combination of well educated general practitioners, a readily available range of contraceptives and

349      *Fertility*

easy and rapid access to early abortion. Sadly, it is rare anywhere and in most of the world this situation is simply impossible.

**4 Infertility and subfertility**

The incidence of infertility is difficult to determine with precision because control populations that lack any social or artificial constraints on fertility are not easy to find. However, a number of studies on populations of women having frequent, unprotected intercourse suggests that pregnancy fails to occur in four out of five menstrual cycles (see Section c). This figure provides, therefore, a baseline of fertility against which to measure sub- or infertility. Failure to conceive within one year is therefore worrying, and after 2 years of failure infertility is clearly indicated. A second consideration is that infertility is a problem for the couple concerned and subfertility in two individuals may intereract to yield the infertile state. Estimates of 10–15% of couples experiencing difficulty with conception and successful delivery of a normal child are probably not wildly inaccurate, and reveal the magnitude of a problem that has been ignored by much of the medical profession until recently. Even now, diagnosis of, and therapy for, infertility is probably more often than not protracted, stressful and ineffective.

The causes of infertility vary greatly with socio-economic and geographic factors. The male and female partners contribute more or less equally to the problem. Psychological factors causing sexual dysfunction (such as impotence, vaginal spasm, premature ejaculation and disorders of sexual identity) are common amongst infertile patients, but it is difficult to determine whether these are responses to, rather than causes of, the infertile state.

Four major classes of disorder account for about 75–80% of all cases of infertility: disorders of the female tract, disorders of ovulation, spontaneous abortion and seminal inadequacy.

*a Disorders of the female tract*

The diagnosis of tubal obstruction is made by visual assessment with a laparoscope and by attempting to insufflate air or dye from the cervix through the tubes under direct vision. Alternatively, it can be made from a hysterosalpingogram using a radio-opaque dye. Tubal obstruction is usually a secondary consequence of tubal infection and fibrosis. The incidence is elevated after frank or asymptomatic gonorrhoea, chlamydial infection, tuberculosis (more especially in developing countries), postabortal or postpregnancy sepsis, and in users of the IUCD. Tubal infection leads to adhesions that restrict oviducal movement and egg pick up, to loss of cilia and thus egg and sperm transport, or to simple physical blockage. At present, surgical treatment for this large group of patients has a very low chance of success, and the only effective therapy offered is *in vitro* fertilization of eggs aspirated from the ovary, with replacement of the conceptuses into the uterus, thereby bypassing the block. Cervical stenosis, infection or malfunction secondary to endocrine disorders are encountered and a common problem is endometriosis in which endometrial tissue grows inappropriately in the oviduct, ovary or peritoneal cavity. It is painful, its origins are unclear and it is difficult to treat.

*b Disorders of ovulation*

This general classification covers a range of disorders. Primary amenorrhoea was discussed earlier. Here we consider only disorders relating to malfunction of the matured reproductive system, namely *absent cycles (secondary amenorrhoea), irregular cycles, anovulatory cycles* and *abbreviated luteal phases.* These conditions are often associated with stress, obesity, strenuous exercise, anorexia nervosa or use of various drugs, such as neuroleptics or tranquilizers and may resolve if the primary cause is removed or alleviated. Indeed, in one study merely taking women into clinical care and giving placebo treatments resulted in 30% of cases in successful pregnancies! Many other patients, although classified as secondarily amenorrhoeic, have never had entirely normal cycles and could therefore represent failures of terminal maturation of the neuroendocrine system at puberty.

The endocrine features of a normal menstrual cycle were described in Chapter 5, and the associated cyclical changes in the woman's anatomy and physiology in Chapter 7. An ideal clinical investigation would examine all of these through at least one cycle, but such a procedure would be time consuming, costly and inconvenient. In practice, therefore, a more limited range of preliminary tests are applied in an attempt to assign the endocrine defect to one of the following categories.

i Hyperprolactinaemia

A common cause of menstrual irregularity, this condition was discussed in Chapter 5 in some detail. Diagnosis is by two or more blood samples, ideally each at the same time of day under non-stressful conditions, and by X-ray tomography for evidence of changes in the sella turcica caused by pituitary tumours which account for 25–30% of cases. Stress-induced hyperprolactinaemia is relieved by bromocriptine therapy.

ii Primary hypothalamic failure

Absence or deficiency of GnRH reaching the pituitary results in depressed gonadotrophin and oestrogen levels and failure of ovulation. Thus, evidence of a menstrual cycle will be lacking, e.g. absence of a basal body temperature shift at midcycle and of changes in consistency and penetrability of cervical mucus by spermatozoa. Two endocrine challenge tests can be used on these patients. In the *progestagen withdrawal* test, patients are given 5–10 days of progestagen treatment after which they are observed for evidence of withdrawal bleeding. Failure to bleed is associated with the prior inadequacy of oestrogen priming (and thereby endometrial progesterone receptor induction—Chapter 7). In contrast, *challenge with exogenous GnRH* should make up for endogenous deficiencies and lead to an LH surge and ovarian stimulation. Patients were traditionally treated therapeutically with exogenous gonadotrophins, but these are expensive and rather difficult to use, since adequate but not excessive doses must be established. More recently, use of pulsatile infusions of GnRH, as described experimentally in Chapters 5 and 6, has achieved a sequence of complete and fertile cycles. Pregnancy can ensue at a high rate and can proceed to a successful delivery.

iii Hypothalamic–pituitary failure

This syndrome is less well defined, gonadotrophin secretion occurring, but at a level insufficient to support a full cycle. Gonadotrophins may be normal or depressed, and both progesterone withdrawal and GnRH challenges are positive. Oestrogen levels may fail to rise fully or be maintained throughout an anovulatory cycle at slightly elevated levels. Ultrasonography of the ovary reveals antral follicles that fail to mature. The syndrome can be treated successfully in some cases by therapy with exogenous gonadotrophins to maintain or improve follicular growth and oestrogen output; if an endogenous LH surge does not occur or is weak it can be supplemented with exogenous LH. However, as with primary hypothalamic failure, correct dosages of gonadotrophins can be difficult to gauge and slight overdosing can lead to hyperstimulation, and to multiple ovulations and implantations. Therefore often *clomiphene therapy* tends to be used. The mode of action of this drug is not entirely clear. The drug has *anti-oestrogenic* properties and is thought to act on the hypothalamus and/or pituitary to compete with endogenous oestrogen and thereby reduce its negative feedback effects. Elevated gonadotrophins would result. However, clomiphene also stimulates aromatase activity in the ovary. Since local ovarian oestrogen stimulates granulosa cell proliferation and development of LH receptors on these cells (Chapter 4), an important action of clomiphene might be to stimulate ovarian responsiveness to gonadotrophins as well as to elevate gonadotrophin levels. Clomiphene tablets are administered on days 5–9 of the cycle and may, if necessary, be followed by a supplementary ovulating injection of LH 3–4 days later.

iv Defects of follicle maturation

This syndrome is poorly understood and is indicated by *elevated* gonadotrophins, *low* urinary oestrogens, failure to detect mature ovarian follicles ultrasonographically, elevation of ovarian androgen output and the absence of good responses to progestagen withdrawal or clomiphene challenge. Follicles appear to be unresponsive to gonadotrophins, yielding immature follicles with little inhibin output, hence elevated FSH, and little aromatization of androgens, hence elevated androgens and depressed oestrogens. Such ovaries are frequently *polycystic,* i.e. have many small follicles. The primary cause of ovarian failure of this kind is not known; clomiphene, gonadotrophin and GnRH therapy are generally of little use. Surgical removal of part of the ovary (*wedge resection*) is sometimes resorted to with some success. This procedure apparently produces an acute reduction in the prevailing high level of follicular circulating androgens (and of the oestrone derived from them) which may reinstate more normal feedback relationships and leave open the possibility of an LH surge and ovulation. However, the precise sequence of events is far from clear.

| | |
|---|---|
| v 'Anovulatory' cycles that are endocrinologically 'normal' | Failure to eject an oocyte, or the ovulation of inherently deficient and therefore unfertilized eggs, will lead to this syndrome. Luteinization can occur with the egg remaining *in situ,* the so-called *luteinized unruptured follicle syndrome (LUF),* and in circumstances such as these the cycle may appear normal but infertile. There is some evidence that oocytes recovered from follicles laparascopically in women classified in this way are deficient when fertilization *in vitro* is attempted. |
| vi Abbreviated luteal phase | Some women with apparently normal ovulations nonetheless show slow or reduced rises in progesterone, and this is associated with infertility. Such a pattern is also observed quite frequently in women that have undergone clomiphene or bromocriptine therapy, and may therefore represent some deficiency in the maturation of granulosa cells that is manifested by poor luteinization; for example, inadequate development of LH or prolactin receptors (see Chapter 4). However, the precise cause of luteal inadequacy remains uncertain and since it is not yet clear whether in women either LH or prolactin are luteotrophic, a deficiency in these cannot be invoked by way of explanation! Treatment with progesterone during the luteal phase often helps. |
| c *Spontaneous abortion* | Loss of a conceptus can occur either because it is inherently deficient or becomes so as a result of environmental insult, or because it is inadequately supported by the mother. |
| i Abnormal conceptuses | We pointed out at the beginning of this Chapter that four out of every five cycles in which frequent, unprotected intercourse occurred nonetheless failed to yield pregnancies. It seems likely that much of this failure can be accounted for by early loss of the conceptus. Thus, one study in which conceptuses were recovered by lavage from human uteri 4.5 days after the detection of ovulation and the insemination of spermatozoa, |

**Table 14.3.** Incidence of chromosomal abnormalities in human embryos lost by spontaneous abortion.

|  | No. per 100 aborted conceptions | % Surviving to birth |
|---|---|---|
| Triploidy (3 sets of chromosomes) | 12–15 | < 0.01 |
| Tetraploidy (4 sets of chromosomes) | 3–5 | < 0.01 |
| Sex chromosome trisomies (3 sex chromosomes) | <1 | >99 |
| Sex chromosome monosomies (1 sex chromosome) | 10 | < 1 |
| Trisomy for one or two autosomes | 20–40 | 3 |
| Monosomy for one or two autosomes | 1 | None |
| Structural rearrangement of chromosomes | 2–3 | 35 |

yielded a blastocyst rate of only 20%. This figure is remarkably similar to that achieved *in vitro* after fertilization of oocytes in the clinical laboratory. Moreover, many of the 20% of conceptuses that do reach the blastocyst stage may die subsequently, since failure by the conceptus to convert the luteal phase into pregnancy will often not be noticed or will be observed as a slightly lengthened cycle. Between 8% and 25% of menstrual cycles are characterized by detectable levels of hCG over a period of 18–30 days after the last menstruation. This hCG is assumed to be of blastocyst origin and thus reflects a very considerable further loss of conceptuses. Such a result could arise from production of abnormal conceptuses that develop to the blastocyst and then fail, or from a failure of the corpus luteum to respond adequately to hCG stimulation. Since there is at present little evidence to suggest that this early failure of pregnancy is a recurrent event in individual women, it may more likely represent early developmental failure of the conceptus. This conclusion is further strengthened by the high incidence of genetic abnormality observed in recognized pregnancies. Thus, identifiable chromosome abnormalities alone have been detected in 0.5% of all live births, 5% of stillbirths, and 40–60% of spontaneous abortions especially those occurring in the first trimester. These figures mean that *around* 10% of recognized pregnancies (or half of spontaneous abortions) are identified as being chromosomally abnormal. The type of each genetic abnormality observed, and its approximate incidence in abortuses, is recorded in Table 14.3. Three major classes of chromosomal abnormality are represented: *translocations,* i.e. structural rearrangements of chromosomes, *errors of ploidy,* i.e. deletions of duplications of a complete set of haploid chromosomes, and *errors of chromosome number*, i.e. loss or gain of a single sex chromosome or autosome. It is clear from Table 14.3 that some types of abnormality are more common in abortuses and others are compatible with survival to birth. It is also clear that some are missing altogether, e.g. haploids, or are under-represented, e.g. autosomal and sex chromosomal monosomies which might be expected to occur with equal frequency to trisomies (since when one nucleus gains a chromosome at division the other will lose one!). It is likely that these types of abnormality are lethal very early in development (lack of genetic material being deleterious earlier than excess) and are therefore not recorded as abortuses. Analysis of early mouse development indeed confirms that most monosomic and haploid conceptuses die at pre-implantation or early postimplantation stages, whereas trisomic, triploid and tetraploid conceptuses survive for longer. It has been calculated both from clinical data, and by extrapolation from data derived from comparative studies, that 50% or more of all human conceptions may result in genetically abnormal embryos, and that in excess of 90% of all human conceptions fail to survive to term.

Clearly this massive failure rate cannot be due entirely to constitutional genetic defects in all the germ cells of one or both parents, and many if not most genetic abnormalities in fetuses have indeed been shown to arise at or shortly after fertilization. For example, disorders of ploidy are likely to result from failure of polar body formation (10–20% of triploids), from polyspermy (80–90% of triploids) or from failure of one early cleavage division (tetraploidy). Some if not most of the mono- and trisomic conceptuses also arise from abnormalities of oocyte meiotic divisions. As we saw in Chapter 8, these events are sensitive to a number of environmental perturbations, in particular the age of the egg in hours postovulation, and exposure of the female to alcohol or anaesthesia around the time of ovulation. However, a few chromosomal abnormalities in the conceptus may also arise from events in the ovary or testis that affect gametes directly, well in advance of the acute events of fertilization and early cleavage. For example, abnormalities may be induced by exposure to X-irradiation or certain chemicals and such exposure correlates with a subsequently increased natural abortion rate. Moreover, an increased incidence of fetal abnormality is observed with increasing maternal age which may reflect the fact that eggs are maintained in a prolonged dictyate stage (Chapter 4) during which they are susceptible to such damage.

In addition to genetic causes of embryonic and fetal loss, the conceptus may commence development normally but become deformed or incompetent as a result of environmental insult; for example, direct exposure of the fetus to X-irradiation, to certain viruses such as *Rubella* (German measles) and cytomegalovirus, and to certain drugs such as thalidomide, dilantin and 6-mercaptopurine. *Teratogens* such as these are frequently active only at restricted periods of development, most during some point in the first or second trimester. Additionally, women who smoke or who drink heavily put fetal development at risk, especially in the third trimester, by impairing intercirculatory exchange within the placenta (Chapter 9). Unfortunately, unlike most genetic disorders, these environmental insults to the fetus often result not in spontaneous abortion but in the birth of deformed children. However, fortunately, since the insults are environmental they are also in principle preventable.

## ii Maternal defects

Of the 40–50% of spontaneous abortions that are not clearly ascribable to genetic or induced defects of the conceptus itself, many are of uncertain origin. Clearly ascribable are abortions resulting from anatomical problems such as cervical incompetence or implantation in eccentric uterine positions or ectopically. In addition haemolytic diseases of the fetus and neonate account for a diminishing proportion of fetal wastage in Western Europe. Immunological incompatibility of mother and fetus at either the ABO or Rhesus blood group loci poses the major problem. However, recognition of incompatibility during the first pregnancy, in which sensitization of the mother to these fetal antigens is most likely around parturition, allows prophylactic administration of antibody passively to the mother. The antibody mops up any antigen released by fetal bleeds late in pregnancy and prevents active immunization of the mother.

Paradoxically, incompatibility of mother and fetus at the major (HLA) histocompatibility loci may *favour* a successful outcome of pregnancy. The reasons for this are obscure. However, compatibility at these loci predisposes the mother to the condition of *pre-eclampsia* (also called *toxaemia* or *gestosis*), characteristically a disease of the last trimester in which maternal diastolic blood pressure is periodically or continuously elevated, oedema occurs and in severe cases albuminuria occurs. Severe pre-eclampsia can lead to the convulsive state of *eclampsia*. Throughout, both mother and fetus are at risk.

## d Oligospermia

Oligospermia is a loose term that includes defects in sperm motility and morphology as well as a less than average number and concentration of spermatozoa. Deficiencies in the seminal plasma volume or composition may also be included under the same heading. A systematic and quantitative assessment of semen quality (either ejaculated or after recovery from the cervix postcoitally) is rarely performed, and these inadequacies of diagnosis are reflected in a therapeutic success rate that is low. These deficiencies in part reflect a traditional reluctance to acknowledge the role of the male in

**Table 14.4.** Characteristics associated with 'normal' and 'subfertile' semen.

| Criterion | Normal | Subfertile |
|---|---|---|
| Volume (ml) | 2–5 | 1 |
| Sperm concentration (no./ml) | $50–150 \times 10^6$ | $<20 \times 10^6$ |
| Total sperm no. | $100–700 \times 10^6$ | $<50 \times 10^6$ |
| Spermatozoa swimming forward vigorously (%) | >60 | <40 |
| Abnormal spermatozoa (%) | <30 | >60 |
| Viscosity after liquefaction (see Chapter 8) | low | high |
| Cellular debris, leucocytes and immature sperm cells | low but variable | high? |

infertility, but also because objective criteria of what constitutes abnormality and normality in a semen sample are lacking. The availability of quantitative morphometric techniques for analysing spermatozoa, plus laparascopically recovered human oocytes on which to assay sperm function, should reduce these inadequacies. Features generally associated with subfertility are indicated in Table 14.4; however, these values serve as very approximate indicators and ideally several samples at different times and after at least 2 days of abstinence should be evaluated. Low sperm counts can arise because of deficient production (*a- or hypospermatogenesis*) or deficient transport (*obstructive azoospermia*).

Characteristically hypospermatogenesis is associated with smaller testes (less than 15 ml volume) of softer consistency. As spermatogenesis is a remarkably sensitive process, environmental insults can lead quite commonly to hypospermia. For example, dietary deficiency, X-irradiation, heating of the testis and exposure to a range of chemicals (notably cadmium, anti-mitotic drugs used in tumour therapy, insecticides and anti-parasitic drugs) all may impair spermatogenesis. In these cases, removal of the offending agent can result in restoration of spermatogenesis from the Ao stem cell population of spermatogonia, since it is mitotic spermatogonia and meiotic spermatocytes that are the principal targets of these agents (see Chapter 3). Untreatable forms of hypospermatogenesis include cryptorchid testes (see Chapter 1), genetic abnormalities such as XXY, XYY and some autosomal translocations, germ cell aplasia of unknown cause often with hyalinization of tubules, and as a sequel to severe orchitis or to prolonged and intense drug therapy for tumours. In all these patients (except those with Kleinfelter's syndrome), testosterone and LH levels may well be in the normal range. FSH levels, however, tend to be elevated, probably due to the absence of inhibin production (see Chapter 5). In general, elevated FSH is associated with a poor prognosis. Hypospermatogenesis due to a primary neuroendocrine deficit is relatively rare in men and is easily recognized. Treatment may be attempted with exogenous GnRH, gonadotrophins or with bromocriptine, but success is not high.

Obstructive azoospermia is not associated with obvious endocrine disorder and testes are of normal size and consistency. Obstruction to sperm transport usually occurs in the epididymides, as a congenital disorder or secondary to infection with, for example, gonorrhoea or tuberculosis. Not surprisingly, attempts to bypass the blockage surgically are of limited success, reflecting the important role the epididymis plays in sperm maturation (Chapter 8).

Abnormal, slow-swimming or dead spermatozoa in the ejaculate, as distinct from low numbers, might also result from suboptimal spermatogenesis and is certainly increased, for example, in cases of *varicocoele* (varicosity of the spermatic vein), genetic abnormality, or maintained elevated scrotal temperatures. Deficiencies in the maturation of spermatozoa, cytotoxic factors or antisperm antibodies in the fluids of the accessory glands and infection in the genital tract also contribute to abnormal sperm samples.

*In vitro* fertilization has provided a particularly useful route to therapy for hypospermatogenesis, the small numbers of recovered viable spermatozoa being concentrated around the oocyte(s). Alternatively, patients may be offered *gamete intrafallopian transfer* or GIFT, in which oocytes and sperm are mixed and transferred laparoscopically into the oviduct, where fertilization occurs. Such a procedure can also be used where sperm transport up the female tract is problematic.

*e Summary*

Even within the majority of cases of infertility classifiable under the main categories that we have considered, there is little room for satisfaction about the potential for diagnosis and the scope for successful therapy. There remain a group of subfertile individuals (about 20% of all cases) for whom even firm diagnosis is often not possible. However, the diagnosis and treatment of subfertility has made enormous advances over the past 20 years to a point where some reasonable hope can now be offered to around 40–50% of patients.

*f The future*

Over the past 10 years, the plight of the infertile has been taken seriously for the first time. There is little doubt that this change of attitude has stemmed largely from the pioneering work of Bob Edwards, Patrick Steptoe and Jean Purdy in the development of *in vitro* fertilization (IVF) techniques. This work has spawned interest in all forms of infertility, the development of many new techniques (such as GIFT) and has opened up the possibility of pre–implantation diagnosis for patients at risk of genetically transmitted disease, thereby avoiding a requirement for late abortion. However, as has happened historically, the infertile are again at risk of becoming the targets of social reaction, since in many countries this work and its application therapeutically are under attack from influential minorities largely motivated by religious considerations.

**5 Conclusion**

Our failure to adequately control fertility concomitant with our success in decreasing neonatal and infant mortality has, for much of humanity, replaced one tragedy by another. The spectre of deprivation, starvation, mass migration and war that haunted Western Europe until recently is still all too real for much of the world's population. One of the greatest indictments of medical science has been the relative neglect of research in reproduction and its associated clinical disciplines. The social and religious prejudices that have delayed or prevented analysis of these subjects means that much of our knowledge is recent or incomplete. Indeed, many of the pioneers in this area of scientific study are still alive and working today. Echoes of these prejudices recur in debates on the ethics of the so-called 'test tube babies', of abortion, of attitudes to sexual behaviour and the response to AIDS. More importantly, however, are the continuing effects of social and religious prejudice on the effective application of the knowledge that has accrued from such a relatively brief period of rigorous scientific and medical study. Deliberate policies to promote reproduction in some countries fearful of racial imbalance, and the frustration of family planning programmes by religious intolerance in others, is sadly as much a feature of the so-called civilized world as of countries struggling to improve the economic circumstances of their inhabitants. Throughout this book we have refrained as far as possible from imposing a personal view on our account of the science of reproduction. However, we do hope that study of this book will help students training to take a place in the medical and scientific professions to realize just how pervasive the reproductive process is, both for the individual and through society at large, and how an adequate understanding and control of it represents a crucial element in the survival and well-being of both.

**Further reading**

Edwards RG. *Conception in the Human Female.* Academic Press, 1980.

Hawkins DF, Elder MG. *Human Fertility Control.* Butterworths, 1979.

Pepperell RJ, Hudson B, Wood C. *The Infertile Couple.* Churchill Livingstone, 1980.

Pike MC *et al.* Breast cancer in young women and use of oral contraceptives: possible modifying effect of formulation and age at use. *The Lancet* 1983; No. 8356, Vol. II: 926–929.

Potts M, Diggory P. *Textbook of Contraceptive Practice.* Cambridge University Press, 1983.

Symposium Report No. 20, The control of follicle development and ovulation. *J Reprod Fertil* 1983; **69**: 325–409.

Vessey MP *et al.* Neoplasia of the cervix uteri and contraception—a possible adverse effect of the pill. *The Lancet* 1983; No. 8356, Vol. II: 930–934.

# Index

Lactogenesis 315–16
Lactose synthetase 315
Lactotrophins *see* Prolactin
Lecithin levels, amniotic fluid 283, 284
Leydig cells, in testis 8, 12, 52–55
  and spermatogenesis 70–3
Libido 20
  increased by testosterone 193
  loss of, and increased age 339
    and hyperprolactinaemia 140
    and oral contraceptives 190–1
  *see also* Sexual arousal
Light
  effect on reproduction 141–7
  effect on puberty 165–7
Liver function, oral contraception 347
Lordosis posture 16–17, 197
Luteal phase 93–4, 238–9
  abbreviated 353
  and decreased sexual activity 187
  duration, various species 94–5
  FSH and LH levels 118–19
  in rabbit 97
  shortening in rat and mouse 96–7
  steroid levels 124–5
Luteinized unruptured follicle syndrome 353
Luteinizing hormone (LH)
  follicular development
    antral phase 81–5
    preovulatory phase 85–7
  and hyperprolactinaemia 149
  levels
    in early pregnancy 252–3
    menstrual cycle 116–18
    oestrous cycle 142
      timing of surges 142–3
    postmenopausal 113
    at puberty 155–64
      daily variations 155, 157
      sleep augmented secretion 155–9
  luteotrophic, for non-pregnant corpus
      luteum 90, 97, 139
  negative feedback regulation 160
    *see also* Feedback regulation
  and oral contraceptives 345, 348
  properties 71
  regulation by GnRH 109–12
  injected, effects of 114, 162–3
  secretion, female regulation
    by
      GnRH 113–114
      oestradiol 113
      progesterone 113–14
  secretion, male, regulation 124–7
    by inhibin 126–7
    and pituitary–Leydig cell axis 70–3,
      124–6

and pituitary–seminiferous tubule
    axis 126–7
  and testosterone 70–3, 125–6
  surge 117
Luteinizing hormone releasing
    hormone/factor (LHRH, LRF)
  *see* Gonadotrophin-releasing hormone
    (GnRH)
Luteolysis 90–2
  prostaglandin F$_{2\alpha}$ 91–2

Magnocellular neurosecretory system 105–6
Mammary glands, anatomy 310–11
  development, hormonal effects on 311–14
    during pregnancy 313
  ductular system 311
  involution 322–3
  *see also* Lactation
Marsupials
  delayed implantation 144, 239
  maternal behaviour 325
Masculinization 21–22
  androgens
    effects 15, 17–18, 39–40
    testosterone 128–130
  oestrogen, effects 129–30
  'penis-at-twelve' syndrome 23, 47
Maternal behaviour 324–35
  altricial young 325
  bonding 333–4
  food supply 325
  nest building 325
  onset and maintenance, non-
      primates 326–7
    olfactory responsiveness 327
    pup-stimulation 326
    sensitization 326
    vaginal stimulation, sheep 327–8
  precocious young 325
  primates, relationship to human 329–31
    communication 330–4
  semi-altricial young 325
  weaning 340
Maternal blood, in pregnancy
  composition 270
  hydrostatic pressure, water
      movement 272–3
  oxygen tension variation 271
  and placental blood flow 249–50
Maternal immune responsiveness 290–2
Maternal metabolism,
  and fetal metabolism 268–78
    amino acid retention 276
    folic acid and vitamin deficiencies
      277–8
    iron deficiency 277

Nesidioblastosis syndrome 289
Neural tube defects 280, 303
Neuroendocrine reflex
  oxytocin relase 299, 320
  prolactin release 317–18
Norgestrel 48
Norethisterone 48
Norethynodrel 48
Norplant 133
Nuclei *see* Hypothalamus
Nutrition, and onset of puberty 166–8

Obstructive azoospermia 357–8
Odours
  and maternal behaviour 327, 334
  and sexual behaviour 188–9
Oestradiol
  administration after ovariectomy, and sexual
      behaviour 186
  aromatization and testosterone 41, 47, 184
  blood levels 38
    clearance 38, 45
  and breast development 313
  and feedback control
    FSH and LH 113, 114, 116–20
    positive, in male monkey 163
    at puberty 158
    sites 118–24
  and follicular development 53
  GnRH action, oestrous cycle, effects
      on 143, 145
  levels 38
    liver 38
    in menstrual cycle 116–18
    oestrous cycle 142
      surges 119, 121
    postmenopause 339
    in pregnancy 260–1
    at puberty 158
  and maintenance of adult sexual
      behaviour 142, 182–93
  and maternal behaviour, non-primate
      318, 325–7
  metabolic derivatives 46
  properties 40
  and prostaglandin release 298
  synthesis 261–2
  valate 323
Oestriol
  blood levels 38
  metabolic derivatives 46
  in pregnancy 259–61
  properties 40
  synthesis 261

Oestrogens
  anti-oestrogens 48, 49
  body tissue, fate in 38
  and blood cholesterol levels 181
  and breast development 311
  and capillary fragility 181
  and castration, administration after 184
  and corpus luteum 89, 90
  epithelial cells, stimulation 239
  feto-placental unit, synthesis 260
  and follicular development 83, 85
  and hypertension 181, 347
  and implantation 238
  in lactogenesis 315, 319
  levels
    menstrual cycle 114–18, 120
    at parturition 297–8
    in pregnancy 259–62
    at puberty 158
  and masculinization 129–30
  maternal circulation, disappearance 315, 319
  menopausal symptoms, treatment 339
  properties and functions 40–1
  regulatory action, female 175
    cardiovascular system 181
    cervix 178–9
    oviduct 175
    uterus 175–8
    vagina 180–1
  inspermatogenesis 70
  synthetic 48
  uterotrophic effect 176
  *see also* Oestrodiol, Oestriol, Oestrone
Oestrone, blood levels 38
  clearance 45
  levels
    postmenopausal women 339
    in pregnancy 261
    at puberty 158
  metabolic derivatives 46
Oestrous cycle 76, 99–100
  effect of coitus 97, 147
  effect of light 141–7
  in seasonal breeders 141–3
Olfactory attractants
  and maternal behaviour 327, 334
  and sexual behaviour 188–9
Oligospermia 356–8
Oocytes
  deficient 220–1, 355
    in follicular development 79, 81
      effects of LH 85, 87
      maturation, meiotic and
          cytoplasmic 86–7
      secondary 85, 86
  meiosis 78